Computer Communications and Networks

Series editors

A. J. Sammes, Cyber Security Centre, Faculty of Technology,
De Montfort University, Leicester, UK

Jacek Rak, Department of Computer Communications, Faculty of Electronics,
Telecommunications and Informatics, Gdansk University of Technology,
Gdansk, Poland

The **Computer Communications and Networks** series is a range of textbooks, monographs and handbooks. It sets out to provide students, researchers, and non-specialists alike with a sure grounding in current knowledge, together with comprehensible access to the latest developments in computer communications and networking.

Emphasis is placed on clear and explanatory styles that support a tutorial approach, so that even the most complex of topics is presented in a lucid and intelligible manner.

More information about this series at http://www.springer.com/series/4198

Zaigham Mahmood
Editor

Guide to Ambient Intelligence in the IoT Environment

Principles, Technologies and Applications

 Springer

Editor
Zaigham Mahmood
Debesis Education
Derby, UK

and

Shijiazhuang Tiedao University
Hebei, China

ISSN 1617-7975   ISSN 2197-8433   (electronic)
Computer Communications and Networks
ISBN 978-3-030-04172-4   ISBN 978-3-030-04173-1   (eBook)
https://doi.org/10.1007/978-3-030-04173-1

Library of Congress Control Number: 2018961204

© Springer Nature Switzerland AG 2019
This work is subject to copyright. All rights are reserved by the Publisher, whether the whole or part of the material is concerned, specifically the rights of translation, reprinting, reuse of illustrations, recitation, broadcasting, reproduction on microfilms or in any other physical way, and transmission or information storage and retrieval, electronic adaptation, computer software, or by similar or dissimilar methodology now known or hereafter developed.
The use of general descriptive names, registered names, trademarks, service marks, etc. in this publication does not imply, even in the absence of a specific statement, that such names are exempt from the relevant protective laws and regulations and therefore free for general use.
The publisher, the authors and the editors are safe to assume that the advice and information in this book are believed to be true and accurate at the date of publication. Neither the publisher nor the authors or the editors give a warranty, express or implied, with respect to the material contained herein or for any errors or omissions that may have been made. The publisher remains neutral with regard to jurisdictional claims in published maps and institutional affiliations.

This Springer imprint is published by the registered company Springer Nature Switzerland AG
The registered company address is: Gewerbestrasse 11, 6330 Cham, Switzerland

To Imran, Zoya, Arif, Hanya and Ozair
For their Love and Support

Preface

Overview

Ambient Intelligence (AmI) is an element of pervasive computing that brings smartness to living and business environments to make them more sensitive, adaptive, autonomous, and personalized to human needs. It refers to intelligent interfaces that recognize human presence and preferences and adjust smart environments to suite their immediate needs and requirements. The key factor is the presence of intelligence and decision-making capabilities in the IoT environments. The underlying technologies include pervasive computing, ubiquitous communication, seamless connectivity of smart devices, sensor networks, artificial intelligence (AI), machine learning (ML), and context-aware human–computer interaction (HCI). AmI applications and scenarios include smart homes, autonomous self-drive vehicles, healthcare systems, smart roads, industry sector, smart facilities management, education sector, emergency services, and many more.

The AmI process comprises three main stages: Sensing, Reasoning, and Acting. Sensing relies on real-world data collected by interconnected smart sensors present in the IoT environment. During the Reasoning process, the data is processed by intelligent algorithms embedded within the networked devices that perceive the environment and apply reasoned approaches based on AI and ML. This, in turn, provides responsiveness and adaptability, for the Acting stage of the process, to react to the ambient conditions and modify the environment as per the human requirements that are already built within the ambient systems.

Core benefits of AmI include customisation and automation of IoT-based living and industrial environments to meet user needs, however, because of automation, often control in taken away from the users when environments perform incorrect actions. This is generally due to the newness of the approaches that embed intelligence in smart devices, and the diversity, and sometimes unreliability, of such devices. There are therefore real issues of operability, reliability, consistency, connectivity, security, and trust. Other challenges refer to device communication protocols, sensor battery life, sensor robustness, self-testing and self-repairing of smart devices, as well as the heterogeneity of devices from diverse vendors. It is also important that relevant devices' costs are low, network bandwidth is high, device connectivity is seamless, and software systems are platform independent. Besides, much work still needs to be done at all levels on topics such as infrastructure, unobtrusive hardware, semantic web, algorithms, network capabilities, machine-to-machine (M2M) interactions, and device communication standards to effectively support the cyber-physical interconnections. All this is necessarily required for reliable, autonomic, and self-governing systems and networks.

As mentioned above, the advantages of AmI in the IoT environment are obviously huge, however, like any new technological paradigm, issues, and limitations are also many. This is the context within which the present book has been developed.

The focus of the book is on the AmI element of the IoT, in particular the relevant principles, frameworks, and technologies, as well as on benefits and inherent limitations. With this background, the present volume, *Guide to Ambient Intelligence in the IoT Environment: Principles, Technologies and Applications*, aims to capture the state of the art on current developments of smart spaces and AmI-based IoT environments. Majority of the contributions in the book focus on device connectivity, pervasive computing, and context modeling including communication, security, interoperability, scalability, and adaptability. Forty-Four researchers and practitioners of international repute have presented latest research, current trends, and case studies, as well as suggestions for further understanding, development, and enhancement of the AmI-IoT vision.

Objectives

The aim of this volume is to present and discuss AmI-based IoT in terms of frameworks and methodologies for connected smart environments. The objectives include

- Capturing the state-of-the-art research and practice with respect to the principles, frameworks, and methodologies of AmI-based IoT environments
- Presenting case studies illustrating challenges of AmI-enabled IoT, best practices, and practical solutions
- Developing a complete reference for students, researchers, and practitioners of pervasive computing and smart context-aware environments
- Identifying further research directions and technologies with respect to Ami-based IoT and distributed computing vision.

Organization

There are 12 chapters in *Guide to Ambient Intelligence in the IoT Environment: Principles, Technologies and Applications*. These are organized in three parts, as follows:

Part I: *Principles and Technologies*
This part has a focus on concepts, principles, and underlying technologies. There are four chapters. The first contribution explores the AmI research linked to the realization of smart city environments from the topology and architecture point of view. The second chapter extends the Internet of Things (IoT) vision to Industrial IoT to present the state of the art and the future of AmI in such environments. The next chapter discusses AmI implementation suitable for applications in business settings and suggests guidelines and strategies for IT business managers. The final contribution of this part has a focus on adaptability of Ambient Intelligence mechanism based on the well-known component-based approach and presents a relevant AmI system.

Part II: *Frameworks and Methodologies*
This part of the book comprises four chapters that focus on frameworks and latest approaches. The first chapter presents a WLAN-based intelligent positioning system for the presence detection and conducts performance analysis through simulation. The second contribution looks into the need for AmI for connected autonomous vehicles in a VANET and smart city scenario. A use-case involving augmented reality is also presented. The third contribution in this part aims to develop a mathematical model for embedded intelligent controllers for the detection of carbon monoxide in smart living environments. The last chapter provides an assessment of the AmI microcontrollers and proposes the design and architecture of a low-powered high-performance microcontroller.

Part III: *Applications and Use Scenarios*
There are four chapters in this part that focus on AmI-embedded applications and use cases. The first chapter examines the tax services provision of revenue administration in the IoT environment and discusses the changing role of tax providers. The next contribution suggests mechanisms to support vehicle drivers and proposes a low-cost IoT-enabled wearable device to improve the wellbeing of drivers in the context of Bangladesh. The third chapter proposes a vision-based posture monitoring system for the elderly using intelligent fall detection technique. The results of the study are also presented. The fourth chapter and the last in the book discusses new approaches to modern smart facilities management that employs AmI for developing smart office and living spaces.

Target Audiences

The current volume is a reference text aimed at supporting a number of potential audiences, including the following:

- *Network Specialists, Hardware Engineers and Software Developers* who wish to adopt the newer approaches to develop smart living and working environments and smart intelligent software applications.
- *Students and Academics* who have an interest in further enhancing the knowledge of technologies, mechanisms, and frameworks relevant to AmI in the IoT environment from a distributed computing perspective.
- *Researchers and Practitioners* in this field who require up-to-date knowledge of the current methodologies and technologies relevant to the AmI-enabled IoT vision, to further enhance the connectivity of smart devices.

Derby, UK/Hebei, China   Zaigham Mahmood

Acknowledgements

The editor acknowledges the help and support of the following colleagues during the review, development and editing phases of this text:

- Prof. Zhengxu Zhao, Shijiazhuang Tiedao University, Hebei, China
- Dr. Alfredo Cuzzocrea, University of Trieste, Trieste, Italy
- Dr. Emre Erturk, Eastern Institute of Technology, New Zealand
- Prof. Jing He, Kennesaw State University, Kennesaw, GA, USA
- Josip Lorincz, FESB-Split, University of Split, Croatia
- Aleksandar Milić, University of Belgrade, Serbia,
- Prof. Sulata Mitra, Indian Institute of Engineering Science and Technology, Shibpur, India
- Dr. S. Parthasarathy, Thiagarajar College of Engineering, Tamil Nadu, India
- Daniel Pop, Institute e-Austria Timisoara, West University of Timisoara, Romania
- Dr. Pethuru Raj, IBM Cloud Center of Excellence, Bangalore, India
- Dr. Muthu Ramachandran, Leeds Becket University, Leeds, UK
- Dr. Lucio Agostinho Rocha, State University of Campinas, Brazil
- Dr. Saqib Saeed, University of Dammam, Saudi Arabia
- Prof. Claudio Sartori, University of Bologna, Bologna, Italy
- Dr. Mahmood Shah, University of Central Lancashire, Preston, UK
- Dr. Fareeha Zafar, GC University, Lahore, Pakistan

I would also like to thank the contributors to this book: 44 authors and co-authors, from academia as well as industry from around the world, who collectively submitted 12 chapters. Without their efforts in developing quality contributions, conforming to the guidelines and meeting often the strict deadlines, this text would not have been possible.

Grateful thanks are also due to the members of my family—Rehana, Zoya, Imran, Hanya, Arif, and Ozair—for their continued support and encouragement. Every good wish, also, for the youngest in our family: Eyaad Imran Rashid Khan and Zayb-un-Nisa Khan.

Derby, UK/Hebei, China   Zaigham Mahmood
November 2018

Other Springer Books by Zaigham Mahmood

Fog Computing: Concepts, Frameworks and Technologies

This reference text describes the state of the art of Fog and Edge computing with a particular focus on development approaches, architectural mechanisms, related technologies, and measurement metrics for building smart adaptable environments. The coverage also includes topics such as device connectivity, security, interoperability, and communication methods. ISBN: 978-3-319-94889-8.

Smart Cities: Development and Governance Frameworks

This text/reference investigates the state of the art in approaches to building, monitoring, managing, and governing smart city environments. A particular focus is placed on the distributed computing environments within the infrastructure of smart cities and smarter living, including issues of device connectivity, communication, security, and interoperability. ISBN: 978-3-319-76668-3.

Data Science and Big Data Computing: Frameworks and Methodologies

This reference text has a focus on data science, and provides practical guidance on big data analytics. Expert perspectives are provided by an authoritative collection of 36 researchers and practitioners, discussing latest developments and emerging trends; presenting frameworks and innovative methodologies; and suggesting best practices for efficient and effective data analytics. ISBN: 978-3-319-31859-2.

Connected Environments for the Internet of Things: Challenges and Solutions

This comprehensive reference presents a broad-ranging overview of device connectivity in distributed computing environments, supporting the vision of IoT. Expert perspectives are provided, covering issues of communication, security, privacy, interoperability, networking, access control, and authentication. Corporate analysis is also offered via several case studies. ISBN: 978-3-319-70101-1.

Connectivity Frameworks for Smart Devices: The Internet of Things from a Distributed Computing Perspective

This is an authoritative reference that focuses on the latest developments on the Internet of Things. It presents state of the art on the current advances in the connectivity of diverse devices; and focuses on the communication, security, privacy, access control, and authentication aspects of the device connectivity in distributed environments. ISBN: 978-3-319-33122-5.

Cloud Computing: Methods and Practical Approaches

The benefits associated with cloud computing are enormous; yet the dynamic, virtualized, and multi-tenant nature of the cloud environment presents many challenges. To help tackle these, this volume provides illuminating viewpoints and case studies to present current research and best practices on approaches and technologies for the emerging cloud paradigm. ISBN: 978-1-4471-5106-7.

Cloud Computing: Challenges, Limitations and R&D Solutions

This reference text reviews the challenging issues that present barriers to greater implementation of the Cloud Computing paradigm, together with the latest research into developing potential solutions. This book presents case studies, and analysis of the implications of the cloud paradigm, from a diverse selection of researchers and practitioners of international repute. ISBN: 978-3-319-10529-1.

Continued Rise of the Cloud: Advances and Trends in Cloud Computing

This reference volume presents latest research and trends in cloud-related technologies, infrastructure, and architecture. Contributed by expert researchers and practitioners in the field, this book presents discussions on current advances and practical approaches including guidance and case studies on the provision of cloud-based services and frameworks. ISBN: 978-1-4471-6451-7.

Software Engineering Frameworks for the Cloud Computing Paradigm

This is an authoritative reference that presents the latest research on software development approaches suitable for distributed computing environments. Contributed by researchers and practitioners of international repute, the book offers practical guidance on enterprise-wide software deployment in the cloud environment. Case studies are also presented. ISBN: 978-1-4471-5030-5.

Cloud Computing for Enterprise Architectures

This reference text, aimed at system architects and business managers, examines the cloud paradigm from the perspective of enterprise architectures. It introduces fundamental concepts, discusses principles, and explores frameworks for the adoption of cloud computing. The book explores the inherent challenges and presents future directions for further research. ISBN: 978-1-4471-2235-7.

Software Project Management for Distributed Computing: Life-Cycle Methods for Developing Scalable and Reliable Tools

This unique volume explores cutting-edge management approaches to developing complex software that is efficient, scalable, sustainable, and suitable for distributed environments. Emphasis is on the use of the latest software technologies and frameworks for life-cycle methods, including design, implementation, and testing stages of software development. ISBN: 978-3-319-54324-6.

Requirements Engineering for Service and Cloud Computing

This text aims to present and discuss the state of the art in terms of methodologies, trends, and future directions for requirements engineering for the service and cloud computing paradigm. Majority of the contributions in the book focus on requirements elicitation; requirements specifications; requirements classification, and requirements validation and evaluation. ISBN: 978-3-319-51309-6.

User Centric E-Government: Challenges and Opportunities

This text presents a citizens-focused approach to the development and implementation of electronic government. The focus is twofold: discussion on challenges of service availability, e-service operability on diverse smart devices; as well as on opportunities for the provision of open, responsive, and transparent functioning of world governments. ISBN: 978-3-319-59441-5.

Contents

Contributors

Nova Ahmed Department of Electrical and Computer Engineering, North South University, Dhaka, Bangladesh

Shahed Al Hasan Department of Electrical and Computer Engineering, North South University, Dhaka, Bangladesh

Sarika Azad Department of Electrical and Computer Engineering, North South University, Dhaka, Bangladesh

Bernard Butler Telecommunications Software and Systems Group (Science Foundation Ireland—CONNECT), Waterford Institute of Technology, Waterford, Republic of Ireland

Güneş Çetin Gerger Manisa Celal Bayar University, Manisa, Turkey

Shailesh Singh Chouhan Embedded Internet Systems Lab, Department of Computer Science, Electrical and Space Engineering, Luleå University of Technology, Luleå, Sweden

Partho Anthony D'Costa Department of Electrical and Computer Engineering, North South University, Dhaka, Bangladesh

Richa Debnath Department of Electrical Engineering, Tripura University, Suryamaninagar, Tripura, India

Pragnaleena Debroy Department of Electrical Engineering, Tripura University, Suryamaninagar, Tripura, India

Kadir Alpaslan Demir Department of Software Development, Turkish Naval Research Center Command, Istanbul, Turkey

Seda Demir Institute of Social Sciences, Gebze Technical University, Kocaeli, Turkey

Ali H. Dogru Department of Computer Engineering, Middle East Technical University, Ankara, Turkey

Wesley Doorsamy Department of Electrical and Electronic Engineering, University of Johannesburg, Johannesburg, South Africa

Tufan Ekin Department of Software Development, Turkish Naval Research Center Command, Istanbul, Turkey

Alperen Eroglu Department of Computer Engineering, Middle East Technical University, Ankara, Turkey

Alea Fairchild Constantia Institute sprl and Faculty of Economics and Business, KU Leuven, Brussels, Belgium

Brendan Jennings Telecommunications Software and Systems Group (Science Foundation Ireland—CONNECT), Waterford Institute of Technology, Waterford, Republic of Ireland

Bwalya Kelvin Joseph Information and Knowledge Management, School of Consumer and Information Systems, University of Johannesburg, Johannesburg, South Africa

Alper Karamanlioglu Department of Computer Engineering, Middle East Technical University, Ankara, Turkey

Muhammed Cagri Kaya Department of Computer Engineering, Middle East Technical University, Ankara, Turkey

Syeda Shabnam Khan Department of Electrical and Computer Engineering, North South University, Dhaka, Bangladesh

Adnan Mahmood Department of Computing, Macquarie University, Sydney, NSW, Australia; Telecommunications Software and Systems Group (Science Foundation Ireland—CONNECT), Waterford Institute of Technology, Waterford, Republic of Ireland

Md. Tanvir Mushfique Department of Electrical and Computer Engineering, North South University, Dhaka, Bangladesh

Champa Nandi Department of Electrical Engineering, Tripura University, Suryamaninagar, Tripura, India

Ivan Nikitin Innopolis University, Innopolis, Russia

Tolga Onel Department of Computer Engineering, Turkish Naval Academy, Istanbul, Turkey

Ertan Onur Department of Computer Engineering, Middle East Technical University, Ankara, Turkey

S. Padmavathi Department of Information Technology, Thiagarajar College of Engineering, Madurai, Tamil Nadu, India

Babu Sena Paul Institute for Intelligent Systems, University of Johannesburg, Johannesburg, South Africa

Md. Majedur Rahman Department of Electrical and Computer Engineering, North South University, Dhaka, Bangladesh

Saad Azmeen Ur Rahman Department of Electrical and Computer Engineering, North South University, Dhaka, Bangladesh

Sheikh Raiyan Department of Electrical and Computer Engineering, North South University, Dhaka, Bangladesh

Balwinder Raj Nanoelectronics Research Lab, Department of Electronics and Communication Engineering, National Institute of Technology (NIT), Jalandhar, Punjab, India

E. Ramanujam Department of Information Technology, Thiagarajar College of Engineering, Madurai, Tamil Nadu, India

Vitaly Romanov Innopolis University, Innopolis, Russia

Rahat Jahangir Rony Department of Electrical and Computer Engineering, North South University, Dhaka, Bangladesh

Quan Z. Sheng Department of Computing, Macquarie University, Sydney, NSW, Australia

Jeetendra Singh Nanoelectronics Research Lab, Department of Electronics and Communication Engineering, National Institute of Technology (NIT), Jalandhar, Punjab, India

Giancarlo Succi Innopolis University, Innopolis, Russia

Nur E. Saba Tahsin Department of Electrical and Computer Engineering, North South University, Dhaka, Bangladesh

Bedir Tekinerdogan Information Technology Group, Wageningen University, Wageningen, The Netherlands

Bugra Turan Department of Electrical and Electronics Engineering, Koc University, Istanbul, Turkey

Santosh Kumar Vishvakarma VLSI Circuit and System Design Lab, Discipline of Electrical Engineering, Indian Institute of Technology (IIT), Indore, Madhya Pradesh, India

Wei Emma Zhang Department of Computing, Macquarie University, Sydney, NSW, Australia

About the Editor

Prof. Dr. Zaigham Mahmood is a published author/editor of 25 books on subjects including Electronic Government, Cloud Computing, Data Science, Big Data, Fog Computing, Internet of Things, Smart Cities, Project Management, and Software Engineering, including: *Cloud Computing: Concepts, Technology & Architecture* which is also published in Korean and Chinese languages. Additionally, he is developing two new books to appear later in 2019. He has also published more than 100 articles and book chapters and organized numerous conference tracks and workshops.

Professor Mahmood is the Editor-in-Chief of *Journal of E-Government Studies and Best Practices* as well as the Series Editor-in-Chief of the IGI book series on *E-Government and Digital Divide*. He is a Senior Technology Consultant at Debesis Education UK and Professor at the Shijiazhuang Tiedao University in Hebei, China. He further holds positions as Foreign Professor at NUST and IIU in Islamabad Pakistan. He has served as a Reader (Associated Professor) at the University of Derby UK, and Professor Extraordinaire at the North West University Potchefstroom South Africa. Professor Mahmood is also a certified cloud computing instructor and a regular speaker at international conferences devoted to Cloud Computing and E-Government. His specialized areas of research include distributed computing, emerging technologies, project management, and e-government.

Part I
Principles and Technologies

Chapter 1
Ambient Intelligence in Smart City Environments: Topologies and Information Architectures

Kelvin Joseph Bwalya

Abstract Many cities around the world have embarked on ambitious programmes towards creating Smart Cities where information, diverse digital opportunities, and collective intelligence can be harnessed ubiquitously. Smart Cities are conceptualized using citywide smart and intelligent architectures informed by the context in which they are implemented. These architectures make it possible to access information and intelligence anywhere and at any time. Information processing and computing is embedded within the urban infrastructures to a point where immovable city entities such as traffic lights are more intelligent to make real-time decisions based on the happenings in the environment in which they are deployed. Advanced development of ambient computing within the realm of Smart Cities will further culminate into possibilities such as vehicle-to-vehicle communication (V2V) and mobile-to-mobile (M2M) communication. Using extensive and critical literature review, this chapter specifically focusses on the design of information architectures that will ultimately support the enshrining of spatial intelligence within Smart City environments hinged on the internet of things (IoT) and cloud/fog computing. The chapter presents latest trends in the research and practice of ambient intelligence (AmI) linked to the realization of the key principles of Smart Cities from the information topology and architecture point of view. A conceptual ambient intelligence architecture that highlights the building blocks of any ambient intelligence architecture as deployed in Smart City environments is also proposed. The proposed conceptual architecture can be used as a blueprint in the design of ambient intelligence topologies and architectures in different contextual settings.

Keywords Ambient intelligence · Architecture · Topology · Smart City Information architecture · AmI basic architecture · Sensing · V2V · M2M Context awareness

K. J. Bwalya (✉)
Information and Knowledge Management, School of Consumer and Information Systems, University of Johannesburg, Johannesburg, South Africa
e-mail: kbwalya@uj.ac.za

© Springer Nature Switzerland AG 2019
Z. Mahmood (ed.), *Guide to Ambient Intelligence in the IoT Environment*, Computer Communications and Networks, https://doi.org/10.1007/978-3-030-04173-1_1

1.1   Introduction

Smart Cities encompass recent innovative applications and orientations that are geared towards improving the quality of life of human beings in the different living spaces of the city. The conceptualization of Smart Cities hinges on the possibility of pervasive information access to a variable of information culminating into intelligence of the city enabling citizens to make decisions in any given context which they find themselves in. Of late, there has been a delve towards incorporation of hybrid disciplines and ensuring that there is an absolute chance for building intelligent living spaces in the city. In this regard, there is a push for encouraging the embedding of topologies and architectures of ambient intelligence (AmI) into the core design aspects of Smart Cities. The embedding of ambient intelligence into Smart City design will culminate into the achieving the desired true dynamism, intelligence, and ubiquity in Smart City environments.

The need for intelligence in cities and the desire for cities to be smart and intelligent cannot be overemphasized given the ever-growing number of people desiring to live in the vicinity of cities, and the growing economic activities and industrialization. As estimated by the United Nations, over 60% of the world's population lives in the cities. This brings a lot of pressure on the existing city resources that are already evident in cities. The negative impact of the anticipated pressure can only be averted when cities become smart and intelligent to the effect that the available resources and opportunities can be efficiently and effectively shared by everyone. For example, smart electricity grid can culminate into efficient and optimal use of electric energy.

One of the applications, in which AmI has been used consistently, is healthcare setups. Ambient technologies thrive where they are built on scalable network and backbone architectures supporting a diverse range of applications such as machine-to-machine (M2M) communication, vehicle-to-vehicle (V2V) communication, etc., where sensors are made to be as responsive as possible. Such kind of networks is able to support a wide range of LTE devices. Although it is apparent that AmI has many potential applications in everyday situations, these are not realized owing to the many glaring challenges attributed to its implementation. Some of the key challenges acting as roadblocks to the advancement of AmI include energy, context awareness, natural interfaces, power, etc. Another pronounced impediment to realizing the full benefits of AmI has been the use of inappropriate network topologies and information architectures [1]. Basing on the foreword to "Pervasive Information Architectures" by [2], authors in [3] raise the question with regards to how digital, physical and cognitive spaces can be bridged. The paths that can bridge the digital spaces are the different desired information architectures that can be designed and be used in a multitude of contextual settings.

This chapter intends to bring out the key building blocks of context-aware ambient intelligence and propose conceptual models that can be used in the design of the different topologies and architecture for contemporary Smart City designs. The models proposed can be used as blueprints and reference points in the design of actual Smart City architectures. The design considerations of ambient intelligence posit that AmI systems need to be developed in such a way that they are able to recognize and utilize spatial, temporal and contextual data for behavioral recognition.

The chapter is arranged as follows: The next three sections present the formulaic definitions and aspects of ambient intelligence. Next, the chapter explores the different network topologies and information architectures used to design Smart Cities in the realm of ambient computing. Then, the different issues encountered in the design and implementation of dynamic AmI topologies and architecture will be explored. The last part will give a recap of the concepts discussed and articulate the future works in this area.

1.2 Ambient Intelligence

Smart Cities designs aim to incorporate, as much as possible, different technologies that bring about dynamism, intelligence and multiple capabilities to city living. The pivot of Smart Cities in the realm of ambient intelligence is the capability of employing requisite technologies that are able to scan the environment to sense the movements and opportunities available in different living spaces of the city. These technologies are preferably smart and intelligent with powerful dynamic sensory capabilities [1]. Ambient intelligence ushers in an environment that is sensitive to the happenings in the environment (e.g. human presence) and is therefore adaptive and responsive to any changes. Considering the different formulaic definitions of ambient intelligence is important owing to the need to understand the different nuances and opportunities brought about by AmI.

In a bid to understand AmI, the first point of call is the understanding of what intelligence is. Intelligence in this context entails the ability of entities (people, technology, etc.) to harness the different capabilities and informational opportunities embedded into the environment in which they live. Intelligence involves recognizing these opportunities, computing them in a given context and making them available through actionable processes to different entities that may be in need of them. Contemporary intelligence posits that recognition, computation, and action are processed in a parallel mode towards one outcome.

In order to obtain a clear understanding of what AmI is and what it entails, there is a need to comprehend the different principles upon which it is hinged and the understanding of the scenario and environment in which AmI can be realized. The ambient intelligence environment presents a scenario where a number of heterogeneous devices (known as agents) are deployed in a given environment and are expected to share network resources so that each of them is able to access the available opportunities. In such a scenario, it is expected that as mobile agents (gadgets such as mobile phones, personal digital assistants (PDAs)) traverse a given environment, they are able to access the different opportunities available in the different environments. In order for such a scenario to be realized, a high-level mobility transparency and system integration is desired so that systems act as "whole"—single integrated information spaces. In order to achieve true pervasiveness, one of the key

Fig. 1.1 Sensing and
intelligence in Smart Cities

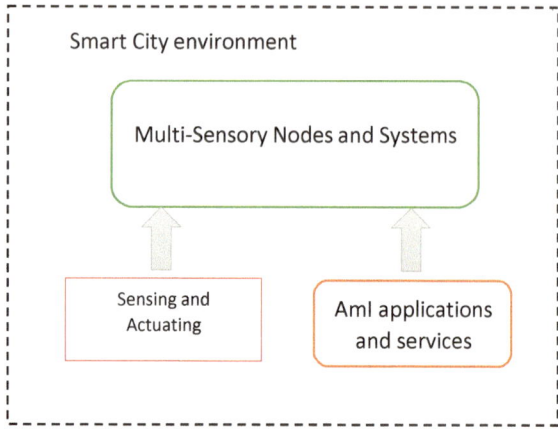

requirements is the need for integrated or distributed information architectures. For
this to be achieved, AmI demands that it uses distributed file systems and databases
which make it possible for multimedia to be shared among agents [4]. Appropriate mobility and migration can be achieved by the continuation of processing of an
application that already started on a machine and moved to another machine. The
application continues from where it was left off on the new machine taking with it
its interface and application context on the new machine. This is an aspect of load
balancing in distributed ambient environments which is so desired in Smart Cities.
Since AmI emphasizes the need for ubiquity so that it can be deployed in any part of
Smart Cities, the UltraWideBand (UWB) technology makes it possible to have high-
speed wireless Internet connection that is able to handle transmission of audio–video
and other data streams which needs high bandwidth. The UWB can work so well in
Smart City environments.

Another principle that AmI is hinged on is ephemeralization which is a concept
meaning "doing more for less." Ephemeralization enables users in the AmI environment to access opportunities unconsciously and automatically using their gadgets
even with limited computing power. Although the conceptual nuances of AmI sound
similar to ubiquitous computing (Ubicomp) , it is clear that AmI is an extension of
Ubicomp to give it more meaning in context and stature. Ambient computing stems
from the convergence of three key technologies, i.e., ubicomp, ubiquitous communication, and intelligent user-friendly interfaces, implying a seamless and integrated
environment of computing enabled by advanced computing and interfaces informed
by the local context. Because of these characteristics, AmI is heavily relied upon by
Smart Cities [5, 6]. The basic setup of AmI as applied in Smart Cities environment
is shown in Fig. 1.1.

The anticipated guaranteed access to distributed resources in a purely dynamic
and ubiquitous environment in ambient intelligence Smart Cities is shown in Fig. 1.2.

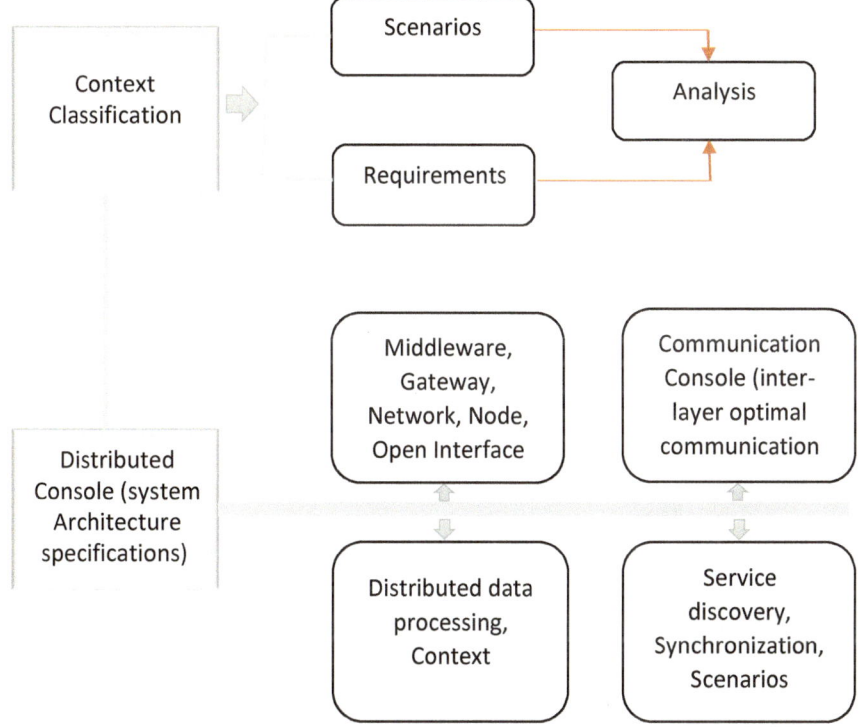

Fig. 1.2 Ambient intelligence in distributed Smart City environments

A clear understanding of ambient intelligence in the realm of Smart Cities in a real-world setup can be achieved by considering the case presented below:

Jac(k) has a chronic disease which needs palliative care and monitoring of his/her condition throughout the day and triggers on his/her watch which remind him when it is time to take medication. On the body, he/she has wearable sensors which monitor his/her heart rate, blood pressure, blood sugar, etc., and if the levels of any one of the measurements vital signs exceed the defined normal, a signal is sent to the doctor or next-of-kin for an intervention who receive the notification through their portal devices (mobile phone, personal digital assistant (PDA) or through e-mail). The wearable sensors are also able to detect the nearby doctors by displaying on Jac(k)'s display unit, on his/her mobile phone the distance to the nearby doctor. Jac(k) can then make a decision on which ones to consult. It happens that as Jac(k) navigates through a given environment, his/her display unit in the area detects that there is a nearby network from the next vehicle which can be accessed for sharing public photos, music, network, etc. The two vehicular networks are now able to share so many network resources such as traffic information, etc., using the open interoperable V2V consoles.

The interactions articulated in the case presented above are made possible by all these interactions between heterogeneous gadgets, made possible by distributed file systems and open user interfaces [7]. The case above has also shown that Smart

City living spaces demand a lot of automation so that there is unhindered sharing of network and information among the different entities deployed in a given environment. Therefore, the gamut of AmI is that it needs to promote unhindered interaction between devices, humans, and smart objects. Therefore, AmI strives towards the transformation of working environments into intelligent spaces which are able to adapt according to the changing user's desires and needs instantaneously [10].

Designing relevant Smart City applications embedded with appreciable degree of ambient intelligence entails incorporating the different principles espoused above [8]. There are a lot of design approaches that have been used to design contextually relevant Smart City applications. Incorporating AmI into Smart City design culminates into change in the design paradigm from that of human-in-the-loop design orientation to a novel design paradigm where the center of design is not the end-user of the technology but the ability of the technology to adapt to the environment in which it is deployed. One of the key challenges in such a scenario refers to the design of interfaces of heterogeneous gadgets so as to optimize on access to the shareable network and the different information resources [9]. Optimization of access to different resources entails that carefully chosen network topologies and information architectures need to be considered. Measuring and modeling the different aspects of spatiotemporal aspects of ambient intelligence requires different mathematical representations (vectors) which is out of the scope of this chapter.

1.3 Ambient Intelligence and Smart Cities

Because of many technological innovations that are being showcased nowadays, it is important to take advantage of the different technology combinations in order to optimally benefits. The design of Smart Cities continuously relies on the different possibilities brought about by ambient intelligence (AmI) and other emerging technological innovations. As a consequence, the ability to integrate two different technologies depends on the level of understanding and interoperability of each of these technologies. Interoperability of different nodes and systems within the wider communication network of Smart Cities will be one of the key issues that need to be addressed in AmI. Ambient intelligence delves well onto the Smart City agenda because its scope is the same as that of Smart Cities which includes the home, hospital, airport, and in general the environment and everything contained therein. AmI enables human beings to take advantage of most of the city environment by unlocking opportunities that were not known or let alone perceived just a few years ago [7].

Lee [11] has proposed that the definition of Smart Cities is dependent upon the context in which it is implemented and the anticipated scope of the Smart City applications. Therefore, the conceptualization of Smart Cities goes towards accomplishing a complex ecosystem geared to improving different aspects of individual citizens and city authorities. A Smart City is a multidimensional and multifaceted conceptualization which can manifest in many aspects of the socioeconomic establishment.

Therefore, technology is not the only requirement for building appropriate Smart Cities and is perceived as one key enabler for Smart City realization. Since the Smart City concept is relatively new, there are no globally accepted formulaic definitions for Smart Cities and consequently, there are no globally accepted principles attributed to Smart Cities. Despite this being the case, the consensus is slowly building around the interplay and balance of smart public administration, smart governance, smart resources, city planning and design, and ICT integration into the urban and regional spaces for a connected and integrated city socioeconomic infrastructure [12]. The smallest unit of a Smart City could be a home. In this scenario, you have a home littered with different types of sensor nodes and systems capturing any movements in the home. AmI works by observing and rightly interpreting the sensor observations to enable various informed decisions

The anticipated benefits of any Smart City project include: an acute contribution towards intelligent public administration and policy-making, improvement of competences of local companies and organizations, and the general empowerment of local community groupings [13, 14]. The implementation of Smart Cities culminates into the improvement of different socioeconomic infrastructures such as having in place smart, adaptive and intelligent university and education hinged on global principles, inspired by local contextual characteristics and desired for future socioeconomic orientation [15].

Achieving the full potential of Smart Cities involves the dovetailing of different aspects of city's design attributes [16]. The dovetailing of the different attributes of the city's socioeconomic infrastructure goes towards achieving spatial intelligence. Spatial intelligence entails the pervasive access to diverse information resources anywhere, anytime, and using any device. Smart Cities design requires the cognizance of community engagement into the governance infrastructure of the city (citizen inclusiveness), overall city administration styles, information integration for ubiquitous information management, data quality, privacy and security, city design taking into consideration sustainability requirements, etc. These requirements are integrated into the overall design of the Smart City concept [17, 18]. For the fact that Smart Cities need a lot of information to be accessed ubiquitously, the need for ambient technologies in Smart City design cannot be overemphasized. In ambient intelligence, especially as applied in Smart Cities, it is important to ensure that transparency, which is the ability of the designers to conceal the design or operational details from the users, needs to be seriously considered in the design and implementation so that the system is perceived as whole. Location-based services tap into the local resources in the area in which the mobile computing agent is [19].

Contemporary cities have embraced ubiquity in many of its establishments to the point that there is access to the Internet anywhere and at any time. Further, agents in highly dynamic environments are able to unobtrusively and seamlessly connect and exchange information [20]. In general, Smart Cities are espoused in context-aware computing paradigms and design architectures. Because of such paradigms, Smart Cities are able to facilitate pervasive information services which enable the achievement of distributed location transparency. Because the components of Smart Cities must continuously change in order to retain their relevance, it is important that the desired requisite network infrastructure be evolving in order to accommodate changing needs. One of the key characteristics of contemporary AmI infrastructures is that they should be open and scalable so that there is no need to redesign the network when there is need to increase its scope. The need for evolutionary network architecture is important because Smart Cities are now designed upon the concept of sustainability and livability [21].

1.4   Trends in Ambient Intelligence

Understanding the different current research endeavors and practice worldwide towards inculcating ambient intelligence in Smart Cities design is important to appreciate the key principles, issues, and models used in contemporary Smart City design. Also, there is a need to understand the full potential usage of ambient intelligence in Smart City environments. Of late, there has been many uses and benefits in everyday life attributed to ambient intelligence.

The conceptualization of AmI is espoused upon the desire to have everything surrounding a man to be smart. This starts from smart and intelligent gadgets in the home to intelligent spaces in the natural environment. For example, AmI will enable the full realization of driverless cars where vehicles will be able to communicate with the environment and making optimal decisions on which routes are optimal. This will be realized by enhanced information capturing, sharing and processing through WSNs, nodes in the different spatial-temporal domains, and the vehicle onboard systems which can make decisions by further passing the processed signals from the onboard systems to the transmission systems and the wheels of the vehicle [5].

AmI is one of the key technology innovative roadmaps that promise to revitalize the health sector. AmI can be used for monitoring the health state of an individual in different environmental conditions [22]. Because of AmI, wearable devices littered with sensor networks can be used on human bodies to collect health information to the unconscious of the user. Wireless devices can be used to continuously monitor the health aspects of the individual. The AmI consoles are embedded with intelligent devices that are able to collect information using numerous sensors from different physical spaces to provide input for various contextual decision-making [23]. In the medical field, AmI is often configured using the star topology where a central node does the coordinating between the sensors and the body area network (BAN). In many aspects, BAN can be used as MobiHealth monitoring systems especially for vital organs and as Ubimon ubiquitous health monitoring systems for wearable sensors, accelerometers, etc. [23]. Further effective application of AmI technologies in healthcare will enable intelligent distributed information processing which will further improve healthcare delivery [23]. It is anticipated that future development in the application of AmI in healthcare will culminate into practical ubiquitous monitoring of patient's status. In this regard, information will be collected from the patient by the wearable medical device embedded with adequate and appropriate sensors to timeously capture information and seamlessly send it over a telecommunications network to the physician, emergency department or to the siblings at home [23]. One of the cardinal problems in the optimization of information access in ambient intelligence environments is the ability of systems to retrieve information stored in different places simultaneously. In healthcare environments, AmI can be used to retrieve information from different health providers using agent-based systems. Research reported in [24] proposed the geriatric ambient intelligence (GerAmI) as an intelligent system that provides up-to-date patient data from all the possible health databases in the country.

Other uses of AmI include the provision of a safer environment by implementing ad hoc environmental sensors and triggers which go off once there is any intrusion, temperature regulation in different contextual settings, and generally supporting individuals in their daily activities so as to enjoy the benefits offered in the surrounding environment. By effectively using AmI, information and knowledge bases in the environment can be accessed using seamless interfaces. AmI can also be used in reverse to observe and monitor the behavior of human beings in different environments. This can be done by monitoring sensor observations. Such a technology can be used in service delivery organizations to observe the posture of the different employees during work hours. Therefore, recognizing the state of a human behavior in a spatiotemporal context is a very important research area in Smart Cities. An example could be understanding the cognitive or physical status of an individual on palliative care and notifying a caregiver if need be [4]. In the same way as ubiquitous computing, cloud and fog computing, ambient intelligence aims to surround human beings with intelligent computing devices embedded in different objects such as furniture, roads, wearables, vehicles, clothes, environment, etc. to aid individuals to make conscious or unconscious decisions anytime anywhere. In this scenario, it is envisaged that there will be object-to-object communication such as vehicle-to-vehicle (V2V) sharing of wireless networks on the roads and ultimate sharing of information, etc. [5]. Together with robotics (extensively used in telemedicine and surgery) and artificial intelligence (AI), there are a lot of uses of ambient intelligence in everyday real-life situations. Many of the new models of the luxury brand vehicles are fitted with some degree of AmI. For example, AmI can be used to alert a sleepy driver that the vehicle has crossed the yellow line and is on course for disaster or can be used to deny the ignition of a vehicle when the driver is extremely drowsy [25].

Realizing the key benefits of AmI, a lot of innovations are being realized in the realm of AmI in different contextual settings in the world. For example, a lot of research is being done worldwide on emergent themes of AmI and the European Union IST advisory group (ISTAG) positions itself as a key advocate for the AmI vision. Many different types of methodologies have been used to design AmI applications or modeling of conceptual AmI structures. In many contexts, prior research has used hidden markov model (HMM) to model the behavior patterns in an environment with heterogeneous gadgets. Authors in [26] proposed SENSOR9k which was deemed a comprehensive architecture that can be used in ambient intelligence environments. The SENSOR9k is a pervasive sensory infrastructure encompassing both hardware and software aspects embedded onto the backbone of local gateways. There are a lot of modeling frameworks that can be applied in modeling the multidimensional aspects of Ambient Computing. These frameworks have been used in different contextual settings and most have been hinged on model-driven engineering. Study reported in [10] focused on the models@run.time paradigm which is able to take care of context-aware and adaptive systems such as the domain-specific modeling languages (DSML) such as the Eclipse Modeling Framework and the Kevoree Modeling Framework. Reference [27] explored the context awareness supported application mobility (CASAM) or "Application Mobility." CASAM enables different applications to be accessed by different devices whilst in motion.

1.5   Context Awareness and Ambient Intelligence

Ambient technologies require context-aware adaptive systems that can alter their behavior or state in reaction to the changes in the environment in which they are deployed [3]. There is undoubted need for context-aware information architectures in the contemporary information management environments. Context awareness has much to do with the context influencing the kind of information or services that are made available in a given situation. Context awareness entails the ability of an information system to reconfigure itself given the environment in which it is deployed. The reconfiguration may involve altering its topologies or interfaces so that it can readily be integrated into the environment in which it is deployed.

Despite the realization of the importance of contextual awareness, there has been no clear consideration of the designers' view regarding contextual information systems. The need for design principles that incorporate contextual awareness by designers cannot be overemphasized. The study in [28] analyzed the concept of context-aware design by analyzing the different practices and artifacts of designers in contextually dissimilar locations and spaces. Their study proved that different designers in different contexts employ different views and approaches to achieving context-aware designs which are commensurate to their particular location.

1.6   AmI Topologies and Information Architectures

In order to achieve the desired intelligence, fluidity, dynamism, and openness, there is need to carefully choose network topologies and architecture upon which AmI is hinged. Architecture entails the different aspects that are concerned with low-level hardware access, the middleware and the different architectural software aspects that together logically collaborate to provide an information service [29]. Topology is the arrangement of the different elements of the network informed by the desire for optimal integration and communication. Appropriate and desired architecture should be designed in such a way that it is able to be reusable, dynamic and interoperable with heterogeneous technology solutions and is able to offer opportunities for rapid development and deployment. Architectures should be flexible to the point that their scope can be increased without having to redesign the entire system [29].

Different network topologies and information architectures are considered for contemporary AmI applications especially with increased requirement for intelligence in network or information sharing environments. As a result, there has been a strong push towards dynamic and smart ecosystems. These ecosystems, which are a departure from static infrastructure and buildings are the hallmark of contemporary city developments. In highly dynamic city infrastructure, ambient intelligence is an undeniable requirement. A very good information architecture and topology is one that can easily be deployed in any given context and self-adapts to suit the conditions of the context in which it is deployed.

Over the years, the way intelligence is being perceived is changing towards more open and interoperable systems which are able to simultaneously process information and guide decisions. The onus of AmI, therefore, is to open the systems and technology interfaces as much as possible so that there is increased technology system collaboration for sharing network and information resources. At the onset, roots of AmI such as artificial intelligence (AI) are exploiting the "physical symbol system hypothesis" which postulates that intelligence is hinged on symbol processing modeled upon input-compute-output model [30]. In order to have appropriate AmI technologies that are able to capture the intelligence from the environment using requisite sensors, Ref. [31] articulates the key attributes of the desired architecture, viz., the technologies able to capture context awareness multi-agent systems are also able to interact with each other and the environment over dynamic and automatic scenarios, capturing movements in the environment with respect to spatial and temporal dimensions. The anticipated technologies for capturing context awareness are dynamic, seamless, and unconstructive technologies such as the radio-frequency identification (RFID) technology. Such technologies have antennas, tags, readers, and software to capture informational instructions. As is common knowledge, nowadays, "lightweight" agents are embedded into mobile devices which support different forms of communication and allow portability with a wide range of devices. The wireless communication console in the consumer electronics is designed upon the ZigBee Alliance designed upon IEEE 802.15.4 protocol. Topologies such as mesh, star, and tree can be used to ensure that there is optimal access to the network and its resources. In the actual environment, the following can be the configuration of the devices using a star topology:

Referring to Fig. 1.3, the Network Coordinator ensures that there is coordinated response to the events in the network and controls all the devices. In certain circumstances, a voting mechanism can be used to reach consensus. The Router or Repeater sends or receives or resends data to and from the coordinator to the End Devices. The End Device sends or receives data to or from the Coordinator. Using this configuration, it is clear that the Router or Repeater is the intermediate player in Ambient Intelligent networks.

Although not the focus of this chapter, it is worth mentioning that architectural configuration of any smart environment is determined by the level of quality of service desired (QoS), i.e., the number of packets accepted to be lost within a given time, and

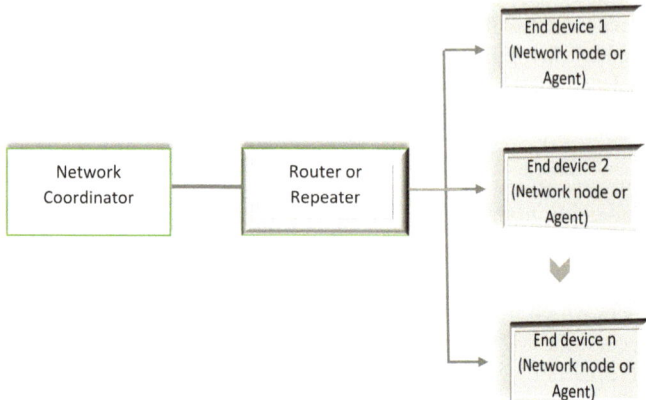

Fig. 1.3 Topology for capturing intelligence in Smart Cities

the routing protocols used. In many dynamic configurations, self-healing networks are used [20]. Based upon the 2011 De Paolo's work, a basic AmI architecture has three layers:

- Physical layer—that has all the different sensory (WSNs) and actuation devices including those for different AmI functionalities and users' applications;
- Application layer—the bedrock upon which the actual different AmI applications are designed and implemented;
- Middleware—that gives the software and hardware abstraction for the applications to be accessed across the heterogeneous applications. It is also a platform upon which virtual databases servicing the AmI network and holding intermittent information are hinged.

In general, the desired architecture for Ambient Computing should have the following themes (as shown in Fig. 1.4):

- Collaborative support for mobile gadget users in ubiquitous environments—mobile users in ubiquitous environments need to enjoy collaboration support so that they have seamless access to devices and artifacts that are available in a given environment; and
- Common middleware used in pervasive computing environments—user tasks need to be distributed and executed in multiple computers using remote task computing execution. Using a requisite middle platform, user tasks are able to be transported from the requesting host computer to a remote or couple of remote computers which have the necessary computing power to process the incoming packets. The middleware is a software or hardware console that allows passage of packets from one mobile device to the other in a highly integrated environment.

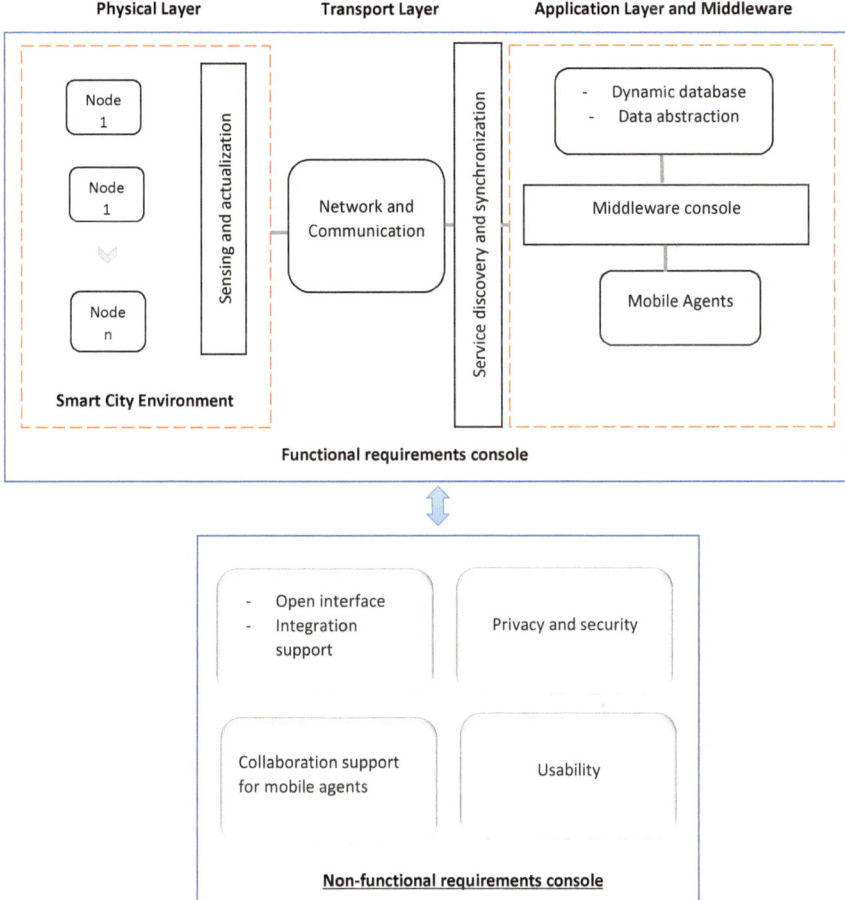

Fig. 1.4 AmI basic architectural layers

Contemporary ambient technologies need architectures that are embedded with concepts intertwined within complex and ever-changing information ecology. In such an environment, information architecture that can stand the test of time is the one that is appropriately informed by the users and the context [3].

Other than context awareness and intelligence, ambient intelligence mainly relies on the peer-to-peer (P2P) communication model to share resources. The P2P model is an asymmetric client–server communication model, endowed with a symmetric one, in which all agents can act as server or client depending on the state at a given time. AmI-P2P model enables agents in the environment to directly communicate with one another. The ambient intelligence model of vehicle-to-vehicle (V2V) is a direct beneficiary to the P2P technology [7]. The different designs of the P2P network configurations are cardinal for providing sustained network coverage and access in nomadic networks characteristics of pervasive AmI systems. One of the common access modes includes the wireless internet access (IEE802.11, Wireless access protocol, etc.) and the wireless broadband access (local multipoint distribution service and multichannel multipoint distribution service). In this network configuration, each of the nodes (gadgets) is recognized as an access point.

Contemporary Smart Cities will be hinged on AmI based on the conceptualization of mobile ad hoc networks (MANETs) in order to achieve access to networks and information anywhere and anytime [9]. Because of continuous advancements in MANETs, different mobile gadgets are able to automatically communicate with one another in Smart City environments. MANETs are designed to be multi-hop wireless networks that do not have fixed infrastructure hinged upon mobile nodes. The concept of Smart City has been perceived with a desire to ensure that information is everywhere and people and machines can access it seamlessly and ubiquitously thereby positively affecting cities' liveability.

Wireless sensor networks (WSNs) are critical dynamic infrastructure to carry data sensed by different sensors to the decision nodes. Generally, WSNs gather environment and personal data continuously and this is used as basis for decision-making. One of the key issues in achieving ambience intelligence in WSNs lies in managing the battery power of the different nodes scattered throughout the WSN technologies (such as ZigBee, radio-frequency identification, etc.) [23]. In the desired network-everywhere aspiration of Smart Cities, the power of the sensor nodes dispersed everywhere in the environment is cardinal for the realization of ubiquitous access to information resources. The prospects for the achievement of a true information-everywhere environment are being made possible by the advancements recorded in the networked infometrical systems (NIMS). The NIMS have been boosted by the advancements recorded in networked embedded sensor and actuator technology which pave the way for contemporary ambient intelligence [32]. Designers of contemporary AmI can use the NIMS as a reference in designing their own AmI systems and applications.

Given the heterogeneous AmI design approaches, several design principles have been utilized in designing contextually relevant AmI services. The following are some of the key approaches:

1. The hallmark of AmI is that people and context are placed at the center of design principles whereas network and other technologies are placed in the background. Due to this realization, many designers of AmI architecture have made their designs based on service oriented architecture (SOA) and agent-based. The SOA and agent-based approaches enable the design of AmI services with the individual at the center of the design.
2. The interconnection of the cloud and sensors is mostly desired in adaptive dynamic Smart City environments which need continued access to heterogeneous information resources. These resources need to be aggregated and appropriately abstracted so that they may be accessed using identical semantics [33].
3. Based on the NIMS (networked infometrical systems) infrastructure, sensing self-awareness can be achieved by having the following: acoustic sensor, image

sensor, and mobile source sensing obstacle. The sensing self-awareness capability enables the management of spatiotemporal uncertainty.

4. It is expected that each of the components of the AmI Space need to have the capability of configuring and reconfiguring itself under both predictable and unpredictable conditions in heterogeneous dynamic environments. The adaptability characteristic is one of the cardinal components of the contemporary Smart City realization.

5. In developing requisite ambient intelligence solutions, ordinary agents and multi-agent systems (MAS) area cardinal for designing dynamic distributed open and interoperable systems.

Of the many approaches which define what stages to include in the design of AmI architecture, the following are some of the key layers desired regardless of the context in which AmI is deployed:

- Sensing and actuation layer—using the agent access application programming interface (API), sensing provides access type devices and actuation provides the necessary functionality to command actuator type devices;
- Device access layer—elements responsible for encapsulating concrete logic to access physical devices;
- Service layer—a layer enabling elements required to solve high-level tasks to be discoverable so that services can be accessed and used by different applications;
- Application layer—a host of elements that represent and implement given functionalities expected by users from the system.
- Context—defining and managing the different elements observed in order to model the current state of the environment. The context defines how the system architecture needs to be configured in order to conform to the state to the environment.
- Middleware—a common platform that brings the system abstraction as a common platform for sharing concepts, vocabulary and ensuring that the AmI system is reusable, scalable, interoperable, and easier to design. The middleware ensures that there are multi-platform compatibility and interoperability among the heterogeneous gadgets in the AmI environment, modularity (loose coupling to enable scalability), distributed operation of the system as a whole and that there is dynamic management of the different components. It is worth mentioning that the middleware needs to be a multi-agent-based that is able to divide a complex AmI system into decoupled components in a multi-agent declarative model.

Given the above, Fig. 1.4 shows the proposed architectural components desired of any AmI design aiming to be embedded onto the Smart City design. Figure 1.4 shows the key attributes which need to be considered in any context desiring ambience intelligence for ubiquitous information access. The proposed conceptual architectural configuration comprises the two main consoles (functional and nonfunctional requirements consoles) and three main layers defining the financial requirements (physical layer, transport layer, and application and middleware) .

In order to design appropriate AmI software and applications, the different possible states of human behaviors need to be mimicked and integrated into the topology and architectural designs.

Figure 1.4 shows multiple modules that need to be integrated in order to achieve true ambient intelligence in any dynamic information management environment. The logical, functional, and technical interoperability enables the different layers of the proposed architecture to sense information from the environment in which it is deployed, contextualized, processed and applied to a given content. Because of the lose coupling of the systems involved, all these processes are executed within the shortest period of time to enable ambient intelligence to positively impact on the quality of citizens.

The proposed architecture is hinged on the ability of the different nodes deployed in the environment to sense correct data and signals based on the capability of multi-faceted and multidimensional WSNs. The WSNs are in the physical layer (PL) which may have advanced technical capabilities to enable to communicate amongst themselves. Further, it is worth mentioning that the sensors are at the center of sensing and actualization which enables ultimate dynamic information to be captured from the environment timeously. The PL integrates the conceptualization of the Smart Cities which aim to ensure that relevant information is littered throughout the environment in order for people to make intelligent decisions. Importantly, the captured information from the environment is transferred to the application layer (AL) which houses three core modules: (1) the dynamic database which offers data abstraction from different highly dynamic information which change as occurrences take place in the environment; (2) the middleware console is the software or hardware abstraction that enable different mobile devices to communicate; and (3) the mobile agents (mobiles phones, PDAs, etc.) enabled by the middleware console. The transfer of information among the different mobile agents is made possible through the transport layer (TL) which has dedicated network and communication. Within the Smart City environment, when mobile agents traverse through different environments littered with WSNs, they are able to discover the services or information available (captured from the dynamic databases) using Services Discovery console. The modules articulated above mainly form the functional requirements of the proposed architecture.

The other part of the proposed architecture is the non-functional requirements console. This console has four modules: (1) collaboration support for mobile agents allowing them to easily discover services around their vicinity; (2) open interface enabling enhanced collaboration and interfacing of heterogeneous gadgets including WSNs ability to exchange information; (3) privacy and security ensuring that no mobile agent is disadvantaged in its quest to participate in the information ecosystem, etc.; and (4) usability requirements module.

The major limitation of this proposed framework is that it has not been used as a reference blueprint in any real-world project. Although the architecture is theoretically and technically sound, there is need for future works to bring out minute technical and operational details of each module to make it stand a chance to be easily adopted in different contextual frameworks.

1.7   Issues in Smart Network Design

Although there are clear and pronounced advancements in the sensor technology design and deployment which sit at the center of smart network design, it is clear that there are several intermittent issues or challenges that need to be addressed. Among these challenges, it can be opined that among the distributed sensor networks there are often calibration errors, as well as white noise in the communication channels culminating into signal degeneration or attenuation. Because of the desired mobility, mobile agents in information environment still find it difficult to correctly propagate signals scanned from the environment or face reduced ability of the sensors embedded onto the environment to correctly capture signals from the mobile agents. Because of the complexity in signal propagation between environmental sensors and mobile agents, there is usually a lot of signal attenuation or distortions from one entity to the other. These limitations have a negative impact on sensing self-awareness which is one of the key desired characteristics for ubiquitous information management [32]. Further, in many cases, one or more of the distributed networks may fail due to failing spatiotemporal sensing ability which in many cases, often than not, are not timely reported to the coordinating node for changing the routing logic tables. This may culminate into communication in the ubiquitous network lost and directly negatively impacting onto the overall feel and degree of the smartness in the city [32]. Given the scenarios above, contemporary adaptable information architectures may incorporate physical adaptation of the sensor network using a robotic mobility which may automatically detect changes in the network and simultaneously alter the network topology to accommodate the change.

Any competent infrastructure needs to be able to handle the different aspects of distributed computing such as:

- masking heterogeneity—various hardware components, communication protocols, operating systems, interfaces, etc. from different applications need to be integrated into the design;
- location or distribution transparency—the existence of an information resource need not necessarily be known with regards to its place of existence or state in which it is stored; etc.

The distributed resources need to be designed using open and standard interfaces which can be used in different instances of distributed computing. These open interfaces can easily be integrated into the ambient design model using standard middleware.

Other challenges needed to be addressed for mobile applications to achieve ambience include limited processing power, energy (battery power of the nodes) , state of the network (connectivity, bandwidth, etc.), user interface, etc.

1.8   Conclusion

Although much research has taken place on the AmI front, it is clear that a lot more needs to be done as AmI research and practice is still considered to be in its nascent stage, especially with respect to the application of AmI in Smart Cities scenrio. Although a lot of ground has been covered in advancing the AmI agenda towards being accepted as one of the core modules of Smart Cities and Smart Environments, a lot still needs to be done. Researchers and practitioners need to acknowledge the different inherent issues that must be resolved if integrating AmI in Smart Cities were to be achieved in any given contextual setting. The following are some of the issues that still need answers before AmI were to advance to greater application in different environments:

(a) Since AmI environment has a lot of sensor nodes and agents exchanging vast amounts of information at any given time, efficient information processing depends on the ability of the agents to simultaneously process information from multiple sources. This information is typically of greater multitudes in size, format, domain, and stature. This scenario entails that because of the generation of Big Data, as the AmI network grows, there will be challenges with regards to data and information storage, segmentation, and processing. The direct implication is that AmI nodes and agents need to be designed with capabilities to handle and manage Big Data. The processing challenge makes the AmI systems somewhat slow and less effective in as far as ubiquitous data and network needs are concerned. Future research needs to find robust mechanism on how to design agents and nodes so that they are able to handle heterogeneous Big Data.

(b) Another issue that is cardinal in realizing the effectiveness of AmI in Smart City environments is the ability of the agents and nodes to ubiquitously process information in a spatiotemporal domain. As the agents move, there are issues of signal propagation and attenuation that need to be addressed. Although significant research has been done in this domain, there is still a lot that needs to be done to achieve a true ambient intelligence environment for Smart Cities. Questions in this research domain would revolve around the understanding of the needed optimum signal sampling rate required for continued access to networks.

(c) Making node and agent designs that take into consideration the dynamic, distributed, and heterogeneous nature of the AmI especially taking into consideration that different agents and nodes have varying storage capacities, computational power, and are generally distributed into the Smart City environment. The different issues revolving around distributed transparency such as location, migration, access, network, etc., need to be considered in the design of the nodes and agents.

(d) Security is one of the key issues in any network. Ambient intelligence needs to be designed and deployed in such a way that any security aspects of the agents or nodes is not compromised.

In the AmI and Smart Cities research, there are a lot of blue oceans that need to be explored. For example, future AmI will study the lifestyle of individuals and learn using learning and optimization algorithms such as artificial neural networks (ANN) and support vector machines (SVMs) to understand the lifestyle preferences of human beings and adjust itself accordingly. It is further worth mentioning that implementing AmI in the future will involve clear definition and realization of the potential of each and every entity in the AmI Space that is in the middle tier, bridging the technological and socioeconomical challenges. This space may be ideally defined by a collection of infrastructure, technologies, applications, and a diverse range of services enabling AmI.

References

1. Augusto JC, Nakashima H, Aghajan H (2010) Ambient Intelligence and smart environments: state of the art. In: Nakashima H et al (eds) Handbook of ambient intelligence and smart environments. https://doi.org/10.1007/978-0-387-93808-0_1, © Springer Science+Business Media, LLC 2010. https://pdfs.semanticscholar.org/f585/8c4ddd20fcf3854508d67bcd8d12d44007e1.pdf
2. Resmini A, Rosati L (2011) Pervasive information architecture: designing cross-channel user experiences. Morgan Kaufmann Publishers, Massachusett
3. Ohlin F (2012) The role of information architecture in context-aware adaptive systems. J Inf Archit 4(1–2):28–37
4. Guesgen HW, Marsland S (2011) Recognising human behaviour in a spatio-temporal context. IGI-Global, USA. http://www.irma-international.org/viewtitle/54670/
5. Punie Y (2003) A social and technological view of Ambient Intelligence in everyday life: what bends the trend? Joint Research Centre (DG JRC), Institute for Prospective Technological Studies. http://www.jrc.es
6. Cook DJ, Augusto JC, Jakkula VR (2009) Ambient Intelligence: technologies, applications, and opportunities. Pervasive Mobile Comput 5(4):277–298
7. Böhlen M, Frei H (2010) Ambient Intelligence in the city overview and new perspectives. In: Nakashima H, Aghajan H, Augusto JC (eds) Handbook of ambient intelligence and smart environments. Springer, Boston
8. Ramos C (2007) Ambient Intelligence—a state of the art from artificial intelligence perspective. In: Proceedings of the 13th Portuguese conference artificial intelligence workshops, LNAI 4874. Springer, pp 285–295
9. Lekova A (2012) Exploiting mobile ad hoc networking and knowledge generation to achieve Ambient Intelligence. Appl Comput Intell Soft Comput 2012, Article ID 262936, 6 pages. https://doi.org/10.1155/2012/262936
10. Moawad A (2016) Towards ambient intelligent applications using models@run.time and machine learning for context-awareness. Unpublished PhD thesis, Université du Luxembourg. https://orbilu.uni.lu/bitstream/10993/23746/1/thesis.pdf
11. Lea R (2017) Smart Cities: an overview of technology trends driving smart cities. https://www.ieee.org/publications_standards/publications/periodicals/ieee-smart-cities-trend-paper-2017.pdf. Accessed 21 Sep 2017
12. Nfuka EN, Rusu L (2010) Critical success factors for effective IT governance in the public sector organization in a developing: the case of Tanzania. In: Proceeding of the 18th European conference on information systems (ECIS), Pretoria, South Africa, June 7–9
13. Asimakopoulou E, Bessis N (2011) Buildings and crowds: forming Smart Cities for more effective disaster management. In: 5th international conference on innovative mobile and internet services in the ubiquitous computing
14. Dlodlo N, Mbecke P, Mofolo M and Mhlanga M (2013), The internet of things in community safety and crime prevention in South Africa. In: International joint conference computers,

information and systems sciences and engineering CISSE 2013. University of Bridgeport, USA

15. Gil-Garcia JR, Pardo TA, Nam T (eds) (2016) Exploring the nature of the smart cities research landscape. In: Smarter as the new urban agenda a comprehensive view of the 21st century city. Springer International Publishing, Switzerland, pp 23–32

16. Paroutis S, Bennett M, Heracleous L (2014) A strategic view on smart city technology: the case of IBM smarter cities during a recession. Technol Forecast Soc Change Int J 89:262–272

17. Hall RE (2000) The vision of a smart city. In: Proceedings of the 2nd international life extension technology workshop. Paris, France

18. Harrison C, Eckman B, Hamilton R, Hatswick P, Kalagnanam J, Paraszczak J, Williams P (2014) Foundations of smarter cities. IBM J Res Dev 54(4):1–16

19. Singhal M, Shukla A (2012) Implementation of location based services in Android using GPS and Web services. IJCSI Int J Comput Sci Issues 9:237–242

20. Lobo P, Acharya S, D'Souza RO (2017) Quality of service for MANET based smart cities. Int J Adv Comput Eng Netw 5(2) ISSN: 2320-2106

21. Aloi G, Bedogni L, Di Felice M, Loscrì V, Molinaro A, Natalizio E, Pace P, Ruggeri G, Trotta A, Zema NR (2014) STEM-Net: an evolutionary network architecture for smart and sustainable cities. Trans Emerg Telecommun Technol 25:21–40

22. Chong N-Y, Mastrogiovanni F (eds) (2011) Handbook of research on Ambient Intelligence and smart environments: trends and perspective. IGI-Global, USA. https://www.igi-global.com/book/handbook-research-ambient-intelligence-smart/41775

23. Dey N, Ashour AS (2017) Ambient Intelligence in healthcare: a state-of-the-art. Glob J Comput Sci Technol 17(3):18–25

24. Kehinde A, Adesina A, Daniel E, Dele S (2012) Agent-based context-aware healthcare information retrieval using dropt approach. Int J Invent Res (IJIR) 5(2):109–118

25. Bosse T, Hoogendoorn M, Klein CA, Treur J (2008) A generic architecture for human-aware ambient computing, Atlantis Press Book −9.75in × 6.5in ABUCbook (Sept 30, 2008 13:18)

26. De Paola A (2011) A cognitive architecture for Ambient Intelligence. Unpublished PhD thesis. http://www.diid.unipa.it/networks/ndslab/pdf/phd/phD-thesis-depaola.pdf

27. Curran K (2015) Recent advances in Ambient Intelligence and context-aware computing. In: Curran K (ed) Advances in computational intelligence and robotics (ACIR) book series. IGI Global, USA

28. Bauer JS, Newman MW, Kientz JA (2014) Thinking about context: design practices for information architecture with context-aware systems. In: iConference 2014 proceedings, pp 398–411. https://doi.org/10.9776/14116

29. Paz-Lopez A, Varela G, Vazquez-Rodriguez S, Becerra J-A, Duro RJ (2010) Integrating Ambient Intelligence technologies using an architectural approach. http://www.gii.udc.es/img/gii/files/integrating_ambient_intelligence_technologies_using_an_architectural_approach.pdf

30. Newell A, Simon HA (1963) GPS, a program that simulates human thought. In: Feigenbaum E, Feldman J (eds) Computers and thought. McGraw-Hill, New York

31. Tapia TI, De Paz Y, Bajo J (2009) Ambient Intelligence based architecture for automated dynamic environments. In: Proceedings of the 10th international work-conference on artificial neural networks: Part 1: bio-inspired systems: computational and Ambient Intelligence, pp 171–180. https://bisite.usal.es/archivos/caepia_context_aware_v.7.pdf

32. Sukhatme GS, Villasenor J, Estrin D (2003) Networked infomechanical systems (NIMS) for Ambient Intelligence. UCLA Center for Embedded Networked Sensing, Technical Report 31, Dec 2003

33. Mitton N, Papavassiliou S, Puliafito A, Trivedi KS (2012) Combining cloud and sensors in a smart city environment. EURASIP J Wirel Commun Netw 247:1–10

Chapter 2
The State and Future of Ambient Intelligence in Industrial IoT Environments

Wesley Doorsamy and Babu Sena Paul

Abstract The advent of the Fourth Industrial Revolution has brought about the drive toward integrating systems and processes through the Internet of Things (IoT) in industrial environments. The major objective of introducing IoT into these environments is to realize dynamic optimization of productivity and efficiency and to mitigate the risks affecting financial loss and safety of personnel. Recent proliferation of sensors and smart embedded devices capable of communicating with each other offers industry the possibility of ubiquitous computing. Other distributed computing paradigms such as cloud computing are also helping to achieve the said objective. The next evolutionary step for ubiquitous computing in relation to industrial environments is the deployment of ambient intelligence (AmI) within the smart devices to bring about seamless integration of the personnel and environmental factors as well as equipment in the workplace. In this chapter, we look at how the application of AmI in the industrial sector has sought progress, and attempt to extricate the past and current challenges in terms to context, architecture, security, and uptake of relevant technologies. A bottom-up approach is taken in reviewing the key technological constituencies of AmI in Industrial IoT (IIoT) from the sensing technologies and communication to overall architectural developments and limitations. Some examples of future application scenarios of AmI in mining, manufacturing, and construction are also presented that offer high-level depictions of how the different aspects of AmI could potentially be brought together to benefit these industrial settings. Ultimately, this work aims to provide stakeholders with an understanding of the possibilities of AmI in IIoT environments through equipping them with knowledge of the state-of-the-art, technological limitations/barriers, and future developments encompassing this application area.

W. Doorsamy (✉)
Department of Electrical and Electronic Engineering, University of Johannesburg, Johannesburg, South Africa
e-mail: wdoorsamy@uj.ac.za

B. S. Paul
Institute for Intelligent Systems, University of Johannesburg, Johannesburg, South Africa

© Springer Nature Switzerland AG 2019
Z. Mahmood (ed.), *Guide to Ambient Intelligence in the IoT Environment*, Computer Communications and Networks, https://doi.org/10.1007/978-3-030-04173-1_2

Keywords Industrial IoT (IIoT) · Internet of Things (IoT)
Ambient Intelligence (AmI) · Context awareness · Architecture · Mining
Manufacturing · Construction

2.1 Introduction

The Internet of Things (IoT) describes a network of smart devices that interact with each other, with relevant compute-related services and with users of devices and such services [1]. The IoT vision is highly successful in many application areas including health care, transportation, homes, etc. The Industrial IoT (IIoT) extends this vision to industrial environments [2] such as manufacturing, mining, engineering, construction, etc. Ambient intelligence (AmI) further extends the support to the activities of human beings in IoT and IIoT environments, by embedded intelligence in smart devices (such as mobile phones, sensors, actuators, etc.) connected via the IoT. More specifically, AmI incorporates the assessment and automatic manipulation of the surrounding environment to provide effective support to the human personnel and businesses where the environment refers to the following [3]:

- Physical environment such as an underground mine, manufacturing shop floor, office space, etc.
- Contextual environments such as the business functions, processes, system flows, etc.

The range of inputs, outputs, and related technologies coupled with the environment, humans, processes, and AmI systems (in the context of an industrial IoT) are presented in more detail in [3].

The AmI paradigm in IoT environments is continually evolving with the rapid extension of industrial applications, as stakeholders aim to align their businesses to the vision of today's Fourth Industrial Revolution. Industry players have realized that the AmI paradigm provides a new avenue of technological innovation whereby objects and people can interact with each other within the environment as part of an industrial system or process.

The mining sector has been at the forefront of industrial applications of AmI mainly to ensure the safety and comfort of personnel. AmI applied in underground mines began more than a decade ago with remote sensing of toxic gases [4]. Despite the early developments in AmI and IIoT, there is still much work to be done in order to advance existing technologies and the AmI paradigm to other safety-related purposes such as position tracking of personnel, and production-related measures such as determination of personnel performance.

Support and features offered by AmI are also constantly evolving with the ongoing advancement of its technological constituencies such as sensors, communication protocols, and machine learning algorithms. Context or situational awareness is an exciting feature of the evolving AmI paradigm. There is a need for this feature in industrial applications as it offers the possibility of automatically tracking of and providing responses to dynamic variables affecting production and safety in the work environment. However, one of the ongoing challenges of successfully deploying such features is the effective collection and processing of data and knowledge management for correct inferences of the context or current situation of a particular process. This feature of AmI is especially relevant to the manufacturing industry, where it has enormous potential to assist with flexible and reconfigurable manufacturing processes. Over the past two decades, the application of intelligent systems in the construction sector has also rapidly expanded [5].

This chapter is organized as follows. First, we look at sensing technologies and data fusion relating to AmI in the IIoT. We also introduce some current and future AmI-IIoT algorithms and computational intelligence techniques that provide contextual and situational awareness, as well as adaptive and anticipatory functions. The next section discusses AmI architecture focusing on the major challenge of harmoniously integrating people, processes, equipment, and environment. We then discuss future application scenarios of AmI in the mining, manufacturing, and construction sectors. Finally, the key points regarding the current state and future direction of AmI in the IIoT environment are discussed before presenting a brief conclusion.

2.2 Sensing Technologies and Data Fusion in Industrial AmI

Sensors and sensor networks in industrial applications form the foundation on which the IoT paradigm can be built. The main requirement of sensor deployment is the extensive level of cabling for power sourcing and data communication, however, luckily, recent advancements in sensor technologies and connectivity are enabling low-powered, ubiquitous, real-time wireless sensing possible even in harsh industrial environments [6], such as mining.

Industrial sensor networks are now primarily moving toward wireless modes of operation due to the inherent flexibility and accessibility. However, major challenges with wireless sensor networks include their susceptibility to transmission failures, time delays, and network breaches in case of harsh environments where reliability, timeliness, and security are mandatory [7]. There are ongoing research and development in the area of wireless sensor networks to mitigate the aforementioned drawbacks with particular progress in improving transmission reliability in environments with high noise and complex electromagnetic interference as discussed in [7]. Another major challenge about wireless sensor networks is the integration of larger heterogeneous networks. Recent developments to address this major challenge include Software-Defined Networks (SDN) and improvements thereof, as presented in [8].

Radio-Frequency IDentification (RFID) technology is also becoming more popular in relation to smart wireless sensor platforms with the development of sensor-integrated antenna, chip-free antenna backscatter, and nanotechnology-enabled sensors [6].

One of the key characteristics of AmI is social interaction to provide security and ambient personalization. This social interaction is achieved through a variety of means including biometric sensors. Such sensors, often required in practice, are dependent on the categories of biometric characteristics to be measured: whether static or dynamic. Static biometrics require identification and recognition of fingerprints, face and iris features etc.; while dynamic biometrics are based on behavioral characteristics e.g. voice [9].

In the era of the fourth industrial revolution, the idea of a network of sensory smart devices is an attractive one and gaining much attention. Artificial intelligence and machine learning techniques are also making visual and audio recording devices more powerful tools for capturing data, processing and extracting business insights. Practical examples include visual-based sensing being utilized for industrial analytics, thermographic imaging for machine condition monitoring [10, 11], and human-assisted production equipment monitoring [12, 13]. Some examples of recent sensor developments include surface acoustic wave sensors for temperature, pressure, strain and torque measurements [14], and polyurethane foam with conductive carbon ink coating for vibration monitoring [15]. There has also been an increase in "software-based sensors" in the industrial environment to predict key operating variables and monitor dynamic behaviors of smart systems [16, 17].

In typical AmI applications in the IIoT environment, numerous parameters need to be monitored in order to make correct inferences for performance improvement and decision-making, however, most sensing devices only provide a measurement of a single parameter per sensor. Therefore, multi-sensor information is required for more effective analytics. Data fusion is the answer. This takes place when raw data from various sensors is collated and combined or preprocessed prior to the collation of the gathered data. The required level of data fusion is dependent on the parameters being monitored and the types of sensors used, as the sampling rates and sensor outputs often vary considerably in format and quality. Data fusion is a critical and intensive process because the required data (e.g., observational data) may not be sourced only from given sensors. Contextual, open source, social media, and ontological data may also be collated and combined with observational information to make the necessary inferences depending on the given application [18].

2.3 Industrial KB Systems and Computational Intelligence

In the past three decades, we have seen the amalgamation and coming together of subject areas such as knowledge-based (KB) systems, computational intelligence, and machine learning for the development of intelligent systems. KB systems in the industry have made a considerable progress and have now evolved to more sophisticated systems. Earlier experiences noted that the development of intelligent systems requires the processing of knowledge. Nowadays, however, modern computational intelligence and machine learning make it possible to develop intelligent systems that only require initial developmental raw data inputs from experienced and skilled workers and then continuously learn automatically to become more accurate and reliable. AmI is strongly associated with these different subject areas and has progressed accordingly. The future of AmI in industrial environments is thus intertwined with the algorithms and techniques offered by modern AI and machine learning.

2.3.1   Fit-for-Purpose Algorithms

The real-world environment in which AmI is deployed is characterized by uncertainty [19]. This is a critical factor in the IIoT vision within which future AmI systems will reside. The intelligence layer of an AmI-based infrastructure provides the decision support in the presence of this uncertainty. The characteristics of future AmI in an industrial environment, as postulated in [20] are as follows:

- Seamless multimodal interaction with human operators;
- Extensive process, human and environmental knowledge;
- Transparency of support and interaction;
- Implicit nature of actions.

In general, computational intelligence refers to a set of computing systems that have the ability to learn and deal with new instances of situations, and typically have one or more reasoning and intelligence attributes [21]. This area is rapidly evolving from the traditional approaches such as swarm optimization, genetic algorithms and simulated annealing to more novel approaches such as bacterial foraging optimization, bees algorithm, glowworm swarm optimization and more [22].

Although there has been extensive progress in advanced computer vision techniques and low-cost sensors, a primary challenge still exists with regards to the processing of extracted features in real-world scenarios (i.e., environmental parameters, dynamic actions of human beings and smart machinery) in a noise tolerant and robust manner [23]. While many current algorithms rely on thresholds to distinguish anomalous instances from regular ones, learning-based paradigms (such as machine learning and neural networks) offer more in terms of knowledge generalization through a set of known training instances and subsequent classification of unknown instances. For instance, unsupervised learning of neural networks offers the possibility of detecting abnormal events by distinguishing between a set of normal known behaviors from those not conforming to this behavior. This is typically carried out by tracking outliers that do not follow the patterns observed with the majority of known instances.

AmI in industry provides not just assistance and guidance in the workplace through wearable sensors and other relevant smart devices, but also helps with better understanding of worker stress, workload issues, information flows, and bottlenecks with the aim of improving worker health, better working conditions, satisfaction, and productivity [24].

Resource management is a basic requirement in industrial operations and there are numerous studies on different techniques on how to best carry out this significant task. AmI is potentially a key enabler for total resource management in industrial operations as it can provide a platform with which to integrate usage of resources, and monitoring of the implementation of operational strategies by evaluating performance indexes [25]. Resources in an industrial environment have remained, previously, unintegrated into management systems. However, these are now becoming more integrated and accessible through AmI either directly or indirectly. For instance, continuous tracking of productivity of personnel in an underground mine was previously difficult, however, this can now be achieved directly through wearable sensors or by inferencing through reasoning from data obtained from a network of sensors in the AmI environment. This extends the process of how the efficiency of an industrial process can be monitored from output-only based analysis to continuous tracking of the entire process stream. This, in turn, enables a better understanding of the management of resources along that stream.

In this context, one of the major challenges relates to upscaling of the AmI technologies to include more interactions from even more devices [26]. There is much development taking place in enabling human–computer interaction in a more natural way such as gesture recognition, gaze tracking, facial-expression recognition, and spoken dialogue understanding. This is particularly important in an industrial environment where AmI is, or will be, concerned about groups of people carrying out a variety tasks in a dynamic way.

2.3.2 Context Awareness

AmI research and development have the considerable task of addressing and overcoming challenges relating to human-inspired applications that require context and situational awareness [27]. This is particularly crucial in the industry domain where there is a need for the seamless assemblage of complex human behavior, human interaction, constrained machine functions and processes, and dynamic environmental parameters.

Context awareness is a required feature of AmI in all scenarios including industrial applications. In [28], the contextual status of crop growth is inferred through ontology via a series of measurements including humidity, temperature, and pH in an intrafactory cultivation environment.

AmI is truly set apart from traditional assistive working technologies with its cognitive inference capabilities and context awareness that links the hardware and software systems to the environment to operate in the best interests of the humans [27], and in the case of industrial AmI, the businesses.

2.3.3 Adaptive and Anticipatory Functions

An area of high interest, and perhaps most significant in AmI in the IIoT environment, is intelligent assistance to workers performing manual tasks. Research and development presented in [29] is an example of the ongoing progress in this area where an intelligent system was designed to assist workers in a stationary manual assembly. The system uses augmented reality together with hand tracking to make the user aware of incorrect picking actions in the assembly process. This application of projection-based reality and hand tracking is not only a progressive development in manual assembly but presents an opportunity for future design of assistive systems in many other areas, e.g., manufacturing, mining, construction, and many more where manual human work is still prominent. For example, a crane operator in construction [30], or manufacturing environment, or equipment operators in mining [31], could be assisted using similar approach where they are assisted or guided, through augmented reality and hand/movement tracking, by being made aware of incorrect or potentially incorrect actions.

The main challenge encumbering the use of augmented reality and action-tracking based assistive systems in the industry is the lack of skills in usage and development. Another challenge is the lack of supporting infrastructure, e.g., the required bandwidth and the speed of transmission for streaming augmented reality video recordings.

2.4 AmI Architecture and IoT in Industrial Environments

Integrating AmI into an existing industrial environment is probably the biggest challenge. The main ingredients such as sensing, communications, and intelligent algorithms are available, but bringing all these together in a harmonious manner is still not completely achievable. A key question to consider with integrating AmI in an industrial environment refers to the fact that if AmI is people-centered then where exactly the people are considered within the architecture.

Traditionally, the development of architectures normally takes either the bottom-up or top-down approach, as follows:

- A top-down approach first considers the desired high-level attributes of the overall system and then develops downwardly in accordance with the high-level framework decisions. This approach is more applicable to a business or industry where processes and systems are still in development.
- A bottom-up approach is typically taken for a business or industry where the processes and systems are already established and technological "retrofitting" of AmI is being considered.

Both these approaches to architecture development are systematically rigid and therefore pose flexibility, scalability and integrability drawbacks.

Ultimately, AmI should be interpreted as a system of cooperative intelligence that necessarily requires more careful planning. The architecture of this system critically affects the system's overall capability, performance, flexibility, scalability, and reliability.

Cooperative intelligence in architectural designs can take on many forms but it is important that it includes viewpoint specifications that consider associations within the system from different viewpoints [32]. In this way, particular concerns of different stakeholders can be addressed as there will be multiple interrelated layers of industrial AmI architecture. As an example, the AmI architecture in a manufacturing plant should include viewpoints from a functional, logistic, communications, and even enterprise perspectives, as it will potentially interact with varying aspects of the business through different process streams.

Figure 2.1 provides a suggested high-level framework for embedding AmI into an industrial business architecture. This generalized framework offers some key design considerations. The proposed architectural design seeks to obtain cooperative intelligence between AmI and business strategy. The reasoning behind the approach is that AmI can feed the same low-level information into multiple areas of the business but for different inferences. For example, tracking of an autonomous vehicle across the shop floor in a manufacturing plant can provide information to supply chain, operations, and maintenance for inferring statuses about material whereabouts, overall production efficiency, and equipment health, respectively. This design also enables flexibility and scalability at the people–equipment–environment level without modification to the cooperative intelligence layer. Flexibility and scalability of the system architecture to include more or different equipment, more personnel and different environments/process is critical in the long term.

2.5   Industrial Applications of AmI

In this section, we present a number of application scenarios, from the mining, manufacturing and construction areas, relating to the Industrial Internet of Things vision.

2.5.1   Smart Mining

The mining sector is continuing to advance its use of AmI, particularly for safety purposes in the mining scenario. Research presented in [33] is an example of how wireless sensor networks can be used to give an accurate measurement of ambient conditions such as dust, gas, humidity, temperature, and air flow in order to determine potential hazards in the working environment. These types of sensor networks are becoming more prevalent in mining environment and are now being expanded to include the sensors worn by the mining personnel and sensors fixed to mobile equipment. In the future, these sensor networks will feature more in industrial applications with the progressive development of technologies such as smart skins [6]. Figure 2.2

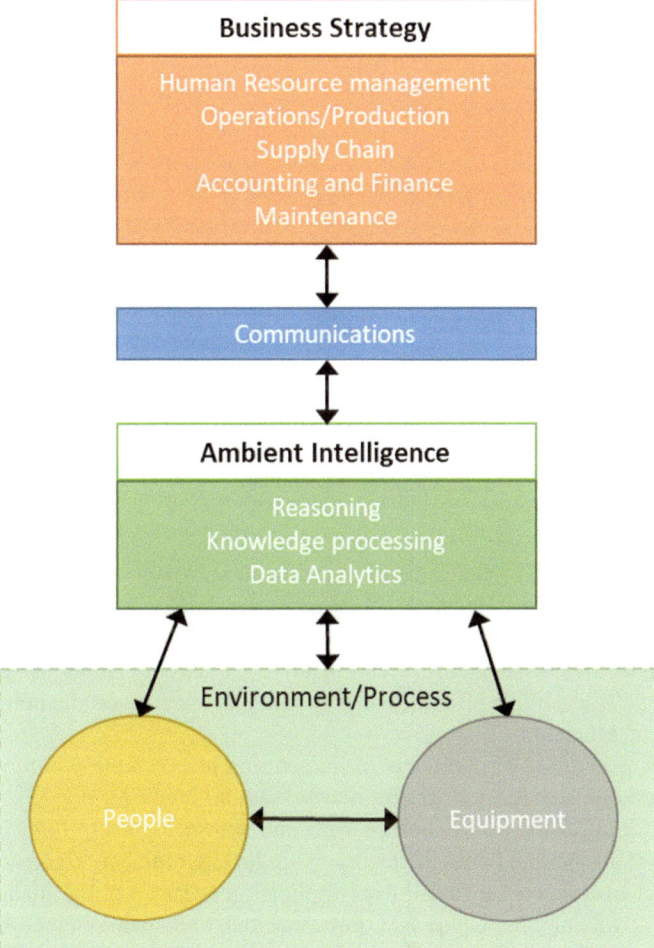

Fig. 2.1 AmI as part of industrial business architecture

presents an example of a future AmI in a smart mining environment whereby the human operator, environment, and machine all seamlessly interact for optimized mining process.

2.5.2   Intelligent Manufacturing

Smart or intelligent manufacturing is a broad concept that aims to optimize production through the advanced use of information and technology [34]. This optimization typically refers to the integration of the entire life cycle of a product including

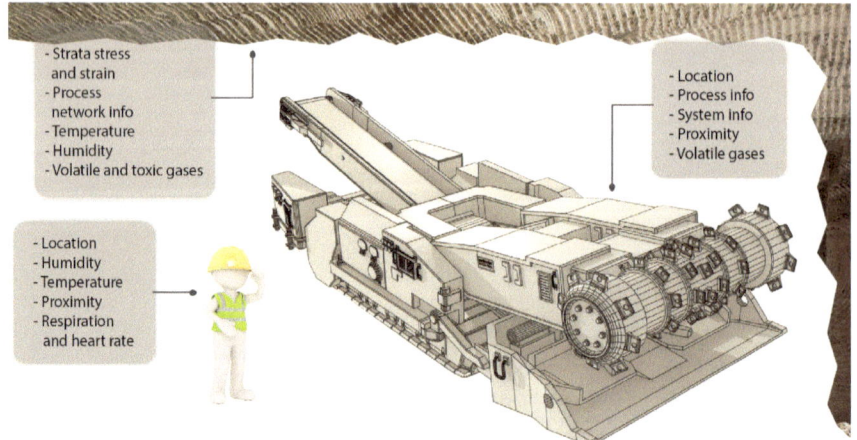

Fig. 2.2 Future application scenario of AmI in smart mining operations

design, production, and management. In this context, AmI is to play a crucial role in smart manufacturing in the form of various smart sensors and intelligent devices together with advanced reasoning systems. The collection and analysis of real-time data pertaining to raw materials, processing, logistics, and staff productivity enhances visibility and traceability offering the possibility of making production decision-making more efficient.

The new generation of intelligent manufacturing has seen the evolution of three specific paradigms, viz,: intelligence, networking, and digitization [35]. These are exactly the aspects served by modern AmI and are analogous to human–cyber—physical systems coupled with interaction with the environment. Figure 2.3 depicts a high-level overview of a future AmI application scenario in the manufacturing environment whereby the equipment, human actors, and environmental factors are monitored and supervised through AmI, in a manner such that the overall process is optimized and made more effective.

2.5.3 Smart Construction

The construction industry has one of the highest accident rates and safety risks globally, as compared to many other industries [36]. Therefore, real-time safety management is essential in these work environments to avoid worker injury or fatality and damage to equipment. Recently, there have been more efforts toward establishing intelligent and automated supervisory/management of activities on construction sites using IoT [37].

Real-time productivity monitoring at construction sites is desirable because processes are relatively complex, resources are dynamic, and deadlines are critical. AmI

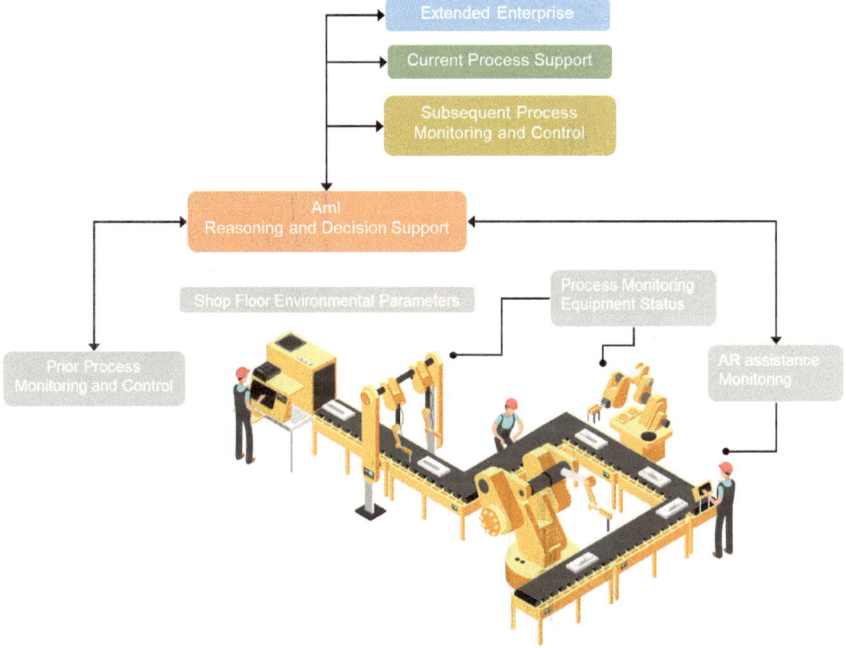

Fig. 2.3 Future application scenario of AmI in intelligent manufacturing environment

has the potential to serve both safety and production monitoring and management functions in the construction industry.

Figure 2.4 depicts a future application scenario of AmI in the construction industry whereby the position and movement of onsite workers and mobile equipment are monitored, including environmental conditions such as temperature and air pollution (e.g., dust). Augmented reality is also used to inspect the site, and track and compare the operational progress.

2.6   Roadmap for AmI in IIoT Environments

The AmI-IIoT paradigm is generally more widely accepted by practitioners of technology and the industrial sector as it is considered as a key enabler and driver of technology-driven business transformation. However, an ongoing challenge to eliciting stakeholder trust is that the AmI-IoT paradigm does pose risks to privacy and security, and also exhibits vulnerabilities in other cases [38]. In summary, the main challenges to widespread implementation and growth of AmI in IIoT are the following:

- Lack of skills and expertise in the related technologies;
- Inadequate existing infrastructure to support the AmI technologies;

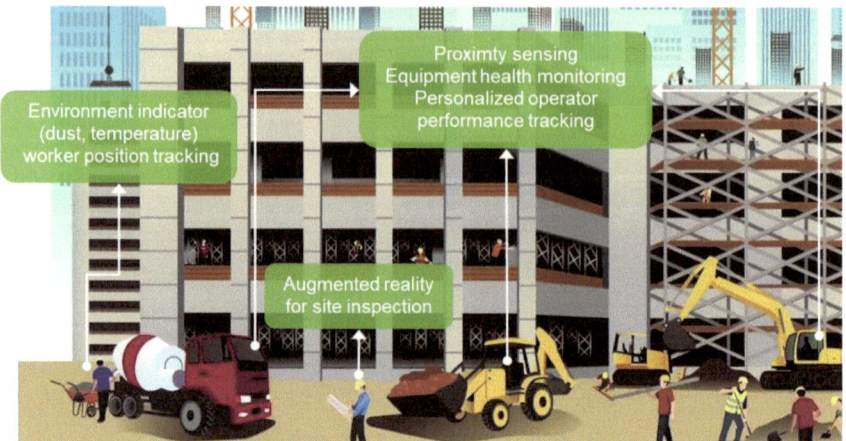

Environment indicator
(dust, temperature)
worker position tracking

Proximty sensing
Equipment health monitoring
Personalized operator
performance tracking

Augmented reality
for site inspection

Fig. 2.4 Future application scenario of AmI in smart construction environment

- Disruption of typical workflow;
- Lack of legal and regulatory framework.

In dealing with the challenge of lack of skills and expertise, industry-based consumers of AmI should seek to develop in-house skills and expertise in the related technologies. This will assist with training and upskilling to be done within the business, but also enable maintenance and further development of the AmI system to be carried out by trained personnel. A benefit of AmI is that it can be developed incrementally. This, in turn, will allow for required support infrastructure to be acquired over a period of time. However, it should be noted that a long-term developmental strategy for the envisaged AmI system should be planned prior to this. In general, integrating technology into existing workflows is a challenge that can only be resolved over a period of time together with the development of necessary training and awareness campaigns. Development of legal and regulatory frameworks for AmI-associated technologies, and use thereof, are the responsibility of all stakeholders and therefore industry, academia, regulatory bodies, and governments all have a role to play. This is also typically an ongoing process and, over the past decade, there have been contributions from working groups, forums, standards organizations and communities of practitioners regarding AmI policy and regulatory frameworks.

The concept of AmI has become more prevalent in real-world scenarios, where focus is much more on ambient-assisted working, which is concerned with advanced AmI infrastructure in the workplace [39]. Industrial work environments, in particular, have a relatively higher degree of process control and automation, with respect to tracking, monitoring, and computer-assisted processes; however, the next step in industrial IoT is true integration of AmI systems with the humans. In an industrial environment, it is essential to provide users with trustworthy automated services.

This is an emerging topic in AmI and will form a crucial research and development area in the coming years, particularly for industrial-based environments. Advanced studies will be required in terms of providing trustworthy and reliable systems, such as those presented in [40] where an ontological approach was used to design trust semantics for ambient services. Similar studies should also be conducted in the industrial environments as this will also assist with developing the aforementioned policy and regulatory frameworks.

2.7 Conclusion

In the recent past, the AmI paradigm has been more prevalent in living and social scenarios such as homes, smart workspaces, and healthcare applications. Nowadays, AmI is seen to be the next technological step in the evolution of IIoT environments as it provides a platform for truly integrated human workers, work equipment, and work environments through state-of-the-art sensing, communications, and intelligent algorithms and technologies. In the IIoT environment, AmI advances the concept of ambient-assisted working and offers the possibility of mitigating safety risks to people and equipment, as well as monitoring and managing productivity. Some of the industrial sectors that have already begun exploiting the benefits of AmI are mining, manufacturing, and construction. A few of the main features of AmI in these industries are the following:

- monitoring of environmental parameters such as air pollution, humidity, temperature;
- measurement of personnel presence and their movement; and
- monitoring of performance and health of equipment, vehicles, smart devices, and machinery, etc.

It is expected that more businesses within these sectors and other industrial sectors will follow suit as they strive to become more competitive in the current Fourth Industrial Revolution.

AmI is an amalgamation of distributed computing technologies and smart devices, and although there has been much progress recently with sensor technologies, machine learning, artificial intelligence, and intelligent algorithms. One of the major challenges is the integration of the advanced technologies into work environment. The approach to developing the AmI architecture is key to ensuring proper integrability, scalability and flexibility. However, AmI architecture in the IIoT environment has not received sufficient attention to date and is an area that is expected to attract more research and development in future. Another difficulty with integration and the stumbling block to uptake of AmI is lack of user trust. Trusting AmI in the workplace requires worker confidence in the underlying technologies and this can be improved through adequate training of personnel and demonstration of the technology's benefits. Ultimately, the true benefits of AmI in the IIoT will only be realized when it is molded into the workplace environment and business.

References

1. Mukhopadhyay SC, Suryadevara NK (2014). Internet of things: challenges and opportunities. In: Internet of things. Springer, Cham, pp 1–17
2. Jeschke S, Brecher C, Meisen T, Özdemir D, Eschert T (2017) Industrial internet of things and cyber manufacturing systems. In: Industrial internet of things. Springer, Cham, pp 3–19
3. Robinson DC, Sanders DA, Mazharsolook E (2015) Ambient intelligence for optimal manufacturing and energy efficiency. Assem Autom 35(3):234–248
4. Sammarco JJ, Paddock R, Fries EF, Karra VK (2007) A technology review of smart sensors with wireless networks for applications in hazardous work environments. Information Circular 9496
5. Irani Z, Kamal MM (2014) Intelligent systems research in the construction industry. Expert Syst Appl 41(4):934–950
6. Cook BS, Le T, Palacios S, Traille A, Tentzeris MM (2013) Only skin deep: Inkjet-printed zero-power sensors for large-scale RFID-integrated smart skins. IEEE Microw Mag 14(3):103–114
7. Liu L, Han G, Chan S, Guizani M (2018) An SNR-assured anti-jamming routing protocol for reliable communication in industrial wireless sensor networks. IEEE Commun Mag 56(2):23–29
8. Silva I, Guedes LA, Portugal P, Vasques F (2012) Reliability and availability evaluation of wireless sensor networks for industrial applications. Sensors 12(1):806–838
9. Henniger O, Damer N, Braun A (2017) Opportunities for biometric technologies in smart environments. In: European conference on ambient intelligence. Springer, Cham, pp 175–182
10. Wang YWA, Hazel T, Hjornevik R, Fjeld O (2017) Equipment monitoring using thermography cameras: remote detection of temperature-related failures. IEEE Ind Appl Mag 23(4):35–44
11. Narain Singh A, Doorsamy W, Cronje WA (2018) Thermal instability analysis of a synchronous generator rotor using direct mapping. SAIEE Afr Res J 109(1):4–14
12. Shimada S, Tanaka H, Hasebe K, Hayashi N, Ochi Y, Matsui T, Nishizaki I, Matsumoto Y, Tanaka Y, Nakamura H, Mizuno Y (2016) Ultrasonic welding of polymer optical fibres onto composite materials. Electron Lett 52(17):1472–1474
13. Wu W, Zheng Y, Chen K, Wang X, Cao N (2018) A visual analytics approach for equipment condition monitoring in smart factories of process industry. In: Pacific visualization symposium (PacificVis). IEEE, pp 140–149
14. Zhgoon SA, Shvetsov AS, Sakharov SA, Elmazria O (2018) High-temperature SAW resonator sensors: electrode design specifics. IEEE Trans Ultrason Ferroelectr Freq Control 65(4):657–664
15. Ajith R, Tewari A, Gupta D, Tallur S (2017) Low-cost vibration sensor for condition-based monitoring manufactured from polyurethane foam. IEEE Sens Lett 1(6):1–4
16. Yuan X, Wang Y, Yang C, Ge Z, Song Z, Gui W (2018) Weighted linear dynamic system for feature representation and soft sensor application in nonlinear dynamic industrial processes. IEEE Trans Ind Electron 65(2):1508–1517
17. Bidar B, Shahraki F, Sadeghi J, Khalilipour MM (2018) Soft sensor modeling based on multi-state-dependent parameter models and application for quality monitoring in industrial sulfur recovery process. IEEE Sens J 18(11):4583–4591
18. Fourati H (2015) Multisensor data fusion: from algorithms and architectural design to applications. CRC Press
19. Ramos C (2007) Ambient intelligence–a state of the art from artificial intelligence perspective. In: Portuguese conference on artificial intelligence. Springer, Heidelberg, pp 285–295
20. Stokic D, Kirchhoff U, Sundmaeker H (2006) Ambient intelligence in manufacturing industry: control system point of view. In: The eighth IASTED international conference on control and applications, pp 24–26
21. Marwala T, Lagazio M (2011) Militarized conflict modeling using computational intelligence. Springer Science & Business Media, London
22. Xing B, Gao WJ (2016) Innovative computational intelligence: a rough guide to clever algorithms. Springer, Dordrecht, pp 105–121

23. Ravulakollu KK, Khan MA, Abraham A (2016) Trends in ambient intelligent systems. Springer, Cham
24. Farringdon J, Nashold S (2005) Continuous body monitoring. In: Ambient intelligence for scientific discovery. Springer, Heidelberg, pp 202–223
25. Chien CF, Hong TY, Guo HZ (2017) A conceptual framework for "Industry 3.5" to empower intelligent manufacturing and case studies. Proc Manuf 11:2009–2017
26. Augusto JC (2009) Past, present and future of ambient intelligence and smart environments. In: International conference on agents and artificial intelligence. Springer, Heidelberg, pp 3–15
27. Bibri SE (2015) The human face of ambient intelligence. Atlantis Press
28. Hwang J, Jeong H, Yoe H (2014) Design and implementation of the intelligent plant factory system based on ubiquitous computing. In: Ambient intelligence-software and applications. Springer, Cham, pp 89–97
29. Wessling F, Gries S, Ollesch J, Hesenius M, Gruhn V (2017) Engineering a cyber-physical inter-section management—an experience report. In: European conference on ambient intelligence. Springer, Cham, pp 17–32
30. Chen YC, Chi HL, Kang SC, Hsieh SH (2011) A smart crane operations assistance system using augmented reality technology. In: Proceedings of 28th international symposium on automation and robotics in construction, Seoul, Korea, pp 643–649
31. Bassan J, Srinivasan V, Tang A (2011) The augmented mine worker—applications of aug-mented reality in mining. In: Proceeding second international future mining conference
32. Sun L, Li Y, Gao J (2016) Architecture and application research of cooperative intelligent transport systems. Proc Eng 137:747–753
33. Henriques V, Malekian R (2016) Mine safety system using wireless sensor network. IEEE Access 4:3511–3521
34. Zhong RY, Xu X, Klotz E, Newman ST (2017) Intelligent manufacturing in the context of industry 4.0: a review. Engineering 3(5):616–630
35. Zhou J, Li P, Zhou Y, Wang B, Zang J, Meng L (2018) Toward new-generation intelligent manufacturing. Engineering 4(1):11–20
36. Li RYM (2017) Smart construction safety in road repairing works. Proc Comput Sci 111:301–307
37. Li RYM (2018) An economic analysis on automated construction safety. Springer, Singapore
38. Bibri SE (2015) The shaping of ambient intelligence and the internet of things: historico-epistemic, socio-cultural, politico-institutional and eco-environmental dimensions, vol 10. Springer
39. Bühler C (2009) Ambient intelligence in working environments. In: International conference on universal access in human-computer interaction. Springer, Heidelberg, pp 143–149
40. Lee OJ, Nguyen HL, Jung JE, Um TW, Lee HW (2017) Towards ontological approach on trust-aware ambient services. IEEE Access 5:1589–1599

Chapter 3
Ambient Intelligence in Business Environments and Internet of Things Transformation Guidelines

Kadir Alpaslan Demir, Bugra Turan, Tolga Onel, Tufan Ekin and Seda Demir

Abstract Ambient intelligence (AmI) is an emerging paradigm bringing intelligence into our lives with the help of intelligent interfaces and smart environments. AmI has the potential to affect our business environments significantly. With the help of AmI, we can find better ways to serve our customers and increase productivity. Internet of things (IoT) is a key enabling technology that provides the necessary infrastructure for ambient intelligence. In addition, ambient intelligence paradigm enhances the use and capabilities of IoT devices. As a result, businesses those want to benefit from this new paradigm and the relevant technologies need to build the necessary IoT infrastructure. In this study, our goal is to help the business and technical managers by developing an AmI enhanced business vision and managing an effective IoT transformation. In this chapter, we discuss an existing implementation of ambient intelligence in the business environment. Furthermore, we envision various future uses of AmI in business environments. We also present issues related to IoT technology transformations. In addition, we provide a set of guidelines, strategies, and best practices for business and IT managers for a successful IoT transformation leading to an ambient intelligence enhanced business environment. We divide the transformation issues into three categories: management issues, technical issues, and social issues. These issues are discussed in detail.

Keywords Ambient intelligence · Smart business environments
Internet of Things (IoT) · Business processes · Enterprise information systems
Management of information systems

K. A. Demir (✉) · T. Ekin
Department of Software Development, Turkish Naval Research Center Command, 34890 Istanbul, Turkey
e-mail: kadiralpaslandemir@gmail.com

B. Turan
Department of Electrical and Electronics Engineering, Koc University, 34450 Istanbul, Turkey

T. Onel
Department of Computer Engineering, Turkish Naval Academy, 34942 Istanbul, Turkey

S. Demir
Institute of Social Sciences, Gebze Technical University, 41400 Kocaeli, Turkey

© Springer Nature Switzerland AG 2019
Z. Mahmood (ed.), *Guide to Ambient Intelligence in the IoT Environment*, Computer Communications and Networks, https://doi.org/10.1007/978-3-030-04173-1_3

3.1 Introduction

Technology is advancing at an enormous pace. Not a single day passes by without the introduction of a new concept, theory, or paradigm. Many applications, services, and technologies are built based on these advancements. These changes affect how we live and how we conduct business. For many organizations and companies, being able to utilize new technologies quickly is a core competitive advantage. Businesses unable to keep up with these changes are in danger of becoming obsolete. According to many, we are on the brink of a new industrial revolution [1, 2]. This new revolution is called Industry 4.0. The goal of the fourth industrial revolution is similar to the goals of previous industrial revolutions. Simply, the goal is achieving mass production with the help of new technologies. The motto of Industry 4.0 is *Smart Manufacturing for the Future*. This motto is promoted by the Germany Trade and Invest (GTAI) [3]. Smart production with the help of recent technological advances is at the core of Industry 4.0 [4]. Naturally, various trend technologies enable this new industrial revolution. Internet of things (IoT) , cloud computing, big data, robotics, and artificial intelligence (AI) are at the core of these trend technologies [1, 5]. However, it should be noted that these technologies only provide the necessary infrastructure for achieving our business goals. In addition to technologies, we also need paradigms to help us in realizing a business vision. Ambient intelligence (AmI) is one of the promising paradigms with its potential to significantly affect how we live and how we operate businesses. AmI provides smart environments that can interact with its users in a natural and intelligent way. For example, customers can walk into a grocery store, take whatever they want, and walk out without paying to a cashier or waiting in a checkout line. The store's smart environment identifies the customer, senses what the customer purchases, and charges the customer's account. This is a good example of an environment enhanced with ambient intelligence.

In this study, we discuss an existing implementation of ambient intelligence in the business environment. Furthermore, we envision various future uses of ambient intelligence in business environments. We investigated and identified issues related to Internet of things technology transformations. We briefly discuss each issue and provide a guideline or a best practice. We divide the IoT technology transformation issues into three categories: management issues, technical issues, and social issues. The management issues consist of business vision, business dynamics, management support, stakeholder management, governance, strategic partnering, teaming, device management, and e-waste and recycling management. The technical issues comprise the technical solution, networking, energy, interoperability, information security, and data analytics. In the social issues category, there are three main issues, viz., user acceptance, privacy and ethics, education and training. Next, we summarize these guidelines. Finally, we conclude the chapter.

3.2 Ambient Intelligence in the Business Environment

There are many application areas for ambient intelligence. Health care, elderly care, security and safety in the workplace, smart homes, smart cities, smart shopping, and smart offices are just some of the application areas [6, 7]. In this section, we focus on the use of AmI in business environments. First, we provide an introduction to ambient intelligence. Then, we discuss how to improve a business with ambient intelligence paradigm. We further provide an example of an actual AmI implementation. Finally, we discuss possible future uses.

3.2.1 Ambient Intelligence

Ambient Intelligence (AmI) refers to *electronic or digital environments that are sensitive and responsive to the presence of people* [8–10]. According to another definition, AmI is *a digital environment that supports people in their daily lives by assisting them in a sensible way* [11]. The terms such as smart environments (SmE) or intelligent environments (IE) are also used for the ambient intelligence concept [12]. Ambient Intelligence is inspired by the idea of the invisible computer [13]. Weiser phrase the idea of *disappearing computer* elegantly as follows [14]: *The most profound technologies are those that disappear. They weave themselves into the fabric of everyday life until they are indistinguishable from it.* In this paradigm, powerful computers are collectively used in the background and end users interact with small appliances [15]. In ambient intelligence, sensors and actuators connected via a network work collectively with a central or distributed powerful computer system. This powerful computer system has enhanced data analysis and decision-making capabilities.

The discussions related to ambient intelligence date back to 1990s [13]. During the 2000s, the European Commission's Information Society Technologies Advisory Group (EU ISTAG) enhanced the AmI vision with a set of reports [8]. Developments in artificial intelligence, ubiquitous computing, deep learning, mobile technologies, and networking technologies such as Internet of things bring life to AmI. Latest research and development on smart homes, smart businesses, and smart cities strengthen the case for AmI.

Ambient intelligence, by design, answers 5 W's of the context quintet: who, where, what, when, and why. "When" and "Where" are considered straightforward, since they include only the time and location information related with the phenomena and can be obtained from the device timekeeper, network time, internal GPS device, or using user localization techniques as mentioned in [16]. "What" refers to the question related to the task of the user. "Who" and "Why" are considered challenging [17]. Answering the "who" question requires authentication, and commonly used today with fingerprint recognition or iris recognition systems, which are available and common with smartphones. More intelligent techniques like authentication from body movements are also possible today with the wearable computing devices [18].

The question of "why" is related to the users' intention and answers the reason for the user's actions. There are studies today to infer user intention from logged user data [19, 20].

Various developments help realize ambient intelligence:

- Increasing communication speeds with lower latency,
- Increasing processing capabilities for massive data analysis,
- Reduced sensor sizes,
- Increased sensor varieties and capabilities through microelectromechanical systems (MEMS), and
- Evolving distributed systems interconnectivity standards.

In a typical ambient intelligence scenario, sensors embedded in the environment collect and send data to central processing units for analysis and decision-making, and the actuators in the environment execute the decisions. The main characteristic of an AmI system is its human-centricity. An AmI system monitors humans and extracts rules from their actions, predicts the needs of its users, and becomes an invisible assistant. ISTAG proposes a holistic user-centric participation view for AmI [8]. Furthermore, ISTAG states that *AmI needs to be driven by humanistic concerns, not technologically determined ones and should be controllable by ordinary people* [8]. As a result, AmI is a *human-first* or *human-centric* paradigm.

There are various factors influencing the adoption of ambient intelligence [21] including usability, technical feasibility, trust and confidence, social and economic impacts. These factors are also applicable to the most Internet of Things (IoT) applications [21]. In the next section, we discuss these and other issues of IoT technology transformations in detail.

The smart environments have the following properties [9, 21]:

- Awareness of the presence of individuals,
- Recognition of the individuals' identities,
- Awareness of the contexts (e.g., weather, traffic, news),
- Recognition of activities, and
- Adaptation to changing needs of individuals.

Figure 3.1 depicts the ambient intelligence vision. Smart homes, patient monitoring, traffic flow control, production control, attendance control, public surveillance, emergency services, and many more emerging applications benefit from ambient intelligence.

Figure 3.2 shows various technologies supporting the realization of ambient intelligence. We list the main ones here. However, there are also other technologies and research areas that further help achieve and improve ambient intelligence. Business intelligence, social studies, and marketing studies are some of the related research areas.

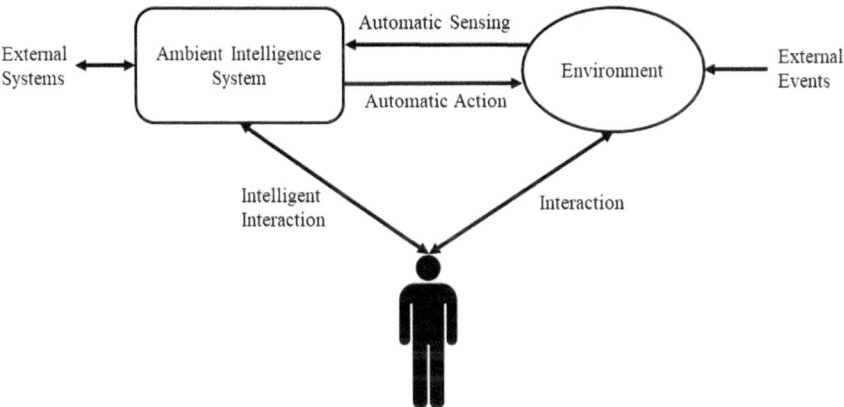

Fig. 3.1 Ambient intelligence vision

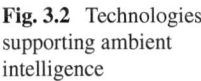
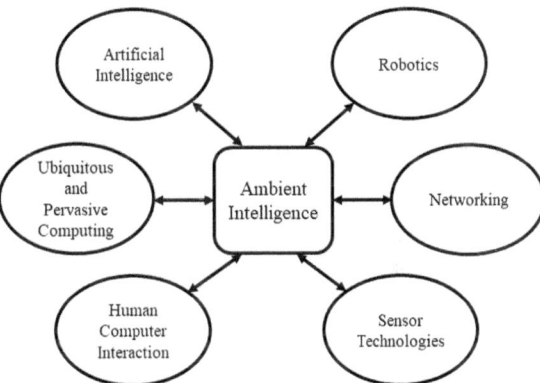

Fig. 3.2 Technologies supporting ambient intelligence

Figure 3.3 presents a conceptual architecture for ambient intelligence (AmI) or smart environments (SmE). In this architecture, sensors, actuators, middleware, artificial intelligence, and system management components are highlighted. These are the essential components in a typical AmI system. Depending on the implementation, other computing and networking devices may also be used.

3.2.2 Improving Business with Ambient Intelligence

Ambient intelligence is expected to create significant changes in the way we live and conduct businesses. Even the introduction of smartphones has affected our lives considerably. We have started to hear concepts such as smart homes, smart vehicles, smart factories, smart offices, and smart cities often. These concepts and related developments are among the main topics of many conferences. Through the use of

Fig. 3.3 A conceptual architecture for AmI/SmE systems

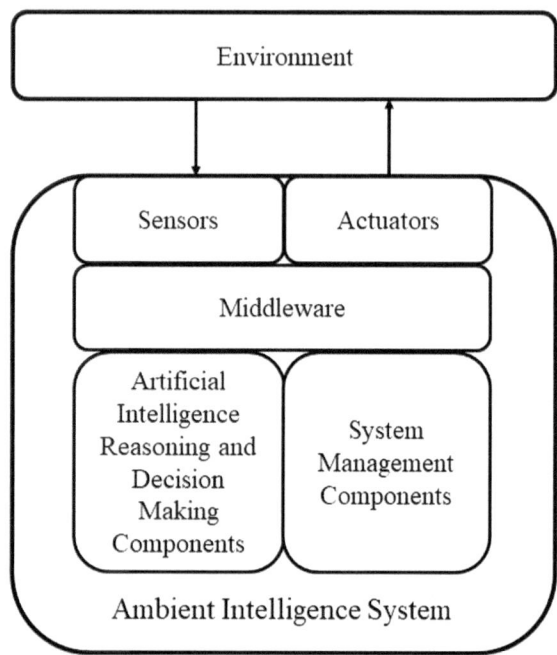

smart and natural human–computer/machine interfaces, we will soon interact with ambient intelligent systems in our homes, offices, and cities, in short everywhere. People becoming comfortable with these systems will expect AmI systems in other areas of their lives. An individual living in a smart home or working in a smart office or having a smart car will expect the smartness in a shopping store as well. Therefore, many businesses will be forced to use the technologies developed within AmI paradigm. Frontier businesses utilizing the AmI systems will develop new competencies and have a chance to increase market share. We believe businesses will have no choice but to use AmI systems in due time. The ones unable to adapt may face closures.

Figure 3.4 shows the concept of business improvement with the ambient intelligence paradigm. Any business venture starts with a business vision offering a value to the society. As long as the value offered stays in demand, the business stays in business [22]. The value offered to the society and the business vision creating this value is dynamic. What constitutes value in a society changes over time. Therefore, businesses revise their business vision to offer new values. Technological developments and paradigm shifts are among the main drivers of new business vision creations and revisions of existing visions. We believe ambient intelligence will lead to the introduction of new businesses and changes in many existing ones. To achieve AmI, a business organization has to undergo a technology transformation. Internet of things lies at the heart of this technology transformation. These transformations often lead to organizational changes, which are often difficult to implement [23].

Fig. 3.4 Business
improvement with ambient
intelligence paradigm

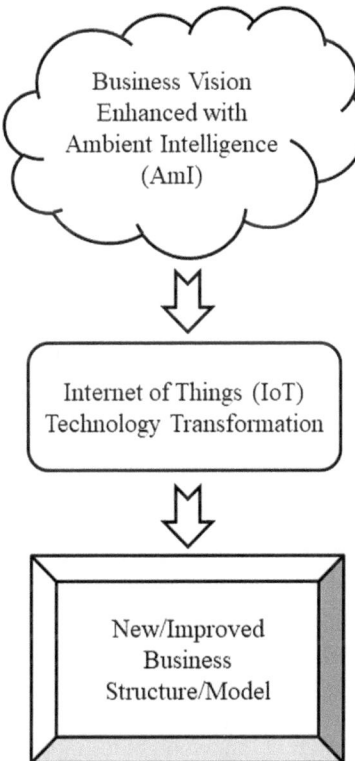

In the following sections, we discuss the issues related to the IoT technology transformation in detail. The business vision is realized by a business model. Thus, to realize the business vision enhanced with the ambient intelligent paradigm, we need a new or improved business model. The IoT technology transformation helps us reach this new or improved business model.

There are various questions that should be answered by business managers before any business improvement effort driven by AmI paradigm, e.g.,

- Is there any value in enhancing business with AmI paradigm?
- Is the technology transformation technically feasible?
- Is the technology transformation economically feasible?
- Is the new/improved business model sound?
- Do my customers adopt my new/improved business model?
- Do my employees adopt my new/improved business model?
- Is the business improvement effort profitable?

The answers to these and other important questions should be carefully analyzed before investing in an AmI technology transformation. Business managers should not follow a hype without adequate investigation. On the other hand, ambient intelligence systems offer great benefits with its natural and seamless use. Such technology utilizations also add value to businesses and brands. We believe utilizing AmI systems will be a great advertisement for frontier businesses.

3.2.3   Current Implementations of AmI

Amazon, one of the world's largest online retail stores, recently introduced a technology that provides a good use case of ambient intelligence in a business environment. It is known as *Amazon Go* [24]. While Amazon is an online retail store, the company acquired a chain of convenience stores. Now, the online retail giant has also brick-and-mortar stores. Amazon has been testing the new shopping technology in one of the stores for some time now. At the beginning of 2018, Amazon opened the Amazon Go store in Seattle. According to the company, *Amazon Go is the world's most advanced shopping technology*. The company advertises it with *No lines, no checkout—just grab and go*. To shop at an Amazon Go store, a customer needs an Amazon account and a smartphone with Amazon Go App installed. The customer walks into the Amazon Go store. In the entrance, the customer checks in with a Quick Response (QR) code generated by the Amazon Go mobile app. Then, the customer takes whatever he/she wants and walks out of the store. The store charges the customer's Amazon account. According to the company's website, computer vision, sensor vision, and deep learning are used in creating the ambient intelligent Amazon Go store. Amazon is not the only company working on cashier-less stores. There are also other companies investing in the concept. Alibaba Tao Cafe is another experimental ambient intelligent store. The way Tao Cafe operates is similar to Amazon Go store. Since it is a café, unlike the Amazon Go store, there are waiters preparing the drinks. However, the purchase is handled by the ambient intelligence system. We expect that these types of stores will increase in the near future.

Figure 3.5 shows the steps for a generic use case of ambient intelligent store shopping. Note the simplicity of the use case. One distinguishing aspect of this use case is the natural and seamless way the user interacts with the system.

3.2.4   Future Opportunities for Businesses

The use of smartphones is expanding at an enormous speed. Consider the mobile phones used 15 years ago. They provided very limited functionality when compared to what our smartphones provide today and they are not considered smart. As technology evolves, our current smartphones will not be considered smart in the future. Touching a smartphone will be considered outdated when voice control is mature enough. Furthermore, smartphones may become our virtual personal assistants. We may choose the brand of our smartphone based on the installed virtual personal assistant software. Moreover, smartphones may evolve to personal assistant devices

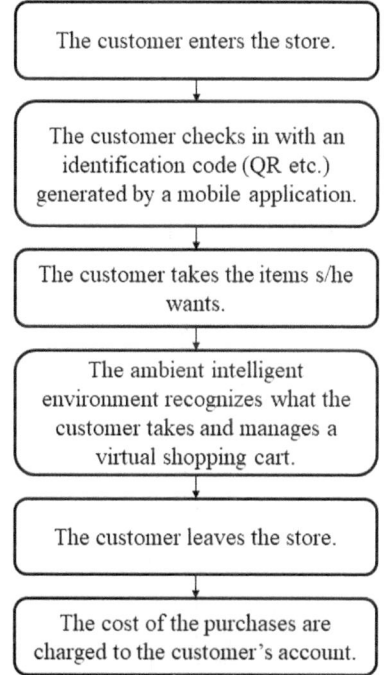

Fig. 3.5 A generic use case steps for ambient intelligent store shopping

The customer enters the store.

The customer checks in with an identification code (QR etc.) generated by a mobile application.

The customer takes the items s/he wants.

The ambient intelligent environment recognizes what the customer takes and manages a virtual shopping cart.

The customer leaves the store.

The cost of the purchases are charged to the customer's account.

with calling functionality. European Commission's Information Society Technologies Advisory Group (EU ISTAG) prepared a set of scenarios for 2010 in 2001 [25]. The focus of the scenarios was mostly related to the daily routines of individuals. One of the scenarios is "Dimitrios and The Digital Me (D-Me)" [25] that describes a virtual personal assistant. This D-Me was thought to be available for 2010; however, we still have a long way to go for the realization of the vision for 2010. We believe the realization of the possible implementations described in other scenarios will also take some time.

EU ISTAG states that their report [25] does not include an industrial scenario. However, they strongly express that AmI technologies offer tremendous opportunities for broad industrial change. They predict significant changes in the industrial baseline [25], e.g.,

- From value chains to value networks,
- Introducing mass customization and improved customer orientation,
- Better prototyping and generic knowledge management capacities,
- Transforming product sales in services revenues,
- End-of-life management, and
- Developing workforce health and safety programs.

As expressed earlier, there are many application areas of ambient intelligence. In this section, we only focus on some of the possible business and commercial application areas. Note that these are just a few scenarios. However, they will provide some useful ideas for business managers and researchers.

3.2.4.1 Ambient Intelligent Stores—Smart Stores

In the previous section, we provide a current implementation of an ambient intelligent store. Amazon Go is one of the first smart stores open to the public. Alibaba Tao Café is currently in the experimental stage. These stores sell packaged food and drink products. We believe there will be other types of stores utilizing ambient intelligence. Fashion stores or clothing stores may implement a different use case other than the one presented in Fig. 3.5.

Consider the use case below:

1. A customer walks into a fashion store.
2. The intelligent environment recognizes the registered customer using the data in the customer loyalty program.
3. The customer allows the ambient intelligent system to view/access the customer's social media accounts.
4. The ambient intelligent system examines the photos of the customer to understand/interpret the fashion choice of the customer.
5. Based on the latest fashion trends, the store's inventory, and the customer's fashion choice, the ambient intelligent system helps the customer to choose clothing that the customer will be highly satisfied with.
6. The customer takes or wears the clothing and leaves the store.
7. The system bills the customer.

There are many ways to extend this use case. For example, the ambient intelligent system may interact with the customer's virtual personal assistant installed on the smartphone. Both systems may work together to help the customer choose the best product.

3.2.4.2 Workplace and Business Process Optimization—Smart Offices

Many companies have open office environments. While various optimizations are used in the initial design of these offices, we may benefit from ambient intelligence systems to optimize the environment further. The AmI system may sense and analyze the movements and interactions of the people in the office. So, from time to time, we may redesign the offices that make the work more efficient and employees more comfortable in the workplace. Furthermore, the AmI system may constantly monitor and adjust various office environmental conditions such as the lightning level, the air conditioning, etc. These optimizations will help businesses achieve better working conditions leading to higher performances.

3.2.4.3 Increasing Workplace Safety—Smart Workplaces

Workplace safety is important. There are many laws and regulations for ensuring safety in the workplaces. Businesses spend resources to achieve a safe workplace. Companies hire specialists to inspect workplace against violations. They train employees to increase awareness. They install systems to increase workplace safety. Still, there are accidents and mishaps leading to deaths, injuries, loss of property, or damage to property. Factories, construction yards, shipyards, mining yards, transportation, or logistics hubs are all among high-risk workplaces. Use of ambient intelligence systems may help detect and prevent unsafe conditions and situations. Furthermore, these systems may quickly activate safeguard measures and systems against fires and other unfortunate events.

3.2.4.4 Increasing Workplace Security—Smart Workplaces

Law enforcement agencies such as police, highway patrol, and gendarmerie are mainly responsible for public security. Ambient intelligence will help to ensure and increase public security in smart cities. Businesses employ private security to ensure security in workplaces. It is currently a need for many businesses. As a result, organizations and companies install security systems and employ private security guards. Therefore, private security is a big industry. Banks, airports, bus terminals, shopping malls, business offices all employ private security. Ambient intelligence systems may help ensure security and reduce associated costs. Criminals and individuals with criminal intents will be demotivated with the presence of ambient intelligence systems that can sense and react to unwanted or unlawful acts. We believe private security industry will be significantly affected by the widespread use of ambient intelligence systems.

3.2.4.5 Increasing Marketing Effectiveness—Smart Marketing

According to American Marketing Association (AMA), *marketing is the activity, set of institutions, and processes for creating, communicating, delivering, and exchanging offerings that have value for customers, clients, partners, and society at large.* Effective marketing requires a deep understanding of consumer behavior as a result of comprehensive and reliable data gathering. The Internet and social media use increased marketing opportunities. Companies have already started utilizing consumer data gathered through Internet and social media use. They become a medium for effective and targeted marketing. As AmI systems become widespread, gathering consumer behavior data will get easier, cheaper, and quicker. Some of this data may also become available in real time. AmI systems may become an important source of consumer behavior data and act as a new medium for marketing.

3.3   IoT Technology Transformation Issues

An IoT transformation process leading to the realization of a business vision enhanced by the ambient intelligence paradigm is a highly challenging information technology/system (IT/IS) project. Consequently, the IoT transformation process is subject to the challenges and dynamics of an IT project implementation. Some of the guidelines are similar to guidelines provided for IT project success.

During IoT transformations, managers need to deal not only with technical issues but also with management and social issues. Figure 3.6 presents the management, technical, and social issues related to IoT transformations. We prepared guidelines for all categories of issues that appear in the following sections.

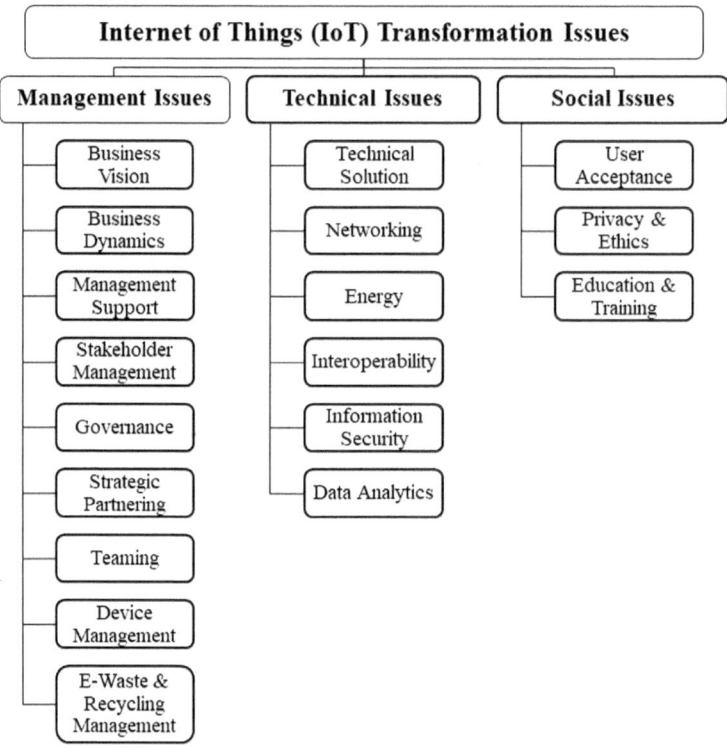

Fig. 3.6 Internet of Things (IoT) technology transformation issues

3.3.1 Management Issues

3.3.1.1 Business Vision

A business vision is *a lively mental image of the business' future state*. Whether it is spelled out clearly or not, an entrepreneur has a business vision that is driven by business goals and value offerings to the public. The vision keeps the business on the right track. It is imperative that the business vision is realistic and achievable. When the vision is new or unique, then the profit potential is high for commercial enterprises. The vision is realized by a business model. For a successful business, the realistic business vision should be supported by a sound business model.

There are many factors affecting the creation of a realistic business vision supported by a sound business model. One of the key factors is the effective and efficient utilization of technology. While this is easy to say, often it is nontrivial to implement. Good business managers continuously seek ways to enhance the business vision and improve the business model. They follow technological developments, trends, and paradigms. They choose the appropriate technologies to invest. While some technological developments lead to limited business improvement, some technological developments driven by trends and paradigms offer great improvement opportunities. We believe ambient intelligence is such a paradigm offering great improvement opportunities. However, following the AmI paradigm requires an IoT technology transformation, which is often challenging as many technology transformations are. These transformations require support from all stakeholders. In addition, there are many issues to consider during these transformations. In the rest of the section, we discuss these issues in detail.

As a result, having a clear business vision enhanced with ambient intelligence paradigm and ensuring that all stakeholders understand this vision is a crucial first step in achieving an IoT transformation.

> *Have a clear business vision and ensure that all stakeholders understand and share this business vision*

3.3.1.2 Business Dynamics

Business dynamics is the desirable or undesirable interaction of dynamic factors affecting a business or an organization. There are two main types of business dynamics: internal and external. Internal business processes and group dynamics within the organization create the internal dynamics. External business processes and relationships with customers and business partners create the external dynamics. Both types of dynamics determine the capabilities and limitations of a business organization. A business improvement effort driven by the ambient intelligence paradigm requires an in-depth investigation of both internal and external dynamics. An IoT technology transformation leading to a deployment of an ambient intelligent system is a big organizational change for many businesses. Business managers and decision-makers should carefully investigate how the internal and external dynamics are affected due to this organizational change. They should try to reduce consequences while maximizing the benefits during this business improvement effort. Most importantly, they should set the scope right and well manage the scope throughout the technology transformation. Development of an ambient intelligence system is, in fact, an IT project. Scope management is quite a challenge in IT projects [26, 27].

As a result, the transformation process should be well-scoped and well-designed considering both the internal and external business dynamics affecting organizational capabilities and limitations.

Study internal and external business dynamics and develop a transformation process based on organizational capabilities and limitations

3.3.1.3 Management Support

Technology management can be defined as combining the disciplines of engineering and management sciences for planning, development, and implementation of technological capabilities. These capabilities help realize strategic and operational objectives of the organization. The key elements of technology management that are related to technological transformations are [28]:

- Identification and evaluation of technological options,
- Integration of technologies into the organization's activities, and
- Implementation of new technologies in products and/or services.

It is obvious from the list that the role of management support in a technological transformation is very important. Sharing a common vision, following a viable strategy, good leadership, and taking responsibility are important factors for a successful IT transition [29]. Managers have the potential and power to positively affect other members of the organization [30–32]. When the necessary support from management is provided, the chances for success in an IT transformation are much higher due to the leadership, responsibility, and decisiveness factors.

Another aspect where management support is necessary to the technological transformation is securing adequate funds and manpower. Without the necessary resources, it is not possible to reach the desired goals.

In summary, management support is a success factor in IT projects [26] including IoT technology transformations for the development of an ambient intelligence system.

Secure management support

3.3.1.4 Stakeholder Management

As expressed earlier, an IoT transformation process leading to the realization of a business vision enhanced by the ambient intelligence paradigm is a challenging IT project. A stakeholder is a person, a group, or an entity that can affect and be affected by a project. According to the association for project management (APM) body of knowledge definitions, *stakeholder management is the systematic identification, analysis, planning, and implementation of actions designed to engage with stakeholders.* Earlier studies indicate that stakeholder management is crucial for any type of information technology/system project [26, 27]. Ambient intelligence aims to create a shift in the existing business models. As a result, it is not easy to predict how different stakeholder will react to this paradigm shift. Some users of the ambient intelligence systems may find the AmI systems comfortable, seamless, and intuitive. Some users may find the idea of invisible computer uncomfortable. Some will object to AmI systems due to privacy and ethics concerns. Many people may be out of jobs due to tasks performed by automated systems. This may create resistance among public and employees of companies trying to achieve an IoT transformation. While AmI systems offer great benefits to many people and businesses, there will also be people and businesses affected undesirably. Business managers and project managers of technology transformations should pay special attention to the concerns of stakeholders those are affected adversely.

Identify all stakeholders and pay attention to stakeholder management

3.3.1.5 Governance

Governance is *all the processes related to governing, whether undertaken by a government, a market or a network, over a social system such as a formal or informal organization and whether through the laws, norms, power or language of an organized society* [33]. According to another definition [34], governance relates to *the processes of interaction and decision-making among the actors involved in a collective problem that lead to the creation, reinforcement, or reproduction of social norms and institutions.* The difference between management and governance lies in the responsibility. In management, the burden of management is on the managers. In governance, the burden of governance is on all stakeholders.

In the context of this study, we focus on the governance of AmI systems and IoT devices. We can talk about two types of governance. The first one is global governance and the second one is organizational governance. An important part of the global governance is the standardization in AmI systems and IoT devices. Without adequate standardization, the market will not mature. The second important aspect of global governance is making the necessary laws and regulations. Currently, an important problem in the IoT domain is the lack of adequate governance [29]. Naturally, there are some initiatives. However, for the IoT domain, we still yet to have a globally recognized governing body to develop international standards, regulate the IoT use, and help develop the AmI market [29]. In addition, we need proper legislation [35].

In the absence of adequate global governance, the cost and effort required for the organizational governance are higher. The organization itself has to fill the gaps due to the lack of laws, regulations, national and international standards, established norms, processes, and procedures. The organizations aiming to benefit from AmI systems have to pay the necessary attention to governance for a successful and uninterrupted business operation.

Develop organizational procedures and standards to achieve IoT governance

3.3.1.6 Strategic Partnering

Development of open, extensible, and sustainable ecosystems between business partners is important to integrate solutions and applications and improve decision-making processes [29]. In the IoT market, these ecosystems will help improve operations, enhance productivity, increase revenues, and reduce risks [36]. Both IoT and AmI markets are in their early stages. Thereby, requirements, use cases, and technological solutions change constantly. The partnership will help develop core competencies for the organizations while outsourcing the management of other competencies to their partners. However, the type of partnership should well define what the organization is partnering for, the partner relationship should define how you work with each other, and the scope of partnership should include the agreement on the technology/solution/intellectual property ownership, the duration of the partnership, and the roles with responsibilities of parties.

Support your partners to initiate an IoT transformation process within their organizations

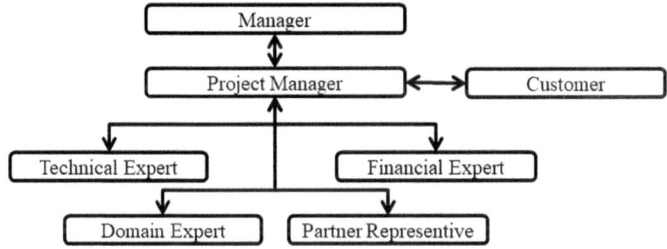

Fig. 3.7 A generic Internet of Things (IoT) technology transformation management team [29]

3.3.1.7 Teaming

Teaming is the formation of a project group responsible for the IoT transformation project. The project manager should always keep in mind that the end result is the realization of a business vision enhanced with ambient intelligence paradigm. This is a big change creating many challenges for the organization. Therefore, development of a good IoT transformation team is crucial [37], just like in a large-scale IT project [29, 26].

Organizations develop, or fail to develop, depending on how well the small groups within those organizations perform [38]. The motivation of the team members with different needs and skills is a must-to-have concept for paving the road to success in a technology transformation.

A management team structure responsible for the IoT transformation is shown in Fig. 3.7. Arrows depict the information flow between team members. Project manager and the customer should not be thought to have a hierarchical relationship. The structure and the size of the team members may be decided depending on the organizational requirements and size.

Build a management team responsible for the IoT transformation process

3.3.1.8 Device Management

We consider the operation, configuration, maintenance, acquisition, and inventory management as part of the device management. While operation, configuration, and maintenance are related to technical management of devices, acquisition and inventory management are related to organizational management of devices. Ambient intelligence systems should work without interruption. Therefore, all aspects of device management may be considered crucial for a reliable AmI system.

Appropriate operational management of the devices comprising ambient intelligence systems and IoT requires efficient interoperable protocols. Tiny sensors and devices lack enough processing power; hence, an interconnection between the sensors and devices with central decision-making and high processing capacity units is necessary. Many small devices are envisioned to be connected to the Internet. There will be 20 billion devices by 2020 [39]. Each device requires to be addressed, localized, and communicated individually, which brings the naming, localization [40], communication, interoperability, process distribution, coordination, fault tolerance, and security problems together. The naming problem deals with addressing each entity. With the inclusion of small smart devices, sensors, and actuators, this problem exacerbates. IPv6-based naming solutions are necessary for device management [41]. After naming each entity, managing their location is the next problem. Locations of the various entities can be managed in a distributed, centralized, or hierarchical manner. Plug and play mechanisms are required for communication and interoperability of devices produced by many different vendors. Middleware-based solutions exist for the communication and interoperability of these heterogeneous systems [42].

There may be many different types and brands of IoT devices comprising ambient intelligence systems. Organizations should be able to monitor and manage them easily. Central management of connected devices helps to achieve an uninterrupted business service [29]. Furthermore, central operational device management enables IT departments to install, monitor, and update the devices and processing units in the system remotely [43]. We advise the use of automation to the maximum extent in the management of these devices.

Configuration management of many different types and brands of devices is a challenge that organizations have to overcome. The sensors and actuators in the ambient intelligence systems may have quite different maintenance requirements and may necessitate a diligent maintenance program. In a promising competitive market such as ambient intelligence systems, there will be many different types and brands of devices with different capabilities. Consequently, acquisition management may not be easy.

Automate management of different types and brands of IoT devices

3.3.1.9 E-Waste and Recycling Management

As the use of ambient intelligence systems and the number of IoT devices comprising these systems increase, the pollution resulting from the disposal of these devices will be a serious threat to our environment [29]. E-waste is one of the fastest growing waste streams in the world in terms of volume and its environmental effect on the world. The benefits of recycling e-waste material can be classified into three categories: economic, environmental, and public health and safety [44]. The best practice for the environment and the economy is to recycle e-waste as much as possible. But this is not an easy goal. There are a lot of obstacles from inadequate regulations to technological challenges for recycling. Currently, there are some regulations on e-waste [45]. An overview of the current e-waste management issues and practices can be found in [46].

In addition to global aims, we must have similar goals at the organizational level. Maintenance and disposal procedures must be well established and implemented. Proper recycling methods for e-waste should be used. The factors affecting successful e-waste management at the organizational level can be summarized as follows [46]:

- Using ecologically designed devices,
- Collecting e-waste properly,
- Recycling material through safe methods,
- Dispose of e-waste by appropriate techniques, and
- Raise awareness on the effect of e-waste.

Develop an e-waste and recycling management program

3.3.2 Technical Issues

3.3.2.1 Technical Solution

IoT devices in AmI systems are embedded into the environment aiming to serve proactively in the anticipation of user needs and only a subset of IoT devices may be appropriate in the context of AmI targeting contextual awareness, recognition of the individuals, and adaptation to user needs. Ambient side consists of sensors, communication, and software whereas intelligence includes data mining, artificial intelligence, machine learning, and data retrieving from backhaul. Networking and energy considerations are normally considered to construct the AmI framework. Furthermore, more complex topics such as machine learning and data mining can be added to the framework. However, without providing reliable connectivity and lifetime accessibility for AmI nodes, practical deployments will not be feasible. With the proliferation of IoT targeting applications and networking/energy technologies, selecting the mature, scalable, and expandable technical solutions will play an important role in the cost and time savings for integration.

Integration of AmI solutions poses challenges mainly in networking, energy, and interoperability. Hence, a technical strategy addressing the integration and a reference framework establishment is the key to manage the interdisciplinary nature of AmI systems. The technical strategy should at least involve detailed phases of AmI system development including defining specifications, implementation, integration, testing, deployment, and operation. Technical strategy phases should be well documented to address future needs of the deployed AmI systems. Furthermore, a reference framework should be defined with the consideration of technology standards to provide repeatable architecture with low cost and reduced risks to ensure interoperability and consistency between multiple AmI projects.

An IoT transformation leading to the deployment of an AmI system is a big change for many organizations. In creating a technical solution, business and technical managers should carefully consider many business and technical aspects starting with the business vision, the business dynamics, and the business processes. Therefore, decision-makers should take their time and seek for the best technical solution.

Take your time and seek the best IoT solution aligned with your business dynamics

3.3.2.2 Networking

AmI systems, sensitive and responsive to the presence of people, aim to support everyday routines with enhanced connectivity and processing features. Interoperability and seamless integration are foreseen to be key enablers of AmI. Furthermore, AmI devices, becoming more interconnected and smart, are further expected to be more hidden in the environment providing only user interface access.

In this respect, we are experiencing a massive connectivity issue and the connectivity is increasing at exponential rates. According to an estimate, 45 trillion networked sensors are believed to be integrated into our daily lives within the next 20 years [47]. These devices will get smarter as artificial intelligence technology advances. They will process big amounts of data effectively and efficiently. Moreover, they will be able to execute decisions with little to no human intervention.

However, according to a UN Report [48], the majority of the world's population lacks Internet access due to lack of infrastructure, affordability, or lack of skills. Hence, AmI devices should not only target certain regions of the world, but should also embrace undeveloped regions of the world. Thereby, when, AmI connectivity is considered, not only the fully Internet connected devices but also the infrastructure independent, machine to machine type connected devices should be regarded.

Many novel technologies in network infrastructures are being introduced due to advancements in IoT [49], fifth generation networks [50], and edge computing [51]. AmI systems will be required to handle large amounts of real-time data. Furthermore, users will expect an uninterrupted service. As a result, organizations may need to review their current network infrastructure and probably modernize the infrastructure to handle this large amount of data.

Modernize your network infrastructure to enable a large amount of real-time data transmission

3.3.2.3 Energy

Providing the necessary power source for IoT sensors of AmI systems is an important research area. AmI sensors will be deployed to our surroundings with almost no additional power supply and maintenance requirements. There are various significant developments in terms of providing power to devices to be used in AmI systems. Hence, energy harvesting (EH) and wireless power transfer technologies (WPT) are believed to be a key enabler for AmI device prevalence. Energy harvesting can be explored under five different categories:

- Solar energy harvesting,
- Strain from Piezoelectric,
- Thermal,
- Magnetic induction from power lines, and
- RF energy harvesting.

Wireless power transfer technologies (WPT) are currently only available via RF energy. In addition to state-of-the-art EH and WPT methods, effective utilization of ultralow-power electronics, flex batteries, and supercapacitors will play an important role for practical realizations of AmI based applications.

On the other hand, power sharing between nodes and access points may also be a key enabler to respond needs with efficient resource utilization. Consider a traffic camera and weather monitoring station sharing a single photovoltaic battery, the camera can be on to monitor rush hour traffic for traffic routing, whereas the weather monitoring station will be only activated to inform drivers about real-time severe weather conditions. We believe redundant energy harvesting methodologies and wireless power transfer technologies are game changers for fast integration of IoT nodes to our AmI universe.

Before an IoT transformation for the deployment of an AmI system, organizations should analyze their current power capacity. They should carefully monitor the existing capacity and improve when needed or beforehand. Furthermore, they should closely monitor the technology developments on energy sources used in IoT devices of the AmI systems.

Analyze and improve the current power capacity based on the current and future requirements of IoT devices to be deployed

3.3.2.4 Interoperability

Interoperability through AmI devices can be built upon standard prevalent IoT technologies. IoT device requirements vary widely in terms of power, connectivity, size, price, processing power, and other dimensions. Hence, there exist tens of standards to fulfill interoperability between IoT devices [52]. However, considering IoT as the infrastructure, providing a vast number of choices for AmI services, a wise selection of the relevant infrastructure subset will ease interoperability.

As the processing and storage capabilities of the AmI devices will be limited, connectivity will be key to realize Artificial Intelligence-based decision and interaction. As more AmI sensors will learn more from each other, more use cases will be covered, and various level of information will be exchanged between distributed AmI devices. Imagine that your car controls your home appliances to make your dinner ready; your electric shaver wakes up your car to heat seats and passenger compartment; and your refrigerator monitors your food inventory and sends your shopping list to your car. To ensure this kind of seamless integration and interoperability, heterogeneous connectivity features and network technology selection, such as discovery and data protocols, transport and communication layers, will be prominent [53].

Due to varying AmI device sizes and power requirements, same network technology usage may not be possible. Thereby, machine learning-based coexistence control mechanisms will be required for reliable AmI connectivity mechanisms. Similar technologies, fulfilling identical QoS requirements (range, data rate, etc.) such as IEEE 802.11ah and IEEE 802.15.4 g should be considered for coexistence control, as they are highly likely to be utilized interchangeably.

> *Standard-driven IoT devices will accelerate the transformation to AmI, providing scalability with seamless integration*

3.3.2.5 Information Security

Gartner predicts that there will be 25 billion connected devices in 2020 [54]. In addition, a significant percentage of these devices will be sensor-based. We envision that an important percentage of these devices will be used for production and business purposes. While the use of these devices offers many opportunities for business, there are likely to create many information security-related concerns.

Heterogonous structure of the IoT makes the communication protocols that connect devices vary in type and capability. RFID, WiMAX, cable broadband are some examples of these connections. This heterogeneity of the devices and variety of communication protocols [55] make conventional security solutions hard to use. Sophisticated authentication and access control mechanisms are required [56].

Another challenge is the low processing capability of the tiny appliances makes them tempting targets for network attacks [57]. Tiny devices cannot perform sophisticated cryptographic protocols. However, they should still be kept secure from the malicious attacks.

Ambient intelligence systems will offer unprecedented human–machine interactions. These interactions should be safe and secure. Furthermore, to sense the individuals these systems will collect and store biometric data from people. This data should be kept secure and users should be able to rely on the ambient intelligence systems. While maintaining security, it should be kept in mind that the usability and the performance of the AmI system should still be within acceptable limits, though developers should consider the tradeoff between security and the efficient usability [58].

> *Balance security and usability in the use of IoT devices*

3.3.2.6   Data Analytics

The deployed sensors within ambient intelligent systems will collect enormous amounts of data. For example, Oracle sponsored the development of a catamaran equipped with 300 sensors and cameras. These devices generate data containing 3000 variables per second and 200 gigabytes of video per day [59]. Such high volumes of data require intelligent data processing techniques. A huge amount of data related to the user behavior and habits will continuously be collected by the devices constituting IoT. These data should be processed by powerful devices and efficient algorithms. Efficient wireless and wired interconnection between the tiny data gathering and actuating devices and the powerful big data processing servers are necessary components. Machine learning techniques are used for big data analysis. Storage for the large data is another problem. Mark up and tagging techniques are used for creating a semantic view of the large data [60]. These semantic views can help to develop privacy and ethics policy implementation for the sensitive data. As the volumes of data collected increase, there will be a crucial need for improved server and storage systems.

> *Improve server and storage systems to analyze data collected by IoT devices*

3.3.3 Social Issues

3.3.3.1 User Acceptance

A successful technology transformation requires user acceptance. There may be resistance to the new technology due to many factors including privacy and security concerns, and personnel downsizing. When the technology transformation is taken into account at the enterprise level, studies show that the number of early adaptors is low. In a technology transformation, there will employees with the fear of losing their jobs as a result of the transformation. Thus, the managers should be clear, fair, and informative about the negative and possible results of any change, so that the employees will be ready for the outcomes of the transformation.

Another important factor in achieving user acceptance is to ensure that the ambient intelligence system meets the expectations of the stakeholders. Information gathering activities such as questionnaires can be used to identify the expectations of the people affected by the technology transformation. Conferences, briefings, and other informative activities may be used to aid in achieving the technology transformation. There are various approaches, methods, and techniques to ease the organizational change process. For example, utilization of social hubs is a prospective effective method for successful organizational change [23].

Seek ways to achieve user acceptance. Pay special attention to conferences, training, and other information sharing activities

3.3.3.2 Privacy and Ethics

IoT devices in ambient intelligent systems will be capable of collecting data from machines, systems, products, employees, and customers [29]. Sensors located everywhere including houses and private areas collect sensitive and private data [35]. The data collected will be directed to storage servers and data centers. Privacy and ethical issues surround this type of aggressive data collection and storage. There are two approaches to ensure ethical data collection. The first approach is to ask the user for permission. This approach is mostly used in smartphones and many users are familiar with it. There are implementations to realize this approach such as a privacy definition language [61]. The second approach is to inform user of the data gathering and let the users adjust their behaviors accordingly [62]. Data processing mechanisms involving different privacy levels are needed. Community habits are expected to change with the involvement of these smart devices in our lives. Some people may lose their jobs due to increased automatization. On the other hand, new business areas emerge. Ethics of intelligent autonomous systems and their interactions with the humans are among challenging issues.

> *Inform users and let them to adjust privacy settings for private data collections using IoT devices*

3.3.3.3 Education and Training

Any experience having a formative effect on the way one thinks, feels, or acts may be considered as educational. Training can be described as maintaining and upgrading skills throughout the working life. Education and training are among the central aspects of achieving success in IoT [35]. After deciding the goals that will be achieved by using ambient intelligence in our organization, the next important step will be educating and training the stakeholders whether they are implementers or users. This will give them a clear view of what will be better, which disadvantages and handicaps will be removed in the new ambient intelligent environment. It will help to eliminate worries of users about data privacy and misuse of information. A help desk will also be beneficial for providing necessary information to the users about the problems they might encounter in the new ambient intelligent environment. Education and training will help prevent misuse of the system. Furthermore, users equipped with the necessary knowledge and skills will use and benefit from this technology to its full potential.

Additionally, education and training help achieve user acceptance.

> *Develop and conduct an effective education and training program for users*

3.4 IoT Technology Transformation Guidelines

The summary of IoT transformation guidelines developed in this research is presented in Table 3.1.

These guidelines and best practices are provided to help business managers in achieving an IoT transformation leading to a realization of ambient intelligence enhanced business vision.

Table 3.1 Internet of things (IoT) transformation guidelines

Management issues	
Business vision	Have a clear business vision and ensure that all stakeholders understand and share this business vision
Business dynamics	Study internal and external business dynamics and develop a transformation process based on organizational capabilities and limitations
Management support	Secure management support
Stakeholder management	Identify all stakeholders and pay attention to stakeholder management
Governance	Develop organizational procedures and standards to achieve IoT governance
Strategic partnering	Support your partners to initiate an IoT transformation process within their organizations
Teaming	Build a management team responsible for the IoT transformation process
Device management	Automate management of different types and brands of IoT devices
E-waste and recycling management	Develop an e-waste and recycling management program
Technical issues	
Technical solution	Take your time and seek the best IoT solution aligned with your business dynamics
Networking	Modernize your network infrastructure to enable a large amount of real-time data transmission
Energy	Analyze and improve the current power capacity based on the current and future requirements of IoT devices to be deployed
Interoperability	Standard-driven IoT devices will accelerate the transformation to AmI, providing scalability with seamless integration
Information security	Balance security and usability in the use of IoT devices
Data analytics	Improve server and storage systems to analyze data collected by IoT devices
Social issues	
User acceptance	Seek ways to achieve user acceptance. Pay special attention to conferences, training, and other types of information sharing activities
Privacy and ethics	Inform users and let them adjust privacy settings for private data collections using IoT devices
Education and training	Develop and conduct an effective education and training program for the users

3.5 Conclusion

In due time, we will live in smart homes, go to work driving through a smart city in smart cars, work in smart offices and factories, and shop in smart stores. Ambient intelligence is the paradigm that will lead to all these developments. As AmI environments increase and people get used to these types of smart environments, they will expect even more and better. Smart environments will be everywhere as these offer a natural and easier way to interact with systems, computers, and machines. As a result, our environments will become smart, the computers will be everywhere and they will be invisible.

In this chapter, we focus on the current and possible future implementations of ambient intelligence in the business environments. We further put forward a conceptual business improvement model with ambient intelligence paradigm. We believe, among other technologies, Internet of things (IoT) will be at the heart of ambient intelligence paradigm. Businesses will need IoT technology transformations to achieve a business vision enhanced with ambient intelligence. Therefore, we discuss management, technical, and social issues dealt with during an IoT technology transformation. We provide a set of guidelines to achieve these transformations. We believe this chapter will be an essential reading for entrepreneurs, business managers, and technical managers willing to benefit from ambient intelligence paradigm in their businesses. Furthermore, researchers will find the chapter valuable as a source of research questions for their future studies.

Disclaimer and Acknowledgements The views and conclusions contained herein are those of the authors and should not be interpreted as necessarily representing the official policies or endorsements, either expressed or implied, of any affiliated organization or government. This work extends our previous studies [29].

References

1. Demir KA, Cicibaş H (2018) The next industrial revolution: industry 5.0 and discussions on industry 4.0, industry 4.0 from the management information systems perspectives, Peter Lang Publishing Group
2. Demir KA, Cicibas H (2017) Industry 5.0 and a critique of industry 4.0. In: 4th international management information systems conference, Istanbul, Turkey, 17–20 Oct 2017
3. MacDougall W (2014) Industrie 4.0: Smart manufacturing for the future. Germany Trade & Invest. https://www.gtai.de/GTAI/Content/EN/Invest/_SharedDocs/Downloads/GTAI/Brochures/Industries/industrie4.0-smart-manufacturing-for-the-future-en.pdf. Accessed 15 June 2017
4. Erkollar E, Oberer B (2016) Endüstri 4.0 Akıllı Üretim İçin Politika ve Programlara Ait Bir Örnek: Alman Akıllı Çözümleri. In: Tecim V, Tarhan Ç, Aydın C (eds) Smart technology & smart management, İzmir, Turkey
5. Gilchrist A (2016) Industry 4.0. Apress, Berkeley, CA, USA
6. Augusto JC (2009) Past, present and future of ambient intelligence and smart environments. In: International conference on agents and artificial intelligence, pp 3–15. Springer, Heidelberg
7. Cook DJ, Augusto JC, Jakkula VR (2009) Ambient intelligence: technologies, applications, and opportunities. Pervasive Mob Comput 5(4):277–298
8. IST Advisory Group (2003) Ambient intelligence: from vision to reality, European Commission

9. Ramos C, Augusto JC, Shapiro D (2008) Ambient intelligence—the next step for artificial intelligence. IEEE Intell Syst 23(2):15–18
10. Aarts E, Encarnação JL (2005) Into ambient intelligence. In: Aarts E, Encarnçao J (eds) True visions: tales on the realization of ambient intelligence. Springer, Heidelberg
11. Augusto JC (2007) Ambient intelligence: the confluence of ubiquitous/pervasive computing and artificial intelligence. Intell Comput Everywhere, 213–234. Springer, London
12. Augusto JC, McCullagh P (2007) Ambient intelligence: concepts and applications. Comput Sci Inf Syst 4(1):1–27
13. Norman D (1998) The invisible computer. The MIT Press, Cambridge
14. Weiser M (1991) The computer for the 21st century. Sci Am 265(3):94–104
15. Remagnino P, Foresti GL (2005) Ambient intelligence: a new multidisciplinary paradigm. IEEE Trans Syst Man Cybern-Part A Syst Hum 35(1):1–6
16. Zhang D, Zhao S, Yang LT, Chen M, Wang Y, Liu H (2015) NextMe: localization using cellular traces in internet of things. IEEE Trans Industr Inf 11(2):302–312
17. Brooks K (2003) The context quintet: narrative elements applied to context awareness. In: Human-computer interaction international proceedings, vol 2003. Erlbaum Associates
18. Parimi GM, Kundu PP, Phoha VV (2018) Analysis of head and torso movements for authentication. In: 2018 IEEE 4th international conference on identity, security, and behavior analysis (ISBA). IEEE, pp 1–8
19. Lin CH, Ho PH, Lin HC (2014) Framework for NFC-based intelligent agents: a context-awareness enabler for social internet of things. Int J Distrib Sens Netw 10(2):978951
20. Schiaffino S, Armentano M, Amandi A (2010) Building respectful interface agents. Int J Hum Comput Stud 68(4):209–222
21. Lee F (2017) Ambient intelligence — the ultimate IoT use cases. https://medium.com/iotforall/ambient-intelligence-the-ultimate-iot-use-cases-5e854485e1e7
22. Drucker PF, Wilson G (2001) The essential Drucker, vol 81. Butter-worth-Heinemann, Oxford
23. Demir KA, Ozkan BE (2015) Organizational change via social hubs: a computer simulation based analysis. Proc Soc Behav Sci 210:105–113
24. Amazon Go. https://www.amazon.com/b?ie=UTF8&node=16008589011
25. European commission's information society technologies advisory group (EU ISTAG) (2001) Scenarios for ambient intelligence in 2010. Office for Official Publications of the European Communities, Luxembourg
26. Demir KA (2008) Measurement of software project management effectiveness. Doctoral dissertation, Naval Postgraduate School, Monterey
27. Demir KA (2009) A survey on challenges of software project management. In: Software engineering research and practice, pp 579–585
28. Task Force on Management of Technology (1987) National research council (U.S.) cross-disciplinary engineering research committee; national research council (U.S.) Manufacturing studies board: management of technology: the hidden competitive advantage. National Academy Press, Washington, Washington, D.C
29. Cicibas H, Demir KA (2016) Integrating Internet of Things (IoT) into enterprises: socio-technical issues and guidelines. Yönetim Bilişim Sist Derg (Management Information Systems Journal) 1(3):105–117
30. Thong JY, Yap CS, Raman KS (1996) Top management support, external expertise and information systems implementation in small businesses. Inf Syst Res 7(2):248–267
31. Thong JY, Yap CS, Raman KS (1997) Environments for information systems implementation in small businesses. J Organ Comput Electron Commer 7(4):253–278
32. Keen PG (1981) Information systems and organizational change. Commun ACM 24(1):24–33
33. Bevir M (2012) Governance: a very short introduction. OUP Oxford, Oxford
34. Hufty M (2011) Investigating policy processes: the governance analytical framework (GAF). In: Wiesmann U, Hurni H et al (eds) Research for sustainable development: foundations, experiences, and perspectives. Geographica Bernensia, Bern, pp 403–424
35. Bassi A, Horn G (2008) Internet of things in 2020: a roadmap for the future. European Commission: Information Society and Media, Brussels, Belgium

36. Hewlett Packard Enterprise (2016) The internet of things: turning ordinary things into extraordinary business outcomes, Number: 4AA6-3316ENN, March 2016
37. Intel (2015) Integrating IoT sensor technology into the enterprise. White Paper, December 2015
38. Edmondson AC (2012) Teaming: how organizations learn, innovate, and compete in the knowledge economy. Wiley, San Francisco
39. Rayes A, Samer S (2017) Internet of things—from hype to reality. The road to digitization, vol 49. River Publisher Series in Communications, Denmark
40. Zhu M, Song F, Xu L, Seo JT, You I (2017) A dependable localization algorithm for survivable belt-type sensor networks. Sensors 17(12):2767
41. Samad F, Memon ZA (2018) The future of internet: IPv6 fulfilling the routing needs in internet of things. Int J Futur Gener Commun Netw 11(1)
42. Chelloug SA, El-Zawawy MA (2017) Middleware for internet of things: survey and challenges. Intell Autom Soft Comput 1–9
43. SAP (2014) Next-generation business and the internet of things. http://go.sap.com/documents/2013/10/02247623-0a7c-0010-82c7-eda71af511fa.html
44. Kumar A, Holuszko M, Espinosa DCR (2017) E-waste: an overview on generation, collection, legislation and recycling practices. Resour Conserv Recycl 122:32–42
45. Zhang K, Schnoor JL, Zeng EY (2012) E-waste recycling: where does it go from here? Environ Sci Technol 46(20):10861–10867
46. Kiddee P, Naidu R, Wong MH (2013) Electronic waste management approaches: an overview. Waste Manag 33(5):1237–1250
47. Janusz B (2014) Trillion sensors movement in support of abundance and internet of everything. In: SensorsCon 2014
48. International telecommunications union—broadband commission (2017) The state of broadband: broadband catalyzing sustainable development
49. Ding M, Pérez DL (2018) Promises and caveats of uplink IoT ultra-dense networks. arXiv preprint arXiv:1801.06623
50. Ericsson.com (2018) Network slicing for IoT service deployment. https://www.ericsson.com/digital-services/trending/economic-study-5g-network-slicing. Accessed 25 Apr 2018
51. IBM scientists team with The Weather Company to bring edge computing to life (2017). https://www.ibm.com/blogs/research/2017/02/bringing-edge-computing-to-life/. Accessed 25 Apr 2018
52. IEEE (2017) Internet of things IEEE standards. http://standards.ieee.org/innovate/iot/stds.html
53. Lee SK, Bae M, Kim H (2017) Future of IoT networks: a survey. Appl Sci 7(10):1072
54. Gartner (2014) http://www.gartner.com/newsroom/id/2905717
55. Marksteiner S, Jimenez VJE, Valiant H, Zeiner H (2017) An overview of wireless IoT protocol security in the smart home domain. In: 2017 Internet of things business models, users, and networks. IEEE, pp 1–8
56. Yang Y, Longfei W, Yin G, Li L, Zhao H (2017) A survey on security and privacy issues in internet-of-things. IEEE Internet Things J 4(5):1250–1258
57. Cranor LF, Garfinkel S (2005) Security and usability: designing secure systems that people can use. O'Reilly Media, Inc., Sebastopol
58. Alrawais A, Alhothaily A, Hu C, Cheng X (2017) Fog computing for the internet of things: security and privacy issues. IEEE Internet Comput 21(2):34–42
59. Burkitt F (2014) A strategist's guide to the internet of things. http://www.strategy-business.com/article/00294?gko=a9303
60. Shi F, Li Q, Zhu T, Ning H (2018) A survey of data semantization in internet of things. Sensors 18(1):313
61. Dehghantanha A, Udzir NI, Mahmod R (2010) Towards a pervasive formal privacy language. In: 2010 IEEE 24th international conference on advanced information networking and applications workshops, 20–23 Apr 2010
62. Weber RH (2010) Internet of things-new security and privacy challenges. Comput Law Secur Rev 26(1):23–30

Chapter 4
Runtime Adaptability of Ambient Intelligence Systems Based on Component-Oriented Approach

Muhammed Cagri Kaya, Alperen Eroglu, Alper Karamanlioglu, Ertan Onur, Bedir Tekinerdogan and Ali H. Dogru

Abstract Technological improvements of the Internet and connected devices cause increased user expectations. People want to be offered different services in nearly every aspect of their lives. It is a key point that these services can be reached seamlessly and should be dynamically available conforming to the active daily life of today's people. This can be achieved by having intelligent environments along with smart appliances and applications. The concept of ambient intelligence arises from this need to react with users at runtime and keep providing real-time services under changing conditions. This chapter introduces a component-oriented ontology-based approach to develop runtime adaptable ambient intelligence systems. In this approach, the adaptability mechanism is enabled through a component-oriented method with variability-related capabilities. The outcome supports the find-and-integrate method from the idea formation to the executable system, and thus reducing the need for heavy processes for development. Intelligence is provided through ontology modeling that supports repeatability of the approach in different domains, especially when used in interaction with component variability. In this context, an example problem exploiting the variability in the density of a smart stadium network is used to illustrate the application of the component-driven approach.

Keywords Ambient intelligence · Component-based software development
Runtime adaptability · Variability modeling · Smart networks · Smart systems

4.1 Introduction

With the introduction of application programs in the early years of computing, the exponential nature of the demand for software was well recognized that has not changed much even today. Various technologies have been developed to support

M. C. Kaya (✉) · A. Eroglu · A. Karamanlioglu · E. Onur · A. H. Dogru
Department of Computer Engineering, Middle East Technical University, Ankara, Turkey
e-mail: mckaya@ceng.metu.edu.tr

B. Tekinerdogan
Information Technology Group, Wageningen University, Wageningen, The Netherlands

© Springer Nature Switzerland AG 2019
Z. Mahmood (ed.), *Guide to Ambient Intelligence in the IoT Environment*, Computer Communications and Networks, https://doi.org/10.1007/978-3-030-04173-1_4

software development in an effort to satisfy this demand. While the related industry was experimenting successfully with compositional approaches such as service orientation and generative approaches such as model-driven development, another dimension was introduced. This new dimension is related to self-adaptation and self-configuration that define the building stones for self-development. Such an ambitious goal is being partially fulfilled by adaptability achievements that are enabled by pre-configured subproblem definitions and their solutions made available for activation at runtime. As smaller scale adaptability, existing sub-solutions can be configured to meet the variability requirements rather than switching for a complete replacement.

Both the component (sub-solution) replacement and component configuration are the implementation level mechanisms for enacting variability. Naturally, many proposals offering adaptability to ambient intelligence (AmI) systems are incorporating variability that is inspired from the software product line (SPL) based techniques [1]. Variability is addressed as the differentiating aspects of a new product that otherwise shares the commonality (complementing variability) with other products in an application domain. Such environments support the modeling of variability, constraint propagation concerning variability from the top-level models to implementation, and resolving the variability at different phases such as design time, load time, runtime, etc. For the purpose of this chapter, runtime variability is exactly what can be utilized because runtime adaptability is necessarily required.

Component-oriented software engineering (COSE) supports the fast development and deployment of software-intensive systems. In addition to the construction-level support through component-based approaches, component orientation further speeds up the earlier analysis and modeling phases by domain-specific guidance and cuts down the work for the declarative modeling that targets code generation such as in object-oriented approaches. Sometimes, component-based approaches also depend on other paradigms for modeling of the problem and the solution spaces. Assuming the existence of a mature domain [2], which corresponds to an advanced work for populating the domain model such as in SPL, the suggested approach can result in significant gains in development time and product dependability. The AmI systems will be developed guided by domain-specific notions and constructs that will quickly map to the existing components. Furthermore, the component set rendering the solution is available for adaptability. The runtime adaptability capability of such an environment, linked to the ontology-based inference that is triggered by runtime events, is exploited for an efficient solution to the runtime adaptability of AmI systems [3].

Ontology-based decision mechanisms contribute through knowledge about optimization parameters for AmI communication systems [4]. This mechanism applies to AmI systems to runtime variations with the help of the variable process model of XCOSEML—extended version of the COSE modeling language (COSEML). The proposed approach is illustrated through a density-aware and density-adaptive AmI network case study for proof of concept.

The rest of the chapter is organized as follows: Sect. 4.2 provides some background general information. Runtime adaptability challenges of AmI systems are discussed in Sect. 4.3 with the help of a case study. Then, component-oriented modeling of an AmI system is explained in Sect. 4.4 with the details of the proposed approach appearing in Sect. 4.5. Ontology-based decision support mechanism is presented in Sect. 4.6, and a discussion of the proposed approach in Sect. 4.7 as well as some open problems in the literature. The chapter is concluded after a summary of related work.

4.2 Background

In this section, we provide some background to our study that includes and refers to AmI, component-oriented software engineering (COSE), and runtime adaptability.

4.2.1 Ambient Intelligence

In the last few years, we have seen a rapid increase in the number and variety of devices connected to the Internet and other networks. We have also noticed the residual user needs for proactively managing the users and ambient environments equipped with these devices. In this regard, ambient intelligence (AmI) is a concept which allows the people to interact with their pervasive environments in a natural way by supporting some autonomous services, and intelligently exploiting some embedded devices and software in response to people demands [5]. AmI concept introduced by the European Community's Information Society Technology (ISTAG) [6], emerging with an attracting significant interest all over the world, brings up a low-cost, dynamic, interconnected, adaptable, sensible, ubiquitous, proactive, and unobtrusive solution via an intelligible decision-making mechanism with the help of effectively monitoring the environment, and users' behaviors and needs [5, 7–9].

An AmI system can include three important layers, viz., the operational, distributed, and intelligent layers [10]. The first layer includes actuators, processors, and other embedded systems used in the environment in order to sense, compute, and control the desired events, data, appliances, and devices providing support and guidance for people in such environment. The second layer has a middleware to properly distribute the data, communication protocols, the related events, databases, and computational resources. The third layer performs artificial intelligence approaches utilizing the combination of the above in order to provide a sensible solution for the environment and the people [9]. Based on the ISTAG report, an AmI system should ease the human life and communication, become cultural and community improvement-oriented, provide robust and confident control, and, if desired, the AmI system could be controlled manually.

In the ISTAG report, adaptive software, sensor technologies, I/O technologies, embedded systems, ubiquitous communications, microelectromechanical systems, smart materials, context awareness, media management, intelligent computing, and natural interaction are all considered as the components of an AmI system [7]. The AmI concept can be exploited in many areas such as industrial applications, smart offices, smart homes, smart agriculture, transportation, intelligent test-beds, tourism, entertainment, safety systems, sports facilities, shopping malls, recommender systems, healthcare applications, and many other smart environments [7, 10, 11]. AmI can be considered as a convergence of the disciplines such as pervasive and ubiquitous computing, embedded systems, context awareness, software engineering, robotics, machine learning, and most importantly artificial intelligence [7, 9, 10].

4.2.2   Component-Oriented Software Engineering

Components are independent software units that are used to build complex systems. The closest approach to component orientation is component-based software development (CBSD). Here, system development refers to using and combining pre-built and pretested components rather than beginning from scratch. Also, development of components conforming to a component model can be considered under the concept of CBSD. In addition to such considerations, component-oriented software engineering (COSE) [12] has also emerged to take advantage of the component notion at all stages of software development. The main motivation is to consider all stages of system development, from requirement specifications to product delivery, but only through the use and reuse of components.

A typical COSE development process starts with the hierarchical decomposition of the system requirements. Next, a search is conducted to find the appropriate components from a set of components (e.g., a library). If there are no suitable components, then existing ones are tried to adapt the requirements with minor modifications. The last option is developing a component from scratch, if necessary.

COSE modeling language (COSEML) [13] is the graphical tool for the COSE approach. This tool lets designers model the system hierarchically with logical (such as packages) and physical (such as components) elements along with their interactions. The tool only shows the static model of a system. Behavioral model and variability modeling capability are added to the textual version of the language—X-COSEML [14]. This version has a choreography-inspired process model called "composition specification". Also, it has a variability model separated from the static and dynamic modeling constructs of the language.

4.2.3   Runtime Adaptability of Software Systems

Runtime adaptability, which has paramount importance in terms of self-healing and self-adaptable systems, is a key concept in many research domains such as AmI applications, pervasive computing, mobile computing, robotics, machine learning, and autonomous computing. Such systems change their behavior, processes, structures, and operations based on the requirements of runtime changes in order to provide a better quality of service, reliability, and better performance. These changes may also result in some new functionalities or enhancements of nonfunctional parameters supporting operational standards and requirements [15, 16].

Inspiring solutions for runtime adaptability are provided in the Software Product Line Engineering (SPLE) domain that provides several variability mechanisms and models in order to increase the reusability of the systems. By using SPL features, dynamic SPLE aims at designing self-adaptive systems, which may include runtime adaption, context-awareness, and self-configuration since especially in recent years with the invention and enhancement of technology, the software systems have a residual complexity. A dynamically configurable and adaptive system senses the users' needs and environmental conditions or limitations at runtime since some changes cannot be detected in advance. Then, it changes the system behavior according to its system variability, features, and architectural models [15, 16].

4.2.4 Ontology-Based Systems

According to Gruber [17], an ontology can be defined as "a formal, explicit specification of a shared conceptualization". Due to the ability to semantically define concepts and their relationships, many domains can be modeled using ontologies. It is preferable to incorporate ontology-based design for numerous systems since ontology can provide analysis and reuse of the domain knowledge. Thus, it is possible to develop specific systems for specific areas in a structured and flexible way.

Ontologies are used for many objectives in software design and development such as specification, reusability, reliability, maintenance, and knowledge acquisition [18]. They are often used specifically for service-oriented and component-based developments.

4.3 Case Study and Problem Description

In this section, we describe the problem statement using an AmI case study on a density-aware and density-adaptive AmI network that is used throughout the chapter. This case study is used to explain both the problem and the solution.

4.3.1   Case Study: A Density-Aware Density-Adaptive AmI
   Network

In this case study, we consider a network management subsystem of a smart sports stadium system. This system is designed to provide a communication service to the stadium users. Base stations, network controller, and user equipment are the basic devices and systems located in this environment. The capacity of the stadium network is sufficient for regular events. However, each type of event has different numbers of participants. For instance, the capacity of the network may not provide adequate quality of service in some rare events that involve more participants than usual, such as derby games and big concerts. In that case, as the number of stadium users increases significantly, namely, the user density increases; the users can share more data such as photos and videos, which reduces the performance of the network and requires more bandwidth and resources. This can be tackled by adding additional base stations. Therefore, the user and base station density of networks change at runtime, which should be monitored and measured at runtime. The network should be adaptable to these changes in order to provide a better quality of service and experience.

As an adaptable network environment, our solution includes base stations, network controller, user equipment, and mobile base stations. All of these appliances are considered as components of the system in our intelligent software that controls the environment and components therein.

The network becomes a dense network because of a derby match or a big concert including more and denser participation. The network controller has an ability to understand this requirement by sensing the environment, monitoring the network traffic or using a knowledge-based method via employing some calculations to estimate the user density. The network controller component can be deployed in base stations or at the edge of the network. In this study, the network controller is a component of mobile edge computing (MEC). By using this intelligent ability, the network controller can enable the required number of mobile base stations in this environment. Any mobile cells added to the system at runtime are also included in the model as components.

As it can be seen in Fig. 4.1, while the first scenario corresponds to a sparse network in terms of the number of participants, during the derby matches or concerts, the density of the stadium users is getting higher. In order to increase coverage and capacity of the network, instead of using some stationary or fixed solutions, mobile base stations can be used for providing a better quality of service to the users.

When the user density is significantly large, network densification, which refers to the addition of more cells in the environment, can provide a solution to enhance the network capacity, coverage, and quality of service. Therefore, nomadic or mobile base stations can be beneficial instead of deploying fixed cells. With the network densification, the density of mobile base stations should also be controlled in order to adapt the network functions and coverage area to the base station density changes. In this case study, when the number of small cells is increased in order to prevent

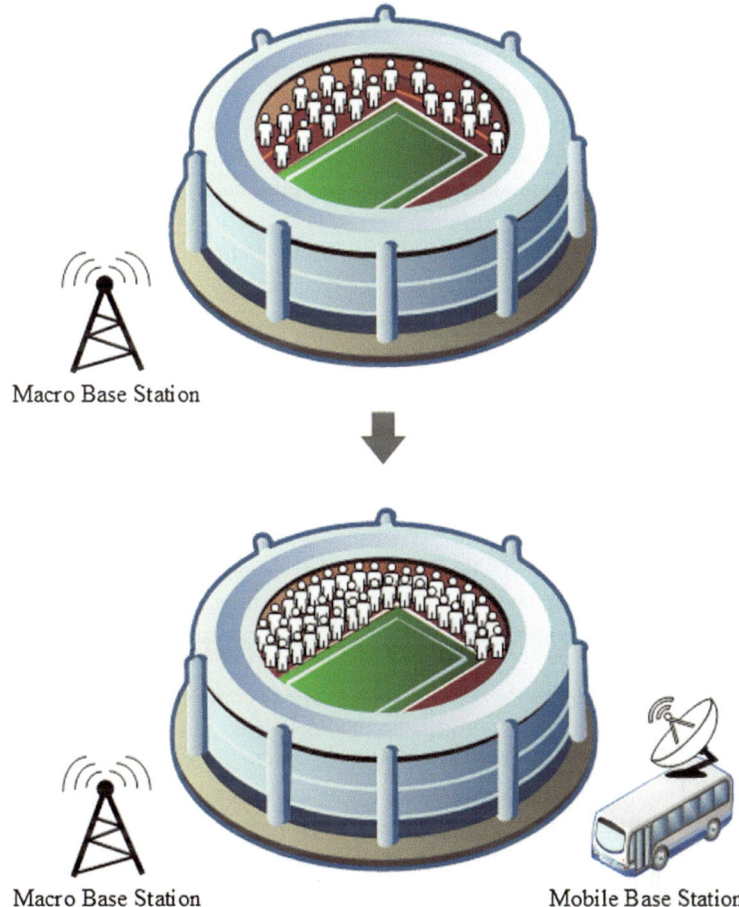

Fig. 4.1 Illustration of sparse and dense networks in a smart stadium

high interference, power consumption and network capacity can be increased. Base stations can change their coverage areas like high or low coverage by altering transmission power which is a possible solution for cell-size adjustment. In this case, we assume that mobile base stations will be connected to another remote macro base station for backhauling. This case is not illustrated in Fig. 4.1 for the sake of simplicity.

4.3.2   Problem Description

With the development of different technologies and smart systems, complexity of systems increases. This makes it harder to meet user requirements for smart systems. AmI systems are expected to meet these requirements in a reliable and unobtrusive way at runtime which can be achieved by self-configurable smart systems. Moreover, AmI systems are used in very different domains. Thus, it is hard to develop a new solution from scratch for each new problem. These challenges can be overcome by configuring AmI systems in a systematic way exploiting commonality and managing the variability of pre-built assets. Also, this variability management must provide runtime support. Therefore, using component technologies along with runtime adaptation capabilities is a sensible solution.

4.4   Modeling the Smart Stadium System

In this section, we use XCOSEML to model the case study described in the previous section. The necessary updates of the smart stadium to deal with the network density are conducted by runtime variability, which is triggered by an ontology-based component.

The variability model of the XCOSEML is called "configuration interface". This model contains variation points, variants, binding times, and constraints. Variation points and variants are shown by variability tags in the process model of the language, namely, the "composition specification". These tags are followed by one interaction or a composite structure that contains a set of interactions. After binding variants to variation points if the condition in the variability tag holds, following interactions are included in the final system. Otherwise, they are removed. This shows how a product is configured using XCOSEML. Components, connectors, and interfaces of that product are included/excluded or configured by selection of the interactions of the composition specification. If the name of the components and connectors are used in a selected interaction, they are included. Interfaces and their corresponding methods are included if they take part in an already included connector. The connector usage and variability binding operation are explained in detail in [14, 19].

Managing variability differs based on the binding times of variation points. In XCOSEML, if the binding time of a variation point is not runtime, all variability tags and their related interactions should be removed from the composition specification of a running system. However, for runtime variability, the selection is made while the product is running. Therefore, variability tags are kept in the composition specification of a product if the binding time is runtime and inclusion/exclusion of system parts or necessary updates are held dynamically.

The configuration interface (SmartStadium_conf) and the composition specification (SmartStadium_cmps) of the smart stadium are provided in Listings 4.1 and 4.2, respectively. In our case study, a macro base station is used to provide services for the smart stadium for a normal density (we called it "sparse"). However, if the network is more intensive than usual (we called it "dense"), then a mobile base station must be included to provide service in support for the macro base station. This decision is taken by the network controller at runtime by estimating the density.

In Listing 4.1, "densityCondition" variation point is defined with its variants "sparse" and "dense". Its binding time is assigned as "runtime". The default variant for this variation point is "sparse". Variability tags in the composition specification use this variation point and its variants as shown in Listing 4.2 at lines 7 and 13.

Listing 4.1 Configuration interface of the Smart Stadium System.

```
1  Configuration SmartStadium_conf
     of Package SmartStadium_pckg
2    externalVP densityCondition:
3    alternative
4      variant sparse
5      variant dense
6    defaultVariant sparse
7    bindingTime runtime
```

The composition specification can be considered as a running program that executes included interactions. Therefore, all unnecessary interactions are removed before runtime. However, runtime variability requires a different approach. As a solution, interactions for possible additions and modifications to the system are kept as passive in the composition specification. Square brackets ("[" and "]") are used to indicate the passive parts of the system at that time. In our example, when the smart stadium system starts, the "densityCondition" variation point is bound with "sparse" as it is the default variant. For this reason, while the sequence of interactions (Listing 4.2, lines 8–12) controlled by the variability tag at line 7 is activated, the other sequence (lines 14–21) becomes passive. Based on the activation of interactions, declarations at the beginning of the composition specification are dynamically updated. In lines 3 and 4, corresponding component and connectors become active when the "dense" variant is selected. To indicate the activation, square brackets are removed from the activated part. To passify the unused parts, square brackets are added.

For the smart stadium system, four basic parts of the system are modeled as components: network controller, macro base station, user equipment, and mobile base station. The network controller component is responsible for the management of the network. It decides when a change is necessary for the system by monitoring the network status with the help of its decision support system. Then, if a change is required, it triggers variant selections in the composition specification. Macro base station is the main service provider of the system. User equipment includes devices that people use to avail services that the system provides. Mobile base station is used to keep providing services seamlessly when the network load is unmanageable. In "SmartStadium_cmps", these components are represented as NC_comp, UE_comp, MacroBS_comp, and MobileBS_comp, respectively. When "sparse" is selected, NC_comp, UE_comp, and MacroBS_comp components are included. In this state, users request different services from the smart stadium. This is handled in the system by connecting user equipment and the macro base station (line 9). Also, the network controller requests density of the network from the macro base station to take action if required (line 10). Moreover, network controller sends directives to the macro base station, in this example to determine the coverage of the macro base station (line 11). The interactions are conducted through connectors. Connectors consist of different parts that we call "connector messages". Each message contains the interfaces used for that communication and corresponding messages. Moreover, messages can indicate operations performed by the connector and communication protocols of the requester and the responder sides of the communication [20]. An example representation of connectors and connector message are as follows: at line 10, "NC-McBS_conn.densityReport" indicates that the NC-McBS_conn connector is used to connect NC_comp and MacroBS_comp by its "densityReport" message.

Network controller estimates the density of the network using the reports sent by the macro base station. If the controller decides the network needs the mobile base station to keep providing service, it binds the "densityCondition" variation point as "dense". Then, because "sparse" and "dense" are alternative variants to each other, the sequence of interactions related to the "sparse" variant becomes passive and the sequence between lines 13 and 21 is activated in "SmartStadium_cmps".

A configurator updates the composition file, and then this change is reflected to the other parts of the system. In this case, the mobile base station is added to the system as a new component (MobileBS_comp). Besides, two connectors are added to the system to communicate with the mobile base station (MbBS-UE_conn, NC-MbBS_conn). Also, in the "sparse" mode, the macro base station covers all the area by high transmission power (line 11). However, when the "dense" variant is set, the network controller begins to use macro and mobile base stations with low transmission power and low coverage area (lines 19 and 20) to apportion the network load of the stadium between two stations and avoid high interference.

Listing 4.2 Composition Specification of the Smart Stadium System

```
1    Composition SmartStadium_cmps
2      import configuration SmartStadium_conf
3      has component NC_comp MacroBS_comp
                      UE_comp [MobilBS_comp]
4      has connector NC-McBS_conn McBS-UE_conn
                      [NC-MbBS_conn] [MbBS-UE_conn]
5
6    Method StadiumProcess:
7     #vp densityCondition ifSelected(sparse)#
8      sequence (
9        UE_comp -> MacroBS_comp {McBS-UE_conn.serviceReq}
10       NC_comp -> MacroBS_comp
           {NC-McBS_conn.densityReport}
11       NC_comp -> MacroBS_comp
           {NC-McBS_conn.coverageHigh}
12     )
13     #vp densityCondition ifSelected(dense)#
14     [sequence (
15       UE_comp -> MacroBS_comp {McBS-UE_conn.serviceReq}
16       UE_comp -> MobileBS_comp {MbBS-UE_conn.serviceReq}
17       NC_comp -> MacroBS_comp
           {NC-McBS_conn.densityReport}
18       NC_comp -> MobileBS_comp
           {NC-MbBS_conn.densityReport}
19       NC_comp -> MacroBS_comp {NC-McBS_conn.coverageLow}
20       NC_comp -> MobileBS_comp
           {NC-MbBS_conn.coverageLow}
21     )]
```

Different products can be derived from managing composition and component variability of XCOSEML. Other assets of the system are determined by inclusion/exclusion of different interactions as explained above. In other words, in XCOSEML, composition specification is changed first, and then other assets change accordingly. Each different configuration of interactions implies variability in composition specification. Other assets can be changed later: components can be added or removed in this way.

Listings 4.3 and 4.4 contain detailed descriptions of connectors "McBS-UE_conn" and "NC-McBS_conn", respectively. In a connector definition, along with connector's service and connector type, connector messages are shown. A connector message is defined to handle different method calls, communication protocols, and connector operations.

Connectors and interfaces are also configurable at runtime. When a runtime change requires usage of a new connector message in the same connector, the passive messages change to active. Active and passive parts of the connectors are shown by using square brackets (Listing 4.4, lines 17–21) as in the composition specification. This representation is also the same for interfaces. Usage of new interfaces or new methods for an existing interface in a new connector message can be handled dynamically.

User equipment represents devices such as smartphones or personal digital assistants (PDAs) used by the visitors of the smart stadium. The connector between a user equipment and a base station can manage communication type of the base station with the device. In Listing 4.3, requester and responder communication protocols are shown in lines 10 and 11 for "McBS-UE_conn" connector. This helps to manage the smart stadium network with different types of devices. For example, for the cellular network, some devices support long-term evolution (LTE) while some of them do not. For LTE-supported devices the "McBS-UE_conn" connector is handy. However, as an example, for 3G compatible devices, requester and responder protocols can be converted to 3G. Connector usage for heterogeneous communication protocols is explained by Kaya et al. [20].

Listing 4.3 The McBS-UE_conn connector

```
1    Connector McBS-UE_conn
2      ServiceType communication
3      ConnectorType stream
4
5      ConnectorMessage serviceReq {
6      RequesterInterface UE_int
7      MethodOut requestService
8      ResponderInterface MacroBS_int
9      MethodIn provideService
10     RequesterProtocol LTE
11     ResponderProtocol LTE }
```

Figure 4.2 shows the model of the smart stadium system when the "sparse" variant is set. In this configuration, there are three components, three interfaces of these components, and two connectors.

In Fig. 4.3, the model of the system is shown when the "dense" variant is set. This change results in the addition of the mobile base station component and two connectors of it and modifications of existing connectors and interfaces. For example, the "coverageLow" message of the "NC-McBS_conn" connector is used instead of the "coverageHigh" message after the runtime change. This change results in another modification in the interfaces of the network controller and macro base station. Before the change, "NC_int" sends a request via the "requestHighCoverage" method and "MacroBS_int" answers via the "provideHighCoverage" method

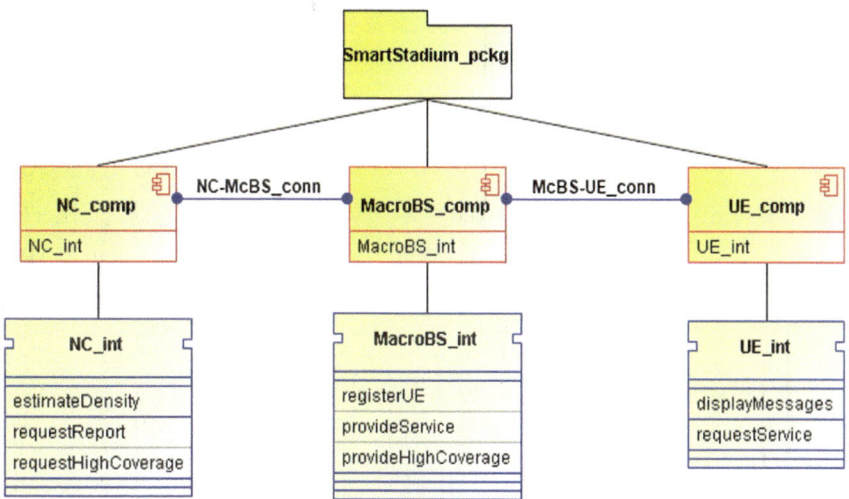

Fig. 4.2 Smart stadium system for the sparse network case

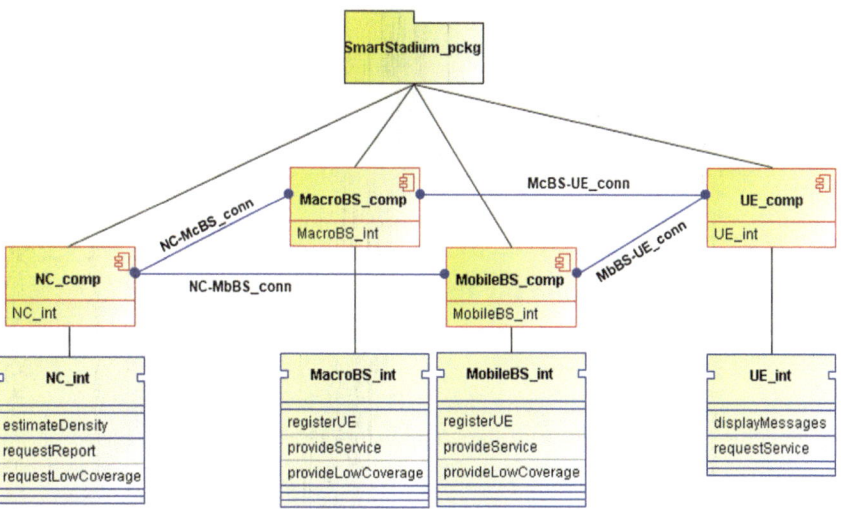

Fig. 4.3 Smart stadium system for the dense network case

(Fig. 4.2). After the update, these methods are replaced by "requestLowCoverage" and "provideLowCoverage".

Listing 4.4 The NC-McBS_conn connector

```
1    Connector NC-McBS_conn
2      ServiceType communication
3      ConnectorType procedureCall
4
5    ConnectorMessage densityReport {
6      RequesterInterface NC_int
7      MethodOut requestReport
8      ResponderInterface MacroBS_int
9      MethodIn provideReport }
10
11   ConnectorMessage coverageHigh {
12     RequesterInterface NC_int
13     MethodOut requestHighCoverage
14     ResponderInterface MacroBS_int
15     MethodIn provideHighCoverage }
16
17   [ConnectorMessage coverageLow {
18     RequesterInterface NC_int
19     MethodOut requestLowCoverage
20     ResponderInterface MacroBS_int
21     MethodIn provideLowCoverage }]
```

4.5   Density Estimation for the Smart Stadium System Network

In this section, the proposed approach is elaborated. A conceptual model is provided for modeling AmI systems with the component-oriented approach. Based on this model, the logic behind the smart stadium system is explained. Moreover, user density estimation in the smart stadium network by using network density ontology is provided in detail.

Figure 4.4 shows the conceptual model for runtime adaptability. Process model picks variability information from the variability model and manages the system. A runtime adapter is employed to monitor the system during runtime and configure the system when necessary. Runtime adapter is a part of the system: in our case, it is represented as a COSEML/XCOSEML component. While a separate component can be employed as a runtime adapter, an existing component can be used as the adapter by adding necessary functionality. A runtime adapter contains a domain ontology, a decision system, and a configurator. It receives the runtime information from the rest of the system, sends queries to ontology to decide the system's state, and configures the system when the conditions hold. Configurator of the runtime adapter changes

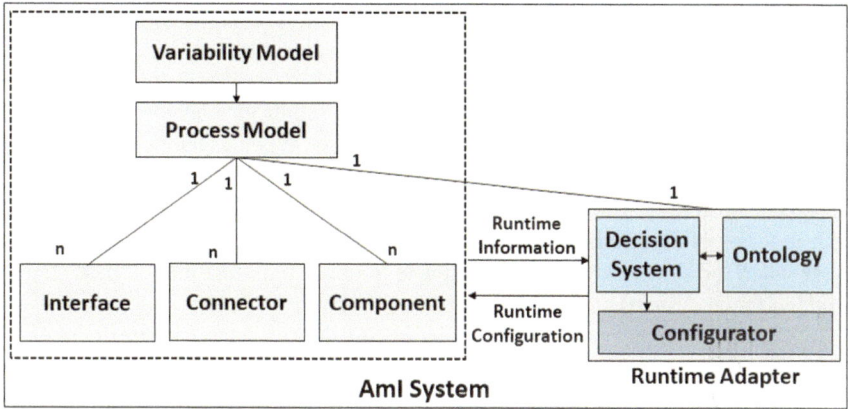

Fig. 4.4 Conceptual model of the runtime adaptable AmI systems

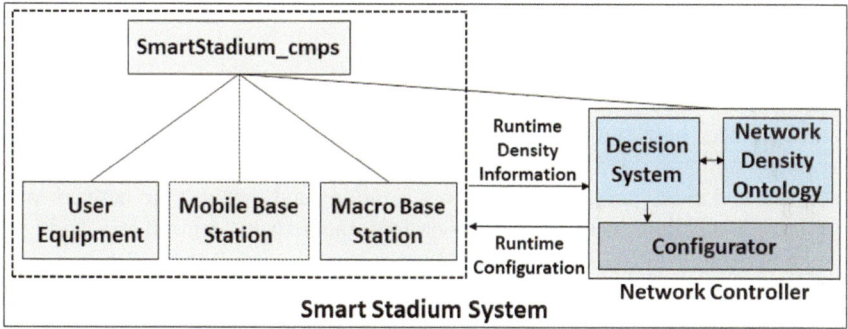

Fig. 4.5 Runtime model of the smart stadium system

the process model by adding and removing square brackets to the process model to indicate active blocks of interactions. Then, component, connector, and interface models are changed based on the change in the process model. After updating all models, the system is updated by changing, adding, or removing software and/or hardware.

Runtime model of the smart stadium system is derived from the conceptual model, and is provided in Fig. 4.5. The process model of the system—SmartStadium_cmps—picks the variability information from the "SmartStadium_conf" configuration interface before runtime. Therefore, "SmartStadium_cmps" has the variability information at runtime.

In Fig. 4.5, we show components of the system along with the process model by omitting connectors and interfaces. The smart stadium system has the network controller, macro base station, and user equipment components. The mobile base station component is added to the system according to network density. Network controller is the runtime adaptor of the system. While Figs. 4.2 and 4.3 show the network functionality of network controller, Fig. 4.5 depicts the role of the component in runtime adaptability. While the system is running, network controller receives the network density information from the rest of the system (usually from the macro base station). The decision system of the network controller decides the network density with the help of the network density ontology. When a change is needed, the configurator affects the SmartStadium_cmps first, and other XCOSEML models are changed accordingly. Then, additions or subtractions or modifications are performed both in terms of system hardware and code. In this manner, necessary changes are made when the system switches from sparse to dense (e.g., the mobile base station is added to the system).

In order to make the network density-aware and density-adaptive, density of users and base stations should be accurately measured at runtime. Therefore, we need robust density estimators. There are different estimation methods such as monitoring network traffic, channel quality indicator level, GPS-based methods, and received signal strength RSS-based approaches [21–23]. In our case study, we select an RSS-based method using received signal strength. RSS is a power measurement of a received radio signal which is a function of distance between a receiver and transmitter. Assume that each user equipment can receive a signal from the macro base station, and each user equipment sends back this sample to the mobile network edge over the macro base station. Mobile edge computing is a key concept promoting computational power and storage at the network edge in order to lessen the network workload and energy consumption of base stations and increase the network performance in a cellular network [24]. Therefore, the network controller is deployed at mobile edge. While performing the density estimation, we need to determine the channel model to convert these RSSs to distances, therefore the network controller should estimate the path-loss exponent which can be used a for path-loss model. After we determine the channel model, the simple path-loss model is performed [23], and then we can obtain estimated distances for each user equipment by using the received signal strengths, the path-loss exponent, and the transmit power. By using these estimated distances, the neighbor relation between the base station and users can be calculated. The division of the summation of neighbor proximity degrees to estimated areas for each RSS measurements collected from k nearest user equipment and consequently effective user density can be found. By considering a threshold value which depends on the participant capacity of the stadium can be determined by the network operators as a design parameter, and the user density can be classified as a sparse or dense network.

In addition to the mentioned approach for calculating user density, since we consider this system as a future cellular network, device-to-device communication is possible. Based on this assumption, we can divide the area into some clusters. In these clusters, we can select a user equipment or a sniffer as a cluster head. Then we can employ different estimators based on distance-matrix or collective RSS measurements [22, 23]. In this case, we can assume that each cluster head can send its measurements to the mobile edge via the macro base station and we can determine the user density of the stadium network. On the other hand, to estimate the base station density, assume that each user equipment collects received signal strengths from at least 2 base stations, and sends these measurements to the mobile edge via its closest base station. Then, by performing the density estimator model in [23], summation of neighbor connectivity degrees, their areas calculated by using estimated distances, and the density of base stations can be acquired.

Although RSS is more vulnerable to small and large-scale fading, it is easy to implement it as a density estimator. The accuracy of the results for our case study depends on the accuracy of the RSS measurements. Here, we assume that the smart stadium network has an intelligent interference management mechanism. Based on these assumptions we can employ this density estimator easily with an ontology-based mechanism.

4.6 Ontology-Based Decision Mechanism

An ontology-based decision mechanism is proposed to make the necessary changes at runtime. Networks of AmI systems are considered as a domain in our case study. Network controller components in the case studies will have this ontology-based support. When a change is required at runtime, this controller informs the process to act accordingly. The decision mechanism is designed as an expert system that uses the ontology model we propose. Thus, it is intended to take advantage of the benefits of ontology constructs.

Jena framework [25] is used while the expert system was being developed. Jena provides a programming interface for ontology application development, independent of the ontology language. This consistent interface provides developers with an environment that integrates the programming language, query languages, ontologies, and reasoning engines. Using Jena rules, inference in OWL and RDFS ontology languages is possible. Jena also offers a complete solution by providing a transitive and generic rule reasoner.

The proposed ontology was created through Protégé tool [26]. Protégé allows to store our ontology in OWL format. Thus, semantic querying and reasoning can be performed on this ontology. There are many factors to consider when performing density estimations in the network management process in smart stadiums. Taking these factors into consideration, the classes and relationships in the created ontology described in Fig. 4.6 have been determined as follows:

- While RSS-based density estimation is performed, the dimension modeling of the network is an essential factor affecting calculations [23]. Density estimation may differ depending on the dimension type. When the dimension type of the network varies at runtime, the estimation method of density will need to be changed accordingly. For this reason, dimension modeling in the constructed ontology should be treated as a separate class.
- "Channel Model" is another important class in ontology. This class helps with how the value of "path-loss exponent" is to be determined. This value can be predicted using a formula or directly fetched from the "Network Controller". "Channel

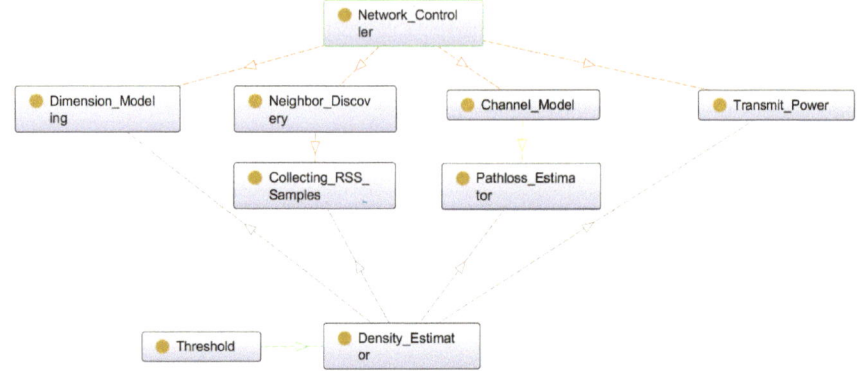

Fig. 4.6 Network density ontology

Model" class dynamically picks one of these two options and sends the resulting choice and value to the "Path-loss Estimator" class [23]. "Path-loss Estimator" class receives the choice and value, and calculates the final path-loss exponent value.

- Another factor that affects the estimation of density is the number of closest neighbors to be taken into account. Methods for efficiently determining the number of closest neighbors to be considered in the density estimation are called "neighbor discovery-based approaches" in the literature [23]. For this reason, we have defined one of the main classes on the ontology as "Neighbor Discovery". This class helps how the extraction of the neighborhood matrix of each node to be determined. This value can be estimated or retrieved directly from the network controller. This selection is fetched by "Collecting RSS Samples" class and accordingly, the summation of nearest neighbor value is calculated.
- Since the transmit power value has an essential role in density estimation, it affects the estimation result. Thus, "Transmit Power" should also be defined as a separate class.
- The task of "Density Estimator" class is to take the calculated values from "Dimension Modeling", "Collecting RSS Samples", "Transmit Power", and "Path-loss Estimator" classes and use it to calculate the final density estimation value. This value is compared with the values in the "Threshold" class to determine whether the density increases or decreases. According to this information, the process model is stimulated. This stimulation will be conducted with inference from Jena rules, taking into account the consequence of the density estimation.

4.7 Discussion

Any solution that modifies an executable software at runtime is liable to conduct necessary adaptations within the software system. After cleaning up the issues with adapting the software, issues about the runtime problems that might occur during this modification should be resolved: system might miss the reception of events during the change process. Furthermore, the process for reacting to an ambient information itself could have been interrupted. So, there can be many related issues that need to be addressed in a domain-specific manner, even at a case-by-case mode identifying what to do in such cases. This measure seems to be solution agnostic, at least for problem-level identification and formulation. The approach proposed in this chapter does not address such issues which are treated as requirements issues that complement the enumerated set of adaptability related event cases. They can be assumed as complex re-initialization procedures to be carried out after a modification. Some general procedures for addressing runtime configurability have already been studied in the Fault Management domain. There, mostly redundant copies of the alternative solutions are considered for runtime replacement. Here, alternative solution parts will be considered for replacement, possibly rendering most of the mechanisms provided by the Fault Management domain usable.

The conceptual model in Fig. 4.4 describes the way how ambient intelligence systems adapt to changes at runtime. This model is faithful to the decomposition principles of the COSE approach. The runtime adapter can be employed as a component of the system. It can be a functional component in that domain (as in our case study) or it can be a separate component whose only responsibility is managing changes of the system at runtime. This modular design is easily adaptable to other domains.

Ontology-based development of the decision mechanism improves extensibility and flexibility. The adaptation of the developed ontology to different domains will be, therefore, greatly facilitated. In this way, runtime adaptation can also be performed in different domains outside the smart stadium. Apart from that, it is also quite practical to be able to modify the constructed ontology for the smart stadium domain. While only RSS-based estimators are used when estimating density in the case study, different estimators can be used in different applications. The ontology will be able to hold common classes and relations between them and easily reconfigure them with new classes and relationships that need to be defined. Beyond that, it will even be possible to combine the results of different estimators. The combined result may have a higher success rate than the single estimator.

4.8 Related Work

In this section, we describe the related work with respect to modeling AmI systems with component-based development; as well as on runtime adaptable AmI systems. Moreover, the literature on ontology usage in smart systems is also reported.

Ambient-intelligent systems provide aggregation of information obtained from various kinds of interconnected smart devices and the environments. The information integration and the human-smart system interaction give rise to two complex problems which should be manageable at runtime in order to make the human-smart system interaction more natural and make the gathered, processed, and reasoned information more distributed, reliable, robust, and reusable. At this point, component-based software engineering (CBSE) eases the interaction between the systems and human being and make software development more adaptable. Vallecillos [27] uses a component-based method considering the relations between the system components and provides a solution to construct the user interface. In comparison to the conventional approaches for developing interactive systems in ambient systems such as user interfaces, the solutions relying on a component-based method attract the attention of researchers as a runtime approach.

Issarny et al. [28] introduce a declarative language for AmI systems that builds upon web services. The language enables automatic retrieval and composition of web services at runtime providing QoS parameters such as security and performance through connector customization. In the proposed approach, required services are sought first from local repositories, then from the web. Services are composed dynamically if needed. Detailed modeling of AmI systems along with variability is not a part of this work.

Dynamic variability modeling and reconfiguration of the feature models for variant addition, selection or removing by using context-aware knowledge requires runtime adaptable approaches [15, 16, 29]. In that sense, a runtime mechanism dynamically changing the structural variability model for a wireless sensor and actuator network (WSAN) including a set of sensor nodes and actuators which sense and give reaction according to the environmental changes at runtime based on the context-aware knowledge is addressed in [16]. For instance, a museum guidance system as an indoor ambient network needs adaptation at runtime in order to be dynamically reconfigured by adding, removing, or modifying its variants of the system's feature model.

Ortiz et al. [16] propose activating some variants in the variability model of this system in order to instantly control whether some ambient measurements (such as humidity, temperature, smoke, and light) are higher than a determined threshold value or not in order to set off some events since this variant called "Event-based" is not pre-activated. Furthermore, the threshold value as an important attribute of this feature must be updated according to the environmental changes. For all nodes, the modification of variabilities in the system feature model requires also proper software updates. In addition to selection variants of system feature model, new hardware and new features can be added into such systems, which requires adding a new variant to the feature model by considering critical and proper variation points and upgrading the system and nodes' software and configuring threshold value regarding the type of corresponding ambient measurements.

Some dynamically adaptive systems include high variabilities in consideration of system design, deployment, and different potential scenarios or application areas such as video surveillance systems requiring runtime configurations. Moisan et al. [30] propose a framework based on model-driven engineering (MDE) perspective for runtime configuration in video surveillance systems. MDE allows developments from a system's abstract level of software to the implementation by using model transformations and software models. They consider an intrusion detection system in a warehouse or a room used as a case scenario for video surveillance system. In the study, feature diagrams are used for the representation of the systems most probable variation points. They consider the system concepts or entities as features such as "Intrusion detection". It is proven that the variability modeling at feature or component level relying on MDE methods and software engineering methods using feature and component models ease the management of the execution of a video surveillance system and the system configuration at runtime. Thus, in this study, we propose a reasoning-based component-oriented modeling in order to provide runtime configuration for an ambient-intelligent dynamic system.

In the literature, there are many studies in which ontologies are used with smart systems. A significant part of these studies is also related to AmI. Besides, it has been determined that the domains where ontologies are used with smart systems are quite diverse. Homola et al. [31] addressed a number of approaches to the area of knowledge representation and reasoning (KR) by associating AmI and KR in their survey. The research in [31] analyses many studies related to ontology evolution, debugging, and reasoning. Stavropoulos et al. [32] introduced an ontology called BOnSAI (Smart Building Ontology for Ambient Intelligence) to enable Ambient Intelligence in a Smart Building. BOnSAI is designed to be domain-dependent and utilizes existing ontologies. Thus, it utilizes these ontologies to model the domain-specific concepts of an AmI application.

Fan et al. [33] proposed an IoT-based smart rehabilitation system that uses an ontology to aid computers with a better understanding of symptoms and medical resources that help to determine a rehabilitation strategy quickly and automatically and reconfigure medical resources according to patient-specific needs. Kim and Park [34] proposed a scalable and flexible ontology model called SHOM for the U-Health Smart Home project. An autonomic system is created using SHOM to make accurate and intelligent decisions.

Teimourikia and Fugini [35] have developed a decision support system aimed at providing runtime safety management in Smart Work Environments. In this ontology, details of the concepts related to the safety domain and how these concepts can be expanded according to a specific use case are shown. It is also proposed to define the constraints of the ontology using logic-based rules.

Karamanlioglu and Alpaslan [36] have developed an ontology-based smart system to automatically detect Service-Level Agreement violations. This system, named SLAVIDES (Service-Level Agreement Violation Detection System), performs an inference by processing the information it receives as a service message. The ontology used by this system is designed generically for use in many different domains.

4.9 Conclusion

In this chapter, a component-oriented approach is employed for realizing runtime adaptability of AmI systems. The key concept in supporting the adaptability has been the variability capability of the approach used in conjunction with an ontology-based inference that supports the portability of the approach for different solution domains. An example case study was provided, to illustrate the approach. The approach was conducted and observed to be fast in development and deployment of a system where adaptations were also straightforward and fast.

One important detail that is not presented in this chapter is the "cleanup" operation after a runtime adaptation task. This operation is related to potentially missed events and interrupted processes. Although such issues are not made more difficult by this approach, future work could prove useful for their inclusion in a solution that presents a more complete view. The components to be included in a domain to support this approach can have specific methods to serve for the necessary operations. Also, some missing work left for future is the testing of the case study in real life, due to the difficulties in exploiting such a large-scale system with thousands of human users.

References

1. Gámez N, Fuentes L (2011) FamiWare: a family of event-based middleware for ambient intelligence. Pers Ubiquitous Comput 15(4):329–339
2. Togay C, Dogru AH, Tanik JU (2008) Systematic component-oriented development with axiomatic design. J Syst Softw 81(11):1803–1815
3. Hansen K, Zang W, Fernandes J, Ingstrup M (2008) Semantic web ontologies for ambient intelligence. In: Proceedings of the 1st international research workshop on the internet of things and services, Sophia-Antipolis, France, pp 1–6
4. Liu Y, Seet BC, Al-Anbuky A (2014) Ambient intelligence context-based cross-layer design in wireless sensor networks. Sensors 14(10):19057–19085
5. Augusto JC (2006) Ambient intelligence: basic concepts and applications. In: International conference on software and data technologies. Springer, Heidelberg, pp 16–26
6. IST Advisory Group (2001) Scenarios for Ambient Intelligence in 2010, European Commission
7. Ramos C, Augusto JC, Shapiro D (2008) Ambient intelligence—the next step for artificial intelligence. IEEE Intell Syst 23(2):15–18
8. Augusto JC (2009) Ambient intelligence: opportunities and consequences of its use in smart classrooms. Innov Teach Learn Inf Comput Sci 8(2):53–63
9. Hornos MJ (2017) Application of software engineering techniques to improve the reliability of intelligent environments
10. Sadri F (2011) Ambient intelligence: a survey. ACM Comput Surv (CSUR) 43(4):1–66
11. Obukata R, Oda T, Barolli L (2016) Design of an ambient intelligence Testbed for improving quality of life. In: Proceedings of the 30th international conference on advanced information networking and applications workshops (WAINA), Crans-Montana, Switzerland. IEEE, pp 714–719
12. Dogru AH, Tanik MM (2003) A process model for component-oriented software engineering. IEEE Softw 2:34–41
13. Dogru AH (1999) Component oriented software engineering language: COSEML, Technical report TR-99-3, Computer Engineering Department, Middle East Technical University, Ankara, Turkey
14. Kaya MC, Suloglu S, Dogru AH (2014) Variability modeling in component oriented system engineering. In: Proceedings of SDPS the 19th international conference on transformative science and engineering, business and social innovation, Kuching Sarawak Malaysia, 15–19 June 2014
15. Bashari M, Bagheri E, Du W (2017) Dynamic software product line engineering: a reference framework. Int J Softw Eng Knowl Eng 191–234

16. Ortiz O, García BA, Capilla A, Bosch J, Hinchey M (2012) Runtime variability for dynamic reconfiguration in wireless sensor network product lines. In: Proceedings of the 16th international software product line conference, vol 2. ACM, New York, pp 143–150

17. Gruber TR (1993) A translation approach to portable ontology specifications. Knowl Acquis 5(2):199–220

18. Ruiz F, Hilera JR (2006) Using ontologies in software engineering and technology. Ontologies for software engineering and software technology. Springer, Heidelberg, pp 49–102

19. Cetinkaya A, Kaya MC, Dogru AH (2016) Enhancing XCOSEML with connector variability for component oriented development. In: Proceedings of SDPS 21st international conference on emerging trends and technologies in designing healthcare systems, Orlando, FL, USA, 4–6 December 2016

20. Kaya MC, Nikoo MS, Suloglu S, Tekinerdogan B, Dogru AH (2017) Managing heterogeneous communication challenges in the internet of things using connector variability. In: Mahmood Z (ed) Connected environments for the internet of things. Computer Communications and Networks. Springer, Cham

21. Basere A, Kostanic I (2017) Spatial sampling requirements for received signal level measurements in cellular networks. In: IEEE 7th annual computing and communication workshop and conference (CCWC), Las Vegas, NV, USA, pp 1–4

22. Locher T, Wattenhofer R, Zollinger A (2005) Received-signal-strength-based logical positioning resilient to signal fluctuation. In: Sixth international conference on software engineering, artificial intelligence, networking and parallel/distributed computing and first ACIS international workshop on self-assembling wireless network, Towson, MD, USA, pp 396–402

23. Eroglu A, Onur E, Turan M (2018) Density-aware outage in clustered ad hoc networks. In: 2018 9th IFIP international conference on new technologies, mobility and security (NTMS). IEEE, pp 1–5

24. Chen L, Zhou S, Xu J (2017) Energy efficient mobile edge computing in dense cellular networks. In: 2017 IEEE international conference on communications (ICC), Paris, France, pp 1–6

25. Apache Jena (2015) A free and open source java framework for building semantic web and linked data applications. https://jena.apache.org/. Accessed 28 Apr 2015

26. Noy NF, Sintek M, Decker S, Crubézy M, Fergerson RW, Musen MA (2001) Creating semantic web contents with protege-2000. IEEE Intell Syst 16(2):60–71

27. Vallecillos J, Criado J, Padilla N, Iribarne L (2014) A component-based user interface approach for Smart TV. In: 2014 9th international conference on software engineering and applications (ICSOFT-EA), pp 455–463. IEEE, Vienna

28. Issarny V, Sacchetti D, Tartanoglu F, Sailhan F, Chibout R, Levy N, Talamona A (2005) Developing ambient intelligence systems: a solution based on web services. Autom Softw Eng 12(1):101–137

29. Floch J, Hallsteinsen S, Stav E, Eliassen F, Lund K, Gjorven E (2006) Using architecture models for runtime adaptability. IEEE Softw 23(2):62–70

30. Moisan S, Rigault JP, Acher M, Collet P, Lahire P (2011) Run time adaptation of video-surveillance systems: A software modeling approach. In: International conference on computer vision systems. Springer, Heidelberg, pp. 203–212

31. Homola M, Patkos T, Flouris G, Šefránek J, Šimko A, Frtús J, Baláž M (2015) Resolving conflicts in knowledge for ambient intelligence. Knowl Eng Rev 30(5):455–513

32. Stavropoulos TG, Vrakas D, Vlachava D, Bassiliades N (2012) Bonsai: a smart building ontology for ambient intelligence. In: Proceedings of the 2nd international conference on web intelligence, mining and semantics, p 30. ACM

33. Fan YJ, Yin YH, Da Xu L, Zeng Y, Wu F (2014) IoT-based smart rehabilitation system. IEEE Trans Ind Inform 10(2):1568–1577

34. Kim J, Park SO (2015) U-health smart system architecture and ontology model. J Supercomput 71(6):2121–2137

35. Teimourikia M, Fugini M (2017) Ontology development for run-time safety management methodology in smart work environments using ambient knowledge. Futur Gener Comput Syst 68:428–441
36. Karamanlioglu A, Alpaslan FN (2018) An ontology-based expert system to detect service level agreement violations. In: Proceedings of the 8th international symposium on business modeling and software design, BMSD

Part II
Frameworks and Methodologies

Chapter 5
Developing WLAN-Based Intelligent Positioning System for Presence Detection with Limited Sensors

Ivan Nikitin, Vitaly Romanov and Giancarlo Succi

Abstract WiFi-Based Positioning Systems (WBPS) play a key role in indoor navigation, but further development of these systems continues to this day. WBPS have been applied to different tasks including mobility tracking and behavior analysis. Mobility tracking allows detecting a user in the environment even if one does not use positioning services. Tracking enables sensing the human presence in different environments, including occupancy detections in smart homes, geofencing, enhanced security and many other scenarios. One of the basic performance criteria of a positioning system is its precision. The general rule states that precision grows with the increase of the number of reference signals used for positioning. However, it is unclear how much information is required to estimate the location of a person reliably. This chapter overviews the current research in the area of Received Signal Strength Indicator (RSSI) based positioning and evaluates a positioning system for localizing a person in an indoor environment, taking into account the number of Access Points (APs) available for estimating the location. We conduct performance analysis of an indoor positioning system based on measurements from a real walk. Additionally, we conduct a simulation, where we analyze the impact of the noise on the positioning quality.

Keywords Indoor localization · WiFi · Mobility tracking · Markov model
Positioning system · WIFI-based positioning system · Simulation · GPS

5.1 Introduction

About two decades have passed since WiFi became a widely used instrument for positioning. It has come a long way from being a conceptual prototype to the de facto localization method in densely populated public places. WiFi-based localization methods allow for positioning in urban areas even through devices such as low-end

I. Nikitin · V. Romanov (✉) · G. Succi
Innopolis University, Innopolis, Russia
e-mail: v.romanov@innopolis.ru

© Springer Nature Switzerland AG 2019   95
Z. Mahmood (ed.), *Guide to Ambient Intelligence in the IoT Environment*, Computer
Communications and Networks, https://doi.org/10.1007/978-3-030-04173-1_5

tablets and laptops that lack both cellular and GPS (Graphical Positioning Systems) capabilities. This became possible through the advent of continuous data collection by smart mobile devices [1]. Once we collect the information about the actual location and the available WiFi signals, we can construct the mapping from the latter to the former. In practice, such an approach proves to work well only in scenarios where we have a plethora of reference signals (known as AP beacons) available. As the number of observed access points (APs) decreases, the positioning accuracy deteriorates quickly [2–6].

Now consider a different problem that arises much more often in the context of ambient awareness and sensing. Assume that we want to learn a relative position of a mobile device and a group of sensors. In this context, a mobile device can act as a sensor, or vice versa, and a group of APs can sense the presence of this mobile device. One of the most useful applications of such sensing system is the ability to determine the location of a mobile device, or, in general, a person [7]. Now, the problem reduces to a well-studied positioning task.

Lately, WiFi has become used for more than just positioning. There were some attempts to utilize it as a behavioral analysis tool. WiFi sensing has also become more available recently after new smart devices started to transmit probe packets that allow them to detect familiar networks quickly. These packets can be used to sense the presence of a device even if it is not associated with an AP. In this respect, there are two major approaches to WiFi sensing:

- The first is based on an AP occupancy count that allows to gather overall statistics about the area business
- The second taps into the meaning of received signal strength indicator (RSSI) value and attempts to use the information about the signal strength to get more granular occupancy data.

By looking at the sequence of visited APs, one can infer the traveling preference of a person. By comparing the signal strengths directly, one can detect the proximity of two people [8]. Chilipirea et al. studied the ways to reduce the noise in the sensed WiFi data [9]. Bellavista et al. [10] compared different experimental datasets that can allow studying human interaction patterns given different measurements, including WiFi signals [10]. Li et al. designed a system that uses WiFi to detect abnormal mass activities [11]. Traunmueller et al. utilized WiFi probe packets to study the common traveling patterns in the city of New York [12]. Basalamah explored an experimental system that allows monitoring the mobility of a crowd using probe packets [13]. Scheuner et al. presented another open source system for WiFi tracking [14]. Vanderhulst et al. presented a sociological based model that uses WiFi to detect spontaneous human encounters using WiFi [15]. Ma et al. studied building occupancy by counting the presence of people in the range of APs and passively collecting probe data [16]. Acer et al. described their experience collecting data about the interaction of people on a large-scale industrial event using WiFi beacons [17].

We can see an increased interest in using WiFi as a tool for crowd behavior analysis. We can expect increased attention to this as the relevant technologies become more accessible and available. The possibility of tracking enables more general awareness about the environment. Moreover, Received Signal Strength Indicator (RSSI) based techniques are transferable to virtually any radio sensor that is capable of measuring the received signal strength.

WiFi gradually becomes a tool for activity recognition and passive tracking. Wu et al. proposed a model to detect three different types of activity using WiFi: stationary, walking, and running [18]. Seifeldin et al. proposed a system for general wireless passive monitoring that allows detecting the position of a person with radio signals, including WiFi [19]. Zhou et al. presented a system for passive human detection [20]. Wang et al. studied the way to identify the activity without an active device [21]. Wu et al. presented a system for non-invasive detection of moving and stationary human with WiFi [22]. Abdelnasser et al. designed a system for gesture recognition using WiFi [23]. Some of these approaches can operate in a device-free manner, allowing to sense activities even for people without any wearable device. We will see later that despite making the appearance, passive tracking methods still have a way to travel before they can be seamlessly deployed.

In this chapter, we aim to look at positioning with WiFi as a tool for ambient sensing, namely sensing human presence. We focus on RSSI based approaches as they offer the simplicity of use. Many techniques for human detection are simply radio-based and are not limited to WiFi. While most of the methods, we are going to discuss later in this chapter, are applicable in a generic radio-based navigation system, WiFi has one significant advantage: its ubiquitous nature. Modern positioning systems rely heavily on the presence of WiFi beacons [24]. The research has shown that the localization precision only grows with the increase in the number of reference signals [2], but it is unclear fundamentally how small this number could be. In a later section of this chapter, we intend to discuss the precision analysis of a positioning system that operates with only a few beacons (APs) available.

The organization of this chapter is as follows. A brief over of existing methods for localization with WiFi is presented in Sect. 5.2. A more detailed formulation of the problem and the requirements for positioning system are given in Sect. 5.3. Section 5.4 overviews state of the art method in RSSI-based localization, including radio propagation models and positioning methods. Section 5.5 discusses how the positioning error is measured, and what are known sources of error for identifying a given location. Section 5.6 touches upon the concept of privacy associated with the topic of WLAN-based positioning. Section 5.7 describes the structure of the proposed prototype of a positioning system, including the description of the radio map estimation method, the mobility model, and precision evaluation. Section 5.8 concludes the chapter.

5.2 Overview of Positioning Approaches

Many different approaches for positioning with WLAN have been developed over the years [7, 24–30]. They include techniques based on RSSI measurements, temporal delays, channel state information, and angle of arrival (AoA). In an ideal situation, we should be able to use WLAN infrastructure for location services in an energy efficient fashion without additional sensors and any changes to the network protocols, but this is not possible for all existing positioning approaches. In this section, we attempt to briefly overview different types of positioning methods, especially, time-based, Received Signal Strength Indicator (RSSI) based and Channel State Information (CSI) based approaches. A more detailed introduction to different methods is provided in Sect. 5.4.

5.2.1 Time-Based Approaches

Most of the time-based approaches for positioning exploit the fact that the signal travels with constant speed. Thus, knowing the time it took for the signal to propagate from one place to another gives us the notion of distance. Time-based approaches are one of the most precise and most difficult at the same time. Makki et al. gave a comprehensive overview of existing techniques [31]. State of the art methods can provide a decimeter level of positioning precision, but it comes with the cost of complexity. In general, time-based approaches are sensitive to such parameters as synchronization error, communication channel bandwidth, sampling rate, and time resolution. Applications of these methods usually require additional layers of protocols on top of IEEE 802.11 standards, and sometimes even modification of hardware [31]. In general, based on proposals in the future IEEE 802.11 standard, this direction of research looks promising, but not readily applicable at the current stage. In this chapter, the focus is on RSSI-based approaches primarily, and time-based techniques are left out of consideration.

5.2.2 Approaches Based on Received Signal Strength Indicator (RSSI)

RSSI based approaches are, by far, the most well-studied ones [1, 32–34]. This group of methods is often referred to as fingerprinting techniques, and the idea is that the value of received signal strength, which is often characterized by the received signal strength indicator (RSSI), is a function of distance or the location in space. RSSI is one of the noisiest sources of information [35], and for this reason it is often combined with other measurements such as angle of arrival [36], device's inertial unit (IMU) readings [37–39], proximity to other people [40, 41], as well as other relevant measurements [42–44].

The advantage of RSSI based approaches is that they are applicable and straightforward even with the modern hardware. More detailed overview of RSSI-based techniques appears in Sect. 5.4.

5.2.3 Approaches Based on Channel State Information (CSI)

A group of methods that made its appearance quite recently uses Channel State Information (CSI) , which provides information about the channel impulse response [33, 45–48]. Refer to Fig. 5.1. The wireless channel suffers from the effect of fading, where the signal can interfere with its copies arrived at different moments of time over different paths. In these approaches, the approximate CSI is always calculated by the WLAN chipset when receiving a new packet as it is a necessary stage in the decoding process. Thus, one can obtain the estimated strength of multiple paths by exposing the physical layer of the WLAN modem [49].

CSI can be utilized in two ways. First, it provides a more granular notion of received signal strength. Two different locations are quite likely to have a similar value of received signal strength, but they are unlikely to have identical multipath profile [50, 51]. Second, CSI can be used to approximate the power of the shortest path between transmitter and receiver. Knowing this power, one can estimate the distance using a propagation model or a simple analytical coverage map [36, 52]. Unfortunately, CSI requires the access to PHY layer of the network protocol. Current standards require the driver modification to grant this access. Moreover, such modified drivers exist only for few selected devices.

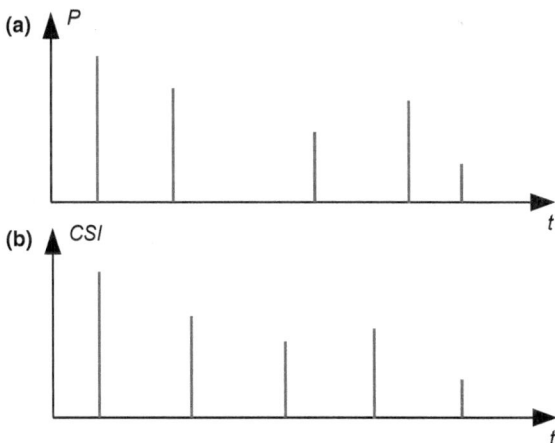

Fig. 5.1 Channel State Information (CSI): **a** wireless channel impulse response (assume impulse signal transmitted)—multiple copies of the same signal arrive at different times with different signal power; **b** resampled version of impulse response obtained by WLAN demodulator

5.3 Problem Formulation

The premise of RSSI based positioning with WiFi (or any wireless sensor network) is the presence of a large number of reference signals (APs). However, that is not the case for environments where dense network infrastructure is simply not needed. The examples of such environments could be warehouses or private residential property. Deployment of a high precision positioning system in such places could be complicated for several reasons. First, the reference ground truth position provided by sensors such as GPS is simply not available indoors, and we cannot solve the problem by creating a mapping from ground truth location to WiFi references. Second, existing alternative approaches often involve an extensive site surveying that aims to determine the relationship between the few available reference signals and the ground truth location. For these reasons the desired ad hoc systems for positioning with WiFi become hardly reachable. Third, even if we are able to construct such a mapping, it might be subject to significant variance. WiFi signals are reported to be significantly affected by noise due to interference and other hardware related biases. For this reason, the common approach to positioning with RSSI often involves many APs to reduce the effect of RSSI noise.

Modern techniques mostly focus on user-side positioning, where the user's device plays the role of a sensor that captures signals from the nearest beacon. Such strategy limits the application of such positioning systems to client-side needs. One could come up with a scenario where positioning should be performed on the side of the network. Examples of such scenarios could be in advanced security systems, geofencing for restricted network access, smart home applications, equipment tracking and other.

In classical RSSI-based positioning, a number of observable APs, together with their average signal strength, create a fingerprint that is later associated with a particular location on the map. Here, we define fingerprint as the observed signal strength of all APs in the area. APs are a perfect source of reference signals since they constantly broadcast their SSID (Service Set Identifier), and a mobile device can estimate their RSSI whenever it wants. One can design a system where the roles of APs and a mobile device are switched. In such a system, the device transmits information, and several of APs estimate the RSSI of the intercepted packet. Although such a system would be able to position a device on the side of the network in a non-invasive fashion, there are several limitations, viz: (1) a mobile device does not constantly transmit signals, and the information about the location of a device is obtained in a rather opportunistic fashion; (2) packets transmitted by a mobile device are hard to intercept. The channel frequency of a mobile device is unpredictable from the standpoint of an unassociated AP, and not any software would be able to detect packets from a protected network; (3) mobile devices due to their small antennas and energy efficient profiles transmit low power signals. As a consequence, devices can be visible only to a few APs in the neighborhood; and (4) there is much more noise in the RSSI measurements from a mobile device due to unpredictable shadowing since they can be worn in different places on a body or in bags.

Furthermore, we attempt to investigate the following questions:

- How can one sense the presence of people in a non-intrusive fashion?
- How the number of APs (sensors) affect the precision of a positioning system?
- How can one reduce the time for site surveying?

To answer the above questions, we try to investigate the state of the art in positioning WLAN-based positioning systems. The next section overviews some existing methods for WLAN-based positioning, including existing methods for analytical radio map estimation, and the positioning techniques that vary from simple KNN-based approaches to calibration free and device-free methods.

5.4 State of RSSI-Based Positioning

The classical pipeline of an RSSI-based positioning, often referred to as fingerprint-based positioning, consists of several steps (Fig. 5.2). First, one has to estimate the signal strength map that associates the physical location with a radio fingerprint. Second, a mapping function is constructed that allows for fast positioning during the localization stage. After this, the system can be deployed for operation.

Next, we are going to elaborate on each of the steps above and discuss how they can help in the presence of the following constraints:

- Positioning is performed on the side of the network
- Few reference signals are used for positioning
- The process of site surveying is simplified

5.4.1 Radio Map Construction

Radio map construction is one of the most important preparatory step in fingerprint-based localization, and the resulting positioning precision heavily depends on its quality. We mentioned before that fingerprint captures the signal strength received

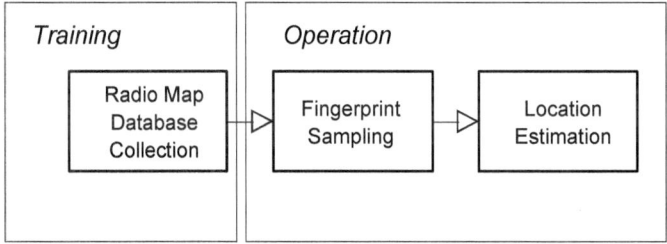

Fig. 5.2 Positioning with RSSI using fingerprints (propagation model-based approaches are not considered)

from all APs in the neighborhood. It is very likely that when many APs are involved, the fingerprint is unique for every location on the map. If we want to build a high precision positioning system, we must ensure the quality of the fingerprints and a dense fingerprint sampling (at least every meter if meter precision is required) during the site survey (Fig. 5.3). There are two ways to associate a location on the map with the signal strength. The first involves analytical model computation, the second resorts to empirical data collection. Now, we are going to explore these two approaches in more details.

5.4.1.1 Analytical Propagation Model

Analytical models were one of the earliest applied to the task of WiFi positioning due to their tight relationship with classical approaches in radio navigation. RSSI based methods in particular often rely on a simple log-domain propagation equation such as:

$$P_{rx} = P_{tx} + G_{tx} + G_{rx} - L$$

where P_{rx} is the received power in dBm, G_{tx} is the gain of transmitter antenna, G_{rx} is the gain of the receiver antenna, and L is the path loss along the wireless channel.

While the rest of the parameters are perfectly deterministic, given the hardware specification, the path loss is an environment dependent variable. The amount of signal power loss depends on the number of factors, and most prominently, the distance from the signal source. There are plenty of models that try to express the relationship of the distance between the transmitter and the receiver to the amount of loss along the transmission channel [53, 54]. These models were developed primarily for the sake of wireless communication and try to capture the properties of radio waves on different frequencies, in different environments. One of the most popular model applied for RSSI positioning is a log-scale attenuation model [55]:

$$P_{rx}(d) = P_0(d_0) - 10 \cdot \alpha \cdot \log_{10}\left(\frac{d}{d_0}\right) \tag{5.1}$$

where $P_{rx}(d)$ is the received power as a function of the distance, $P_0(d_0)$ is a received power measured on the reference distance d_0 from the transmitter, α is an attenuation exponent.

This model provides a very simple way to estimate the received signal strength, but it was shown to exhibit some uncertainty [56]. The problem with this model is that it takes into account only overall environment properties (e.g., attenuation exponent), and therefore the resulting accuracy of RSSI estimates is often low. Wireless signal decays as it naturally propagates through the environment. Log-distance attenuation model accounts for additional sources of signal power loss, such as numerous obstacles, such as furniture, plants, through attenuation exponent. The higher its val-

(a)

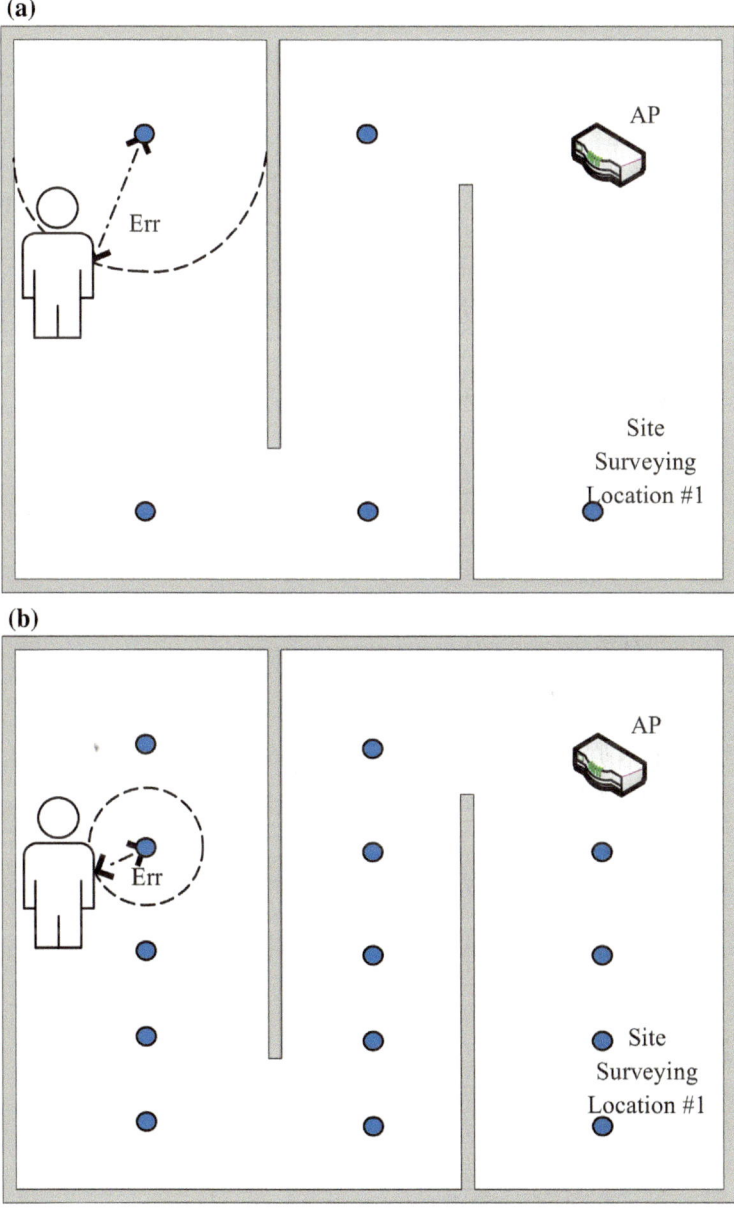

(b)

Fig. 5.3 Example of fingerprint sampling scheme: **a** fingerprint collection is seamless, but positioning quality may suffer; **b** denser sampling allows achieving higher precision

ue—the faster signal decays. However, this approach is unable to capture the granular information about the environment and prone to systematic errors.

5.4.1.2   Ray Tracing

Ray tracing is a viable alternative to analytical propagation models. It is a technique that was extensively applied in the field of computer graphics. It models the propagation of light waves, which in their nature are radio waves. Ray tracing approaches attracted much attention in the field of coverage prediction [55, 57–62]. These approaches include both deterministic and Monte-Carlo estimation of the signal strength distribution. The error between different methods varies around 3–9 dBm [59, 63–66] with the best result of 3 dB achieved in [67].

Techniques for coverage map estimation were also applied to the problem of mobile device localization [55, 58, 68]. Most of these methods rely on a multipath signal propagation model and do not use probabilistic estimation techniques. Before stating the multipath model, let us introduce a number of assumptions:

- The radio signal propagates in straight lines
- The radio signal is partially reflected from the walls of the buildings. $\gamma, \beta \in$ [0, 1] are reflection and transmission coefficients correspondingly, where each one denotes the fraction of the original power
- The angle of reflection is equal to incidence angle; no scattering is assumed.

One of the properties of wireless signal is that it partially reflects from obstacles on its way. When a reflected signal arrives to the receiver, it can cause interference, which results in attenuation. The multipath propagation model is formalized by the following equation [55]:

$$P(l) = 10\log\left(\sum_{i=1}^{K} P_{tx} \cdot L_i(l) \cdot \gamma^{N_{ref}} \cdot \beta^{N_{tr}}\right)$$

where P_{tx} is the transmitted power in watts, L_i is the path loss attenuation factor of a ith path, γ and β are reflection and transmission coefficients of walls respectively, N denotes either number of reflections or transmissions, and K is the number of possible paths between the transmitter and the receiver. The value of L_i is calculated using standard propagation equation [53]:

$$L_i^{(dB)} = 10\alpha \log d_i + 20\log f - 147.55$$

Here, α is an attenuation exponent, f is the WiFi operating frequency, d_i is the distance traveled by the signal along ith path. Scattering is not considered in this model as it was suggested to result in insignificant gains [55].

Needless to say, the application of these approaches requires the knowledge of the floor plan and wall materials. In most of the cases, this information is hardly available. The floor plan can often be inaccurate, and the knowledge of wall material can be idealistic. The piping and wiring in the walls can significantly affect their reflective properties. For this reason, numerical optimization of hyperparameters is employed. The details of optimization procedure will be discussed in later paragraphs in this chapter.

Estimation of the radio map is a clusial stem and there are not many ways to make it labor-intensive. We have presented generic analytical approaches to estimating the average value of the signal in a given location on the map. However, the positioning task tries to solve the opposite problem of estimating the location from a few measurements. We are now going to overview existing approaches to constructing such a mapping.

5.4.2 Localization Function

In order to properly introduce different localization methods, let us formally define the positioning problem. Let $\mathbf{a} = \{a_1, \ldots, a_M\}$ be the set of APs at the disposal of the service provider, where M is the number of available APs. For each network user, a history of observed signal strength is defined with a matrix $\mathbf{R} \in \mathbb{R}^{N \times M}$ with each element being $r_n^{a_m}$, $n = 1 \ldots N$ the nth sample of signal strength received at the AP a_m. Such representation is redundant since most of the time only one associated AP can have the access to RSSI, with rare exceptions. Alternatively, this information can be represented as the sequence of tuples $< r, a_m >_n$, where r is the observed RSSI and a_m is the AP associated at the moment of time n.

The reasonable approach to define the mapping from RSSI to the location on the map is through signal coverage. First, we define the mapping from location to the average signal strength as

$$\mathcal{M} : \mathbf{l} \to \mathbf{r}, \mathbf{l} \in \mathbb{R}^2, \mathbf{r} \in \mathbb{R}$$

where l is the vector for location on the map, and r is the value of RSSI at this location. Evidently, the mapping $\mathcal{M}$, unlike its inverse $\mathcal{M}^{-1}$, is unique. Further, we are going to refer to the inverse $\mathcal{M}^{-1}$ as to localization (positioning) function. Moreover, the observations $\mathbf{r}$ are affected by noise at the receiver, which makes the inverse mapping even less straightforward and more ambiguous. For this reason, additional constraints should be imposed on the inverse mapping procedure. This problem can be interpreted as receiving a corrupted message through a noisy channel (Fig. 5.4).

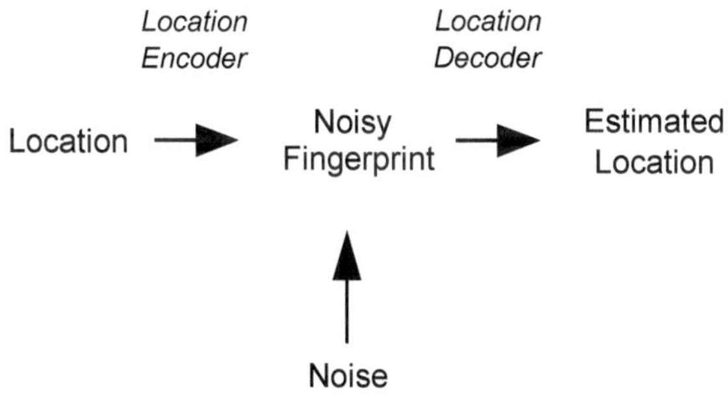

Fig. 5.4 Positioning as a noisy channel estimation problem

Location Encoder

Location Decoder

Location → Noisy Fingerprint → Estimated Location

↑

Noise

In the following sections, we now present brief overview of some existing positioning techniques.

5.4.2.1 Approaches to Localization

In the literature, many positioning approaches have been proposed. These can be categorized into two classes: deterministic and probabilistic.

Deterministic functions create a distinct mapping from the fingerprint to the location on the floor plan that is based on similarity measures. The idea of similarity can be expressed in the following equation:

$$\hat{l} = \mathrm{argmax}_{l \in S}\, p_{det}(l, \mathbf{f}),$$

where $\hat{l}$ is the location estimate, S is the set of all possible location, $\mathbf{f}$ is the fingerprint vector, and p_{det} is a deterministic positioning function. Such methods often use variations of KNN with different distance measures [35, 46, 69–72].

The main difference between the probabilistic approach and positioning is that instead of a unique location, positioning function creates a joint probability distribution over possible locations and fingerprint space. Formally, this is expressed as the following equation:

$$\hat{l} = \mathrm{argmax}_{l \in S}\, p(l, \mathbf{f})$$
$$= \mathrm{argmax}_{l \in S}\, p(l|\mathbf{f}) p(\mathbf{f}),$$

where p is the joint probability density of a location l and an observed fingerprint $\mathbf{f}$. One can reduce this problem to MLE estimator by assuming uniform distribution in the fingerprint space $p(\mathbf{f})$.

Probabilistic approaches allow for more precise positioning as they tend to capture more diverse information about the environment. Another advantage of probabilistic approaches is their ability to model coherent sequences, such as trajectories. Instead of performing pointwise location estimate, one can maximize the probability of a whole path as:

$$\hat{\mathbf{l}} = \mathrm{argmax}_{\mathbf{l} \in S^k}\, p(\mathbf{l}, \bar{\mathbf{f}}),$$

where $\mathbf{l}$ is the path of length k, and $\bar{\mathbf{f}}$ is the sequence of fingerprint samples. The variations of probabilistic positioning functions include Naive Bayes [46, 52], Markov models [42, 43, 68], and models with latent space [73–77].

Methods presented so far work with an existing database of fingerprint measurements. However, very often one wants to alleviate the need for collecting the database in the first place, or to be exact, spend significant resources on its collection.

Next, we are going to introduce techniques that allow deploying a positioning system that can work without such database or constructs the database on the fly.

5.4.2.2   Calibration-Free Methods

Given the information as presented above about the methods for radio map estimation, it is worth mentioning several calibration-free approaches as well. The goal of these methods is to alleviate the need for radio map construction at all. There are several ways to achieve this.

The condition of radio environment is known to be non-stationary. The path loss parameters defined in (5.1) can change over time, and, in general, can be different for different APs. To alleviate the need for manual site surveying, Chintalapudi et al. proposed a method that relies on APs calibrating each other's parameters using non-convex optimization [78]. The positioning is later performed using trilateration. Although such system is easy to deploy, the quality of positioning can suffer due to simplistic propagation model. Lim et al. proposed a similar scheme where the inter-AP measurements are used for online calibration.

An alternative approach to alleviating the site surveying effort is crowdsourcing. Yang et al. proposed to transform conventional floor plans into a high dimensional space where Euclidean distance corresponds to the traveling distance between two points [79]. The need for such transformation comes from the fact that two points are not necessarily connected with a direct path due to the presence of obstacles, such as walls. The crowdsourcing based on measuring the distance between transient positions (step counts) allows the system to collect and adapt the positioning model on the fly.

Wang et al. proposed a system that can automatically detect different landmarks, such as corridors, elevators, and others [80]. Knowing the moment when the user passes the landmark will allow aligning the rest of the user's path with the floor plan. The sensors of a mobile device can be used for path estimation. This system would allow to fill in the gaps with references to unknown locations automatically. Shen et al. propose a similar idea of utilizing sensor fusion and prior knowledge to enhance positioning and facilitate data collection [81].

There are also a number of Simultaneous Localization and Mapping (SLAM) methods for localization with WiFi [74, 82–86]. They are different in a way that they allow to create an internal representation of the environment that is fully suitable for positioning. The challenge is to align the internal map with the actual one. Many of these techniques require the information from Inertial Measurement Unit (IMU) of the mobile device, which is hurting both the battery life of the devices and the possibility of easy deployment.

Thus far, we have considered methods that can use a mobile device as a beacon to locate the bearer. However, some techniques do not require any active device to ensure positioning and work in a passive manner. Next, we are going to briefly discuss how these methods work.

5.4.2.3 Device-Free Approaches

There have been many attempts to bring a device-free positioning into life [87]. The approaches involve many different mediums of information, including pressure, thermal, and sound. In the scope of this chapter, we are interested in wireless-based positioning approaches. Device-free localization implies the tracking system where the object being positioned participates in the process passively. We were able to observe two major trends in Device-Free Positioning (DFP): radio tomography and received signal strength (RSS) fingerprint-based.

Radio tomography is a method that is based on received signal strength along multiple communication links [88, 89]. The idea is to place a significant number of wireless sensors along the perimeter that constrains the positioning area (Fig. 5.5). Sensors communicate with each other, and these communication channels create a set of links. Whenever an object shadows a link, the RSS of several sensors drop. By knowing the location of sensors and the direction of the link, one can restore the position of an object. The limitation of this approach is in difficulty to infer when the changes in RSS occur due to the presence of a human, or due to the changes in the environment.

There have been improvements for the radio tomography approach. Some of them deal with the non-stationary nature of the links [90], some with tracking multiple targets [91], some address multipath problems [92]. The best achieved positioning

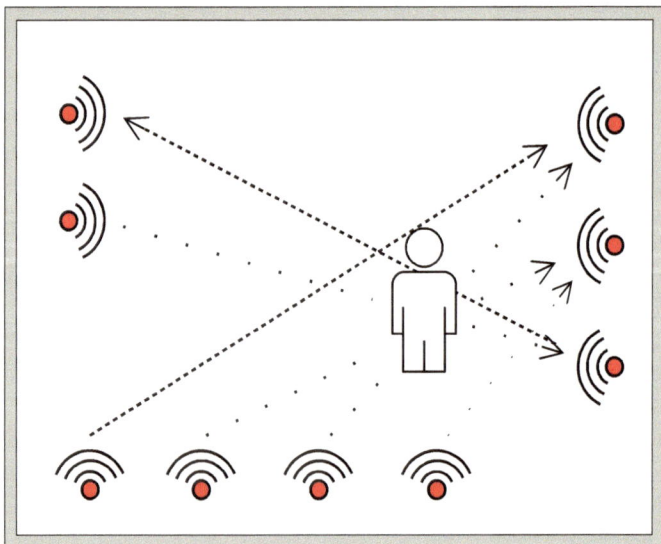

Fig. 5.5 Radio tomography: there are several sensors in the room; some sensor pairs create a link; when an obstacle is present the link is broken; one can estimate the position of an obstacle when several links are broken

accuracy in on decimeter level. Despite these attempt, the most important problem—the number of sensors required for positioning—remain.

An alternative to radio tomography is RSS-based device-free positioning. Unlike previous approaches that created a closed area for tracking, RSS DFP relies on multipath to detect the location of a person. When the wireless signal propagates through the environment, it passes through and reflects from walls and other objects. The presence of a person in the area also impacts the way signal propagates. This fact is used to create RSS DFP. These methods are similar to conventional positioning with fingerprints in a way that they require to creates a mapping from received RSSI to the location where a person is present at the current moment of time. An attractive benefit of this type of approaches is that they can work with as few as one single WLAN AP. The probing of the person's location is performed using the links between the AP and several associated devices. The value of RSSI received by one of the devices in a link changes when the person changes his/her location.

A positioning system like this can be build using a commodity WiFi [19, 93] and other types of RF transmitter–receiver pair [94]. Even though these approaches can work with a few wireless sensors, the most evident drawback is the necessity to create the mapping from person's location to RSSI. Needless to say, the mapping cannot be reused for tracking multiple people.

DFP provides an interesting way to sense the presence of a human, or other activity in the environment, but it currently suffers from several major drawbacks.

5.4.2.4 Limited Information Performance

It was shown before that proper selection of reference signals is crucial to the positioning quality [95]. Often, it is recommended to discard the unreliable measurements with the goal to increase the localization performance. However, a consequent question may arise of how few APs are needed for positioning. Traditional methods of localization based on triangulation and trilateration established this number to be equal to three. For indoor environments, this number is no longer valid due to uneven distribution of wireless signal strength. It does not necessarily lead to a worse outcome, because the presence of the obstacles on the way of wireless signal propagation creates a signal distribution pattern that can be later utilized as an additional source of information.

Most of the existing RSSI based methods are applicable for positioning with one or a few APs, but the performance with such a low number of APs can be discouraging. Nevertheless, several research papers studied the effect of the number of APs. Figuera et al. performed a study where they applied the bootstrap technique for obtaining confidence on positioning [4]. They compared the strength of their method by changing the number of APs from 1 to 4. The results show that the most significant gain is achieved when the number of APs increases from 1 to 2, which corresponds to almost doubling the positioning quality. Adding the third AP reduced the mean error by 25%, but further increase of APs did not lead to statistically significant improvement. Mean error for positioning with a single AP constituted 5.18 m.

Positioning systems can benefit from various sources of information. This includes different priors, such as prior for visited locations, prior for transitions between locations, and so on. Youssef et al. set themselves a goal of positioning with the optimal strategy [3] where the influence of prior user profile on the positioning quality was studied. The main hypothesis of this work is that the prior knowledge about the locations where a particular user can be located greatly improves the positioning performance. Indeed, the results of this study show that in some cases the average positioning error can drop from 20 to under 5 m when using a single AP. But the obvious question of how to estimate these prior remains.

Another kind of prior was used by Zàruba et al. [42]. They modeled the movement of a person in an indoor environment as a random process, and generative model for this process played a role of mobility model prior. The difficulty of this approach is in the temporal correlation of consecutive samples and increased computational pressure on the positioning system. The reported average error of their method constitutes 2.3 m.

The main idea behind previous approaches was in increasing complexity of the model, without introducing any new source of additional information. Sen et al. also studied the influence of the number of APs on the positioning quality [6] while combining RSSI measurements with IMU readings. Their sensor fusion system CUPID achieves single AP positioning error within 8 m, which is halved when five APs are available. It is worth noting that overall low positioning error is achieved due to more sophisticated approaches comparing to previous methods.

It is in practice possible to achieve positioning with signal strength measurements only, but the precision will highly depend on the quality of measurements and noise. One can use additional information about the environment topology to reduce the search space. In the following, we are going to consider how a mobility model can help in position estimation.

5.4.3   Mobility Model Assistance

The approaches that we considered before were primarily based on RSSI readings and pre-recorded coverage maps. Although these systems have proved to be reliable and precise, they consider only a static positioning problem, and one can further improve the positioning quality by harnessing another type of information—spatial proximity. When a positioning system tries to identify a person's location, the fingerprint value is compared with the pre-recorded database, and the location is estimated. In the next moment of time, the process is repeated. One can note that the set of possible locations shrinks if the previous location was identified correctly. Thus, the temporal coherency of locations can help to increase positioning accuracy. The most natural approach would be to filter out the locations that are too far from the current to travel within a constrained time period [39, 55]. However, this solution is too simplistic and definitely can be improved. There are two major trends in incorporating the information about spatial proximity into positioning system: model-based and measurement-based approaches.

Model-based approaches assume the relationship between consecutive locations. This relation is often expressed as a conditional probability function:

$$p(\mathbf{l}) = \prod_{i=1}^{N} p(l_i | \mathbf{l}_{i-1..1}),$$

where $\mathbf{l}$ is a vector of consecutive locations. In a general probabilistic approach, the probability of the current location is a function of all previous locations. To simplify the model estimation, many adopt first- or second-order Markov model [42, 43, 68].

The next question appears to be how to parameterize the transition distribution $p(l_i | \mathbf{l}_{i-1..1})$. One of the simplest approaches is to model it as a normal distribution. In this case, the model is assumed first- or second-order Markov model and can be written as the following:

$$p(l_i | l_{i-1..i-2}) \sim \mathcal{N}\left(\mu_{i-1,i-2}, \Sigma_{i-1,i-2}\right)$$

Note that the transitions are modeled as a normal distribution, locations l are defined in 2D space. The mean and variance can be based on the assumption of zero mean velocity ($V \sim \mathcal{N}(0, \sigma_V^2)$) or acceleration ($A \sim= \mathcal{N}(0, \sigma_A^2)$). The computational complexities for such models are often reduced by using Kalman or Particle filtering [36, 88]. The distribution function for transition probabilities can also take time into account. For example, they can assume that the sample rate is fixed, then we can place constraints on how far a person can go and incorporate environment topology such as the location of walls [2]. A more sophisticated model can also impose the rotational dependency where the person tends to continue moving in the same direction [91].

Other mobility models include a waypoint model with resting [44, 96]. In this model, a person is assumed to pick a random location on the map and start walking towards this direction. Once reaches the destination, the person decides to rest for a duration that is modeled with an exponential distribution.

Wu et al. suggested to map the floor plan into the higher dimensional space where the distance between two locations is equal to traveling distance [39]. Prediction of the next possible position can also be made with a sub-optimal approach where the previous locations are considered correct, and the region of next possible locations is estimated from the current location and the traveling speed [69].

Measurement-based approaches tend to use the same kind of models for filtering but heavily rely on the IMU readings [29]. The access to IMU can significantly reduce the positioning error, but also reduces the battery life of the mobile device [97]. Needless to say, mobility models with measurements are not applicable for DFP or when positioning is performed solely by the network.

5.5 Error Analysis

A general problem with comparing different positioning algorithm is the absence of a unified benchmark. The performance of an RSSI based positioning system is usually evaluated only empirically since it is a function of an intractable number of parameters. Empirical evaluations depend heavily on the specific test site. For this reason, most of the comparison between fingerprint-based positioning techniques should be done carefully.

5.5.1 Error Measurement

The choice of the error metric for positioning system is important as it defines the strength of the approach. Standard approaches to measuring positioning quality are:

- Average error
- Root mean square error
- Error CDF

Average error is one of the most straightforward metrics and can be calculated as

$$err_{avg} = \frac{1}{N} \sum_{i=1}^{N} e_i,$$

where e_i is the error between estimated location and ground truth measurement for ith sample.

Root mean square error (RMS) is a commonly used alternative and it is able to capture the positioning error variance

$$err_{RMS} = \sqrt{\frac{1}{N} \sum_{i=1}^{N} e_i^2}.$$

Error CDF is also a commonly applied metric that is capable of capturing the error distribution of a positioning system

$$err_{CDF}(x) = P(error < x).$$

Besides conventional metrics, more advanced tools are also available. Thus, Youssef et al. used an error function that captures the expected error value [3]. Although it gives more insight, it is harder to estimate the error value in general.

$$\mathbf{E}(Err) = \sum_{l \in L} \mathbf{E}(Err|l)P(l)$$

where l is the correct location, and $P(l)$ is some location prior.

5.5.2 Error Sources

A WiFi positioning system can suffer a significant precision loss due to a number of uncontrolled variables. The inherent source of error is the unexplained variance. The research of the best positioning method based on WLAN is an ongoing process, and some work is dedicated to investigating the influence of different parameters on the robustness and stability of localization with WiFi [95, 98]. Here, we are going to focus on RSSI based positioning.

Modern indoor positioning systems rely heavily on WiFi, and other smaller range beacons. Existing systems tend to utilize all possible resources they have to provide the most reliable location estimate. However, to truly understand the fundamental limitations of WLAN-based positioning, we need to learn its potential pitfalls.

The idea of positioning with fingerprints is based on the assumption that the expected value of RSSI the received from devices located at the same position in space is also the same. Liu et al. have found that this is not entirely true [95]. They report that different WLAN chipsets exhibit different relationship trend between the strength of received signal and the distance between transmitter and receiver, and a thorough calibration is needed before the system can adequately handle a new device. A gap between the observed RSSI value received from different chipsets located at the same location can be as large as 30 dB. In general, the RSSI is not a standardized measure of power, and its interpretation is up to a particular manufacturer [98]. Moreover, some devices showed to be unusable for RSSI-based positioning at all as their power is loosely correlated with the log-distance.

A source of error different from noisy or misinterpreted measurements is the divergence from the model. A set of assumptions that is often utilized by naive fingerprint-based positioning algorithms is

- the distribution of RSSI is lognormal,
- RSSI distribution has spatial correlation,
- the process that generates RSSI value is stationary.

Some research reports mention the inconsistency of these assumptions with reality. Kaemarungsi et al. [98] found that the lognormal assumption does not hold in general. They reported observing different distributions of RSSI conditioning on received power and the presence of line-of-sight (LOS) signal path. Moreover, they have found that the presence of LOS is also associated with higher RSSI variance, and the change in distribution parameters over time indicate the non-stationary noise process. The impact of RSSI variance on the localization is so important, that it is recommended to record it on the site surveying stage, and use it for estimating the position at processing time.

One way the variance can be utilized in the positioning is by selecting a set of reliable signals. Farshad et al. studied the influence of signal strength and variance on the positioning precision [99]. They found that the definition of the fingerprint is a crucial factor, and by considering only the signals from the strongest and the most constant sources for a given location in space, we can significantly increase the precision. This finding suggests that the unstable signals references are one of the big impactors in the final location service quality (especially when the survey takes a brief period of time). The findings are also supported by the other research where it was shown that eliminating some not confident measurements can lead to the increase of accuracy [100, 101].

We see that the final location estimate is subject to many sources of noise. For this reason, it is essential to have as much information as possible. It is reported that the best positioning accuracy is achieved when over three reference signals are present [2, 3, 98]. Whenever these many reference signals are not available, the system can benefit from some prior knowledge such as user location prior distribution [3].

5.6   Privacy with Respect to Positioning

The question of privacy is rarely raised when talking about positioning [2]. However, WLAN-based localization provides the ground for several concerns. Unlike with GPS where the location is determined on the side of the user, the most popular method of positioning in WLAN with WiFi fingerprints requires the user to share the information about his/her location with the server. Research has shown that the information about the sequence of visited locations is highly sensitive as it can reveal the identity of a mobile user. This knowledge hints on two possible modes of attack. In the first, the positioning service provider memorizes the location of the client and sells gathered data. In the second, an attacker intercepts the query and infers the location of a user. These types of attack are of crucial importance as they can allow for tracking people.

The second type of attack allows a malicious user to obtain the intellectual property of a positioning service provider illegally. When reporting user's location, the provider reveals the mapping between WiFi fingerprints and the true location. A malicious user can download an almost identical copy of the original database by sampling the fingerprint space.

The first vector of attack can be dealt with using asymmetric cipher algorithm such as Paillier Cryptosystem. The server calculates the data transmitted to the server in an encrypted way, and the distances on encrypted data. The distances are further transmitted to the client where they are decrypted, and the true position is identified.

The fingerprint database can be protected by randomizing the queries. For this, the server requests the information only about a random subsample of available APs, which complicates the reconstruction of the original database.

The described approach to providing privacy for both mobile user and the positioning service provider comes with a cost of excessive calculations on the side of the mobile device, and the negative impact on the autonomous time. Further details can be found in [2].

5.7   Structure of Proposed Prototype Positioning System

In this section, we are presenting the structure of our positioning system. For any system, the first stage of design is creating the set of satisfiable requirements. As we mentioned at the beginning of this chapter, positioning can be a part of ambient intelligence system. That is, the goal is to create a sensor network that can detect the presence of humans in a closed environment and identify their location. The extension of this task can be activity recognition, which is a function of the semantics of the quarters. We leave the last part out of our consideration and focus only on the ability to estimate person's location reliably. The requirements for our system are

- Non-invasive monitoring: Since the goal of the system is to detect human presence, despite the intent, the system should be able to do this without explicit collaboration
- Operate with only limited information available; We consider the scenario of the system operating in places like private premises or warehouses where it is not reasonable to establish a dense sensor network architecture. In other words, the design that can operate with minimal infrastructural investment is preferred
- Deployment investments are low: It is possible to create a good RSSI based positioning system by creating a thorough empirical fingerprint database. On the other hand, site surveying is a time consuming and complicated process that requires some level of proficiency and preparation, and systems where such expenses unnecessary are preferred.

In a previous section on "device free approaches", we overviewed existing device-free approaches for human presence and location detection. As we have already elaborated, these approaches are not able to satisfy all requirements simultaneously. Solutions that are efficient for location estimation require many sensors installed; which is not possible for large areas. Solutions that do not require many sensors tend to have increased demand for pre-deployment site survey. Moreover, they are limited to specific operation scenarios. As per our observation, this is a general tradeoff for such systems.

Device-Free Positioning (DFP) systems solve a passive localization problem, which is inherently complex. The estimation complexity can be alleviated by allowing for active positioning, which limits the system to detect only users that have a wearable wireless device. Although this would partially contradict the requirement about the non-invasive monitoring, it opens up promising possibilities for the positioning system. DFP CSI based systems require a thorough site survey to capture how the presence of the human body affects CSI of the wireless channel, which is effectively equivalent to capturing all possible positions of a person in the room and later to compare the fingerprint with the database. Even if we can do this for one person, scaling up to detect more people simultaneously will result in a combinatorial explosion of the storage requirements. Such problems do not exist for active positioning because the main carrier of information is the RSSI from the mobile (wearable) device, and this value can be analytically estimated with relatively low error (discussed in Sect. 5.7.1) and is relatively unaffected by the presence of other people in the room.

It is worth noting that active localization system does not necessarily require the explicit collaboration of the person to be detected. Modern mobile devices are known to transmit probe packets that are detected by APs, and thus an AP can estimate the position of a person with a mobile device by observing the signal strength of the probe packet. Alternatively, when the user is connected to the network, associated AP can actively ping for the RSSI. Thus, it is possible to build a sensing system that can detect the presence of a mobile device, which is often an equivalent to detecting a person. The only application that is not possible is security.

Given the reasoning above, we decided to implement an active RSSI based positioning system that operates in an indoor environment and utilized only a few available APs for positioning. The system does not require installation of any special application on the mobile device and is based solely on the principals of existing wireless networks and thus is easily deployed. As any other RSSI-based positioning system, a fingerprint database is needed, and we show below in Sect. 5.7.1 that a relatively good estimation of radio map can be obtained analytically without the site survey. Since only a few APs (even one) are available, the system should incorporate all possible knowledge to achieve the decent accuracy. We incorporate one of the mobility models into our system, which is described in Sect. 5.7.2. Finally, we evaluate the quality of our system using simulation in Sect. 5.7.4.

5.7.1 Radio Map

RSSI based positioning with fingerprints depends heavily on the availability of an accurate radio map. In a previous subsection of Sect. 5.4.1, we discussed a simplistic view how to estimate the received signal strength from a distance between transmitter and receiver using analytical propagation models, and gave explanations why this approach is prone to significant errors in an indoor environment. The alternative approach is to use ray tracing techniques described in a subsection of Sect. 5.4.1. From the various proposed techniques, we decided to use the method of images described in [102] due to its simplicity and optimality of the solution.

The solutions for calculating signal strength map has been studied before but were not widely adopted due to computational complexity. On the other hand, significant advances in the 3D rendering were recently achieved, with sophisticated algorithms executed on commodity graphics cards. Radio waves and light are the same physical phenomenon, and the most recent advances could be adopted from the area of 3D rendering to wireless signal strength estimation. In short, computational capabilities of modern hardware are sufficient for solving this problem when appropriately approached.

The method of images solves the following problem. Consider a complex indoor environment with radio transparent and reflective walls. The transmitter and the receiver are located apart from each other, and the line-of-sight (LOS) propagation channel is not necessarily available. The signal can take different paths when traveling from the transmitter to the receiver (Fig. 5.6). Upon arrival at the receiver, the signals

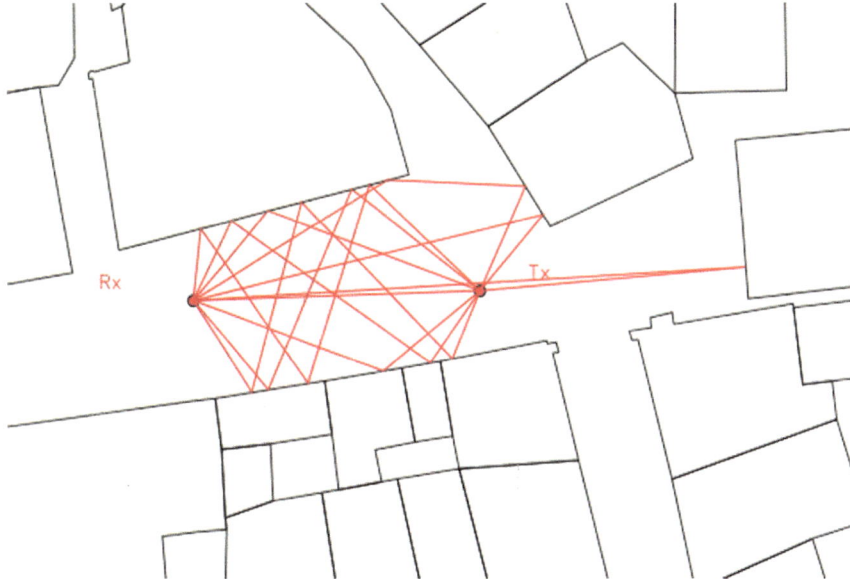

Fig. 5.6 Paths between the transmitter (Tx) and receiver (Rx)

from different paths are superimposed. The arrived energy is distributed over time, but the receiver accumulates the signal arrived within a time frame T. Let us define RSSI as the signal strength arrived at the receiver within the time frame T:

$$RSSI = \sum_{i=1}^{K} P_i^{(rx)},$$

where $P_i^{(rx)}$ is the energy of the signal receiver over path i. We also know that the parameters that define the signal strength at the receiver are the following:

- distance traveled by the signal
- number of times signal reflected from walls
- number of times signal passed through walls.

In the next section, we show that the necessary information can be collected using the method of images.

5.7.1.1 Method of Images

This method is based on the geometrical properties of reflections, that is the angle of incidence is equal to the angle of reflection. The overall rules can be described as follows:

Fig. 5.7 Algorithm
describing the method of
images in pseudocode

```
1: queue ← transmitter location
2: while queue do
3:     im ← object from queue
4:     for wall in W do
5:         new_im ← calculate_image(im, w)
6:         im.children.append(new_im)
7:         if image valid then
8:             queue.append(new_im)
9:         end if
10:    end for
11: end while
```

- At every step of the process we generate an image of the transmitter by finding its reflection on the other side of the wall;
- The signal propagates from the image to the receiver in a straight line;
- The distance between the image and the receiver is the actual distance traveled by the signal;
- The images created in a breadth-first search (BFS) manner;
- If the path between the image and the receiver does not cross the wall that creates that image, the image is invalid;
- One can calculate the number of reflection by counting the breadth order of an image;
- The number of times the signal passes through walls should be calculated on every segment of the signal path.

The method of images is described in Fig. 5.7. The first image is the transmitter itself. At the next step, we pick a wall and find the reflection of the current image and add this new image to the queue. Then, we calculate the image based on other walls. If all walls are processed, we pass onto the next image in the queue, and the process continues. One can notice that this can continue up to infinity. Indeed, the process builds a reflection tree (as shown on the right in Fig. 5.10), where all possible combinations of reflection walls are listed. However, only a few of the combinations are valid. The process continues forever unless the stopping rule is defined. In our case, it could be the path loss of the current path, i.e., we stop exploring the branch of the tree of the path loss is too high. The process is illustrated in Fig. 5.8.

In the reflection tree, every node represents a potential path. The path can be evaluated for validity only in backtrace manner, so pruning the tree is impossible unless additional heuristics introduced. The validity of paths is evaluated with algorithm as shown in Fig. 5.9.

The main obstacle in the way of radio wave propagation in our test environment is walls. The signal achieves the receiver by either direct line of sight, or by bouncing from wall to wall. A fair approach to reducing the computational pressure is providing the information about wall adjacency to eliminate impossible reflections even before thoroughly inspecting them. The simplest implementation evalu-

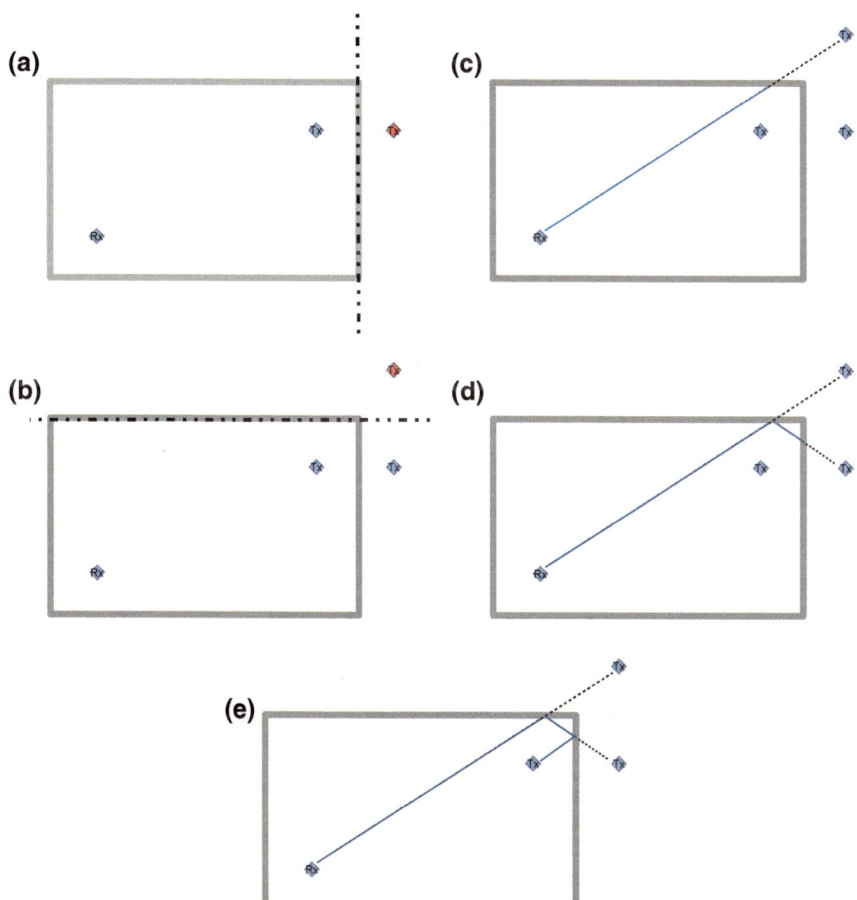

Fig. 5.8 Method of images: **a** first generated image; **b** second generated image; **c** starting backtrace - first intersection with a wall is found; **d** second intersection with a wall is found; **e** path is complete

Fig. 5.9 Algorithm for checking the validity of a generated path

```
1:  rx ← receiver_location
2:  image ← current_image
3:  while image ≠ root do
4:      wall ← image.spanning_wall
5:      path ← line_segment(rx, image)
6:      if path_intersects_wall(path, wall) then
7:          image ← image.parent
8:      else return False
9:      end if
10: end while
```

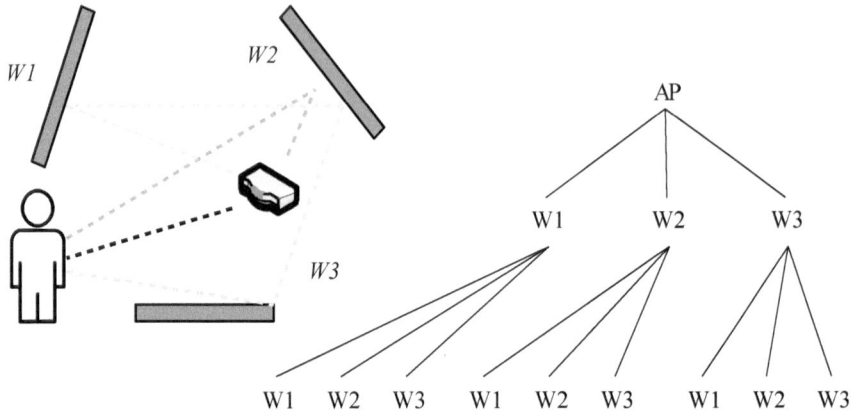

Fig. 5.10 An example environment configuration (left) and reflection tree to describe signal path to the recipient (right). Signal can reach the recipient through many different paths

ates $\frac{N_w!}{n_r! \cdot (N_w - n_r)!}$ (where N_w is the total number of walls in the environment, n_r is the number of possible reflections) possible paths when no such information provided. It is clear that the number of possible combinations will grow very fast with both the number of walls and the number of reflections. We address this problem by introducing the visibility matrix V where the element V_{jk} indicates whether the wall j is visible from the standpoint of the wall k.

Referring to Fig. 5.10, the number of paths in the reflection tree generated using the method of images is much higher than the number of valid paths. One can notice that the number of possible combinations of reflection paths grows exponentially with the number of walls. Several heuristics can be introduced to reduce this number, e.g.:

- Paths that are reflected from the same wall in a row are invalid
- New images should be built only for walls on the opposite side of the current image
- The model can be simplified by prohibiting passing rays, and employing the visibility index, where only walls visible from the image are considered for building new images on the next step
- Special surfaces like walls and ceilings should be handled carefully

Checking the validity of the path is an essential operation for counting the number of collisions of rays with walls. It is crucial to employ spatial index to accelerate this process. We used radix tree as a variant of efficient spatial data structure for spatial join. The details of the implementation are out of scope of this chapter.

5.7.1.2 Determination of Signal Strength

We have defined the signal strength as the energy arrived at the receiver from multiple paths. The equation for RSSI is

$$P(l) = 10\log\left(\sum_{i=1}^{K} P_{tx} \cdot L_i(l) \cdot \gamma^{N_{ref}} \cdot \beta^{N_{tr}}\right)$$

This was discussed in more details in an earlier subsection of Sect. 5.4.1. The path loss is calculated as

$$L_i^{(dB)} = 10\alpha \log d_i + 20\log f - 147.55.$$

Here we have seven parameters

1. Attenuation exponent $\alpha \in [2, \infty]$: It is a parameter that specifies the speed of radio decay in the environment. It is greater in places with numerous obstacles and dense crowds. The value of this parameter should be inferred from the environment
2. Reflection coefficient $\gamma \in [0, 1]$: Specifies the portion of the energy of the incident wave that reflects from the wall. Material properties specify the value.
3. Transmission coefficient $\beta \in [0, 1]$: Specifies the portion of the energy of the incident wave that passes through the wall. Material properties specify the value.
4. The distance between transmitter and receiver d: Estimated for every path between the transmitter and the receiver and can be obtained using the method of images.
5. Radiofrequency f: In case of WiFi, the value is either 2.4 or 5 GHz
6. A number of reflection walls N_{ref}: Estimated for every path between the transmitter and the receiver and can be obtained using the method of images.
7. A number of transmission walls N_{tr}: Estimated for every path between the transmitter and the receiver and can be obtained using the method of images.

The values of α, γ, and β are specific to the environment and are unknown. Moreover, different types of walls have different values of reflection and transmission coefficients. One can try to estimate these values using empirical measurements. We applied standard least squares minimization to optimize for the unknown parameters. The procedure is the following:

1. Pick N location on the map where empirical measurements will be collected
2. For each location collect M measurement. These measurements will be used to estimate statistical parameters of the current location. In our case, it is the expected value of received signal strength $\overline{RSSI}$
3. Calculate the analytical signal strength using the method of images $rssi^{(comp)}$. Apply some initial values to unknown parameters
4. Calculate square error of the analytical estimator

The cost function is

$$J = \frac{1}{N}\sum_{i=1}^{N}\left(\overline{RSSI}_i - rssi_i^{(comp)}\right)^2.$$

We choose to apply constrained gradient descent optimization. Since the optimization function is not convex, we perform several iterations with random initialization. In our case, the values of parameters converged between different iterations.

5.7.1.3 Signal Model

It is often found that wireless signal experiences lognormal shadowing. Thus, the actual RSSI at the receiver is described by the equation

$$\tilde{r}_l = \bar{r}_l + e; e \sim \mathcal{N}\left(0, \sigma_l^t\right).$$

Here σ_l^t is parameterized by location and time. However, neither such information is usually available, nor it is possible to estimate with only RSSI data. Thus, we resort to the simpler noise distribution $e \sim \mathcal{N}(0, \sigma)$ that is invariant in space and time.

5.7.2 Mobility Model

Given the distribution of the received signal strength, we see that the only parameter that changes over time is the average value, which is the function of person's trajectory. Now, not all possible transitions in the trajectory are possible due to physical limitations of person's movements. The simplest way to describe the mobility model of a person is through the distribution of the velocity. However, this approach is not necessarily robust, since speed is usually coherent in time. One of better ways to describe mobility is through the distribution of acceleration. This approach is intuitively more sensible as people also have some inertia. The dynamics model where acceleration is drawn from i.i.d. distribution $a \sim \mathcal{N}(0, 1)$ we can estimate the distribution the next possible location of a person, given the history of previous locations.

Consider the problem of calculating the location of a body in free space given the initial location s_0, initial speed v_0 and the series of accelerations $\mathbf{a} = [a_1 \ldots a_n]^T$. The task is to calculate the series of locations $\mathbf{s} = [s_1 \ldots s_n]^T$, $s_i \in L, \forall i = 1 \ldots n$ occupied by the body in the discrete moments of time $t = t_1, \ldots, t_n$, where the equality $t_i - t_{i-1} = t_j - t_{j-1}$ is not guaranteed for $i \neq j$. The motion equation is given by $s_i = s_{i-1} + \left(v_0 + \sum_{j=0}^{i-1} a_j t_j\right) t_i + \frac{a_i t_i^2}{2}$ where a_n is the acceleration that person processes at the moment of time t_n. The term in brackets describes the velocity accumulated by the moment of time t_{i-1}. After simple transformations, assuming $V_0 = 0$, this equation can be brought to recursive form such as:

$$s_i = s_{i-1} + \frac{s_{i-1} - s_{i-2}}{t_{i-1}} t_i + \sum_{j=i-1}^{i} \frac{a_j t_j t_i}{2} \tag{5.2}$$

Thus, the distribution of s_i given the history of previous locations $s_1, \ldots, s_{i-1}$ and timestamps $t_1, \ldots, t_i$ is

$$\mathbf{p}(s_i | s_{i-2}, s_{i-1}, t_{i-1}, t_i) \sim \mathcal{N}\left(\mu_{s_i}, \sigma_{s_i}^2\right);$$

$$\mu_{s_i} = s_{i-1} + \frac{s_{i-1} - s_{i-2}}{t_{i-1}} t_i;$$

$$\sigma_{s_i}^2 = \frac{(t_{i-1}^2 + t_i^2) t_i^2}{4}.$$

5.7.3 Trajectory Estimation

The ultimate goal of our analysis is to estimate the location of a network user given the history of RSSI measurements. One of the ways to achieve that is by considering the joint conditional probability density $\mathbf{p}(\mathbf{r}, \mathbf{s}|\mathbf{t}) = \mathbf{p}(r_1, \ldots r_n, s_1, \ldots, s_n | t_1, \ldots, t_n)$. The latter can be factorized using the distribution as

$$\mathbf{p}(\mathbf{r}, \mathbf{s}|\mathbf{t}) = \prod_{i=1}^{n} \mathbf{p}(r_i | s_i \ldots s_1) \mathbf{p}(s_i | s_{i-1} \ldots s_1, t_i \ldots t_1)$$

$$= \prod_{i=1}^{n} \mathbf{p}(r_i | s_i) \mathbf{p}(s_i | s_{i-1}, s_{i-2}, t_i, t_{i-1}). \tag{5.3}$$

The form of the factorization stems from the fact that the probability density of RSSI $\mathbf{p}(r_i | s_i)$ depends only on the current location and independent from both previous locations and time intervals. It is worth noting that for the purpose of localization $s_i \in \mathbb{R}^2$ is a 2D vector that describes the location on the map. Since the coordinate components are orthogonal, the density $\mathbf{p}(s_i | s_{i-1}, t_i)$ is decomposed as

$$\mathbf{p}(s_i | s_{i-1}, t_i) = \mathbf{p}(s_i^h | s_{i-1}^h, t_i) \mathbf{p}(s_i^v | s_{i-1}^v, t_i),$$

where s_i^v and s_i^h are vertical and horizontal components of the current coordinate respectively, and $\mathbf{g}_k = [g_1, \ldots, g_k]^T$ is a generic vector with k elements.

Given the continuous nature of s_i, the ways to reduce the computational pressure on the path reconstruction procedure should be considered. Since the precise position estimation is both infeasible due to noise in RSSI measurements and irrelevant due to limited applicability of high precision results on the macro scale, we propose to discretize the environment, where measurements are being taken. Thus, the mapping from RSSI to location will take form $\mathcal{M} : \mathbf{l}^* \rightarrow r, \mathbf{l}^* \in \mathbf{L}, r \in \mathbb{R}$ where $|L|$ is the number of sectors that represent the environment under consideration. In this discrete space $s_i^* = [s_i^{v*} s_i^{h*}]^T \in L$. For the conciseness of the notations in this section, we use s_i to refer to locations in discrete space L.

The meaningful approach to path reconstruction is maximizing the probability $\mathbf{p}(\mathbf{s}_n | \mathbf{r}_n, \mathbf{t}_n)$ using a convolutional Viterbi decoder, i.e.,

$$\hat{s}_n = \text{argmax}_{s_n \in L^n}(\mathbf{p}(\mathbf{s}_n | \mathbf{r}_n, \mathbf{t}_n))$$

The number of possible states for such decoder is $|L|$. As it follows from (5.3), the likelihood of the current state and the probability of transitions are given by $\mathbf{p}(r_i | s_i)$ and $\mathbf{p}(s_i | s_{i-1}, s_{i-2}, t_i, t_{i-1})$ respectively.

5.7.4 Evaluation

To check the performance of our positioning system, we conducted empirical evaluation along with simulation. There were the following sample environments:

- A floor of the university building: This environment allows to test the intended operation mode where the multipath profoundly affects the signal strength distribution, but the system can benefit from diverse signal gradients and temporal correlation of consecutive samples (Figs. 5.11 and 5.12). In Fig. 5.12: in the corridors there are both concrete and hollow walls; rooms are furnished with desks, windows, glasses doors partitioned by glass and metal boards, computers, laptops, chairs, and other wooden, plastic and in some case metallic furniture; during the measurements, doors are kept in some case opened in other cases closed; green line outlines the ground truth path; estimated locations are shown with red diamonds.
- Outdoor environment between two buildings: This environment tests the paths obtained by walking between two dormitory buildings. The distribution of RSSI

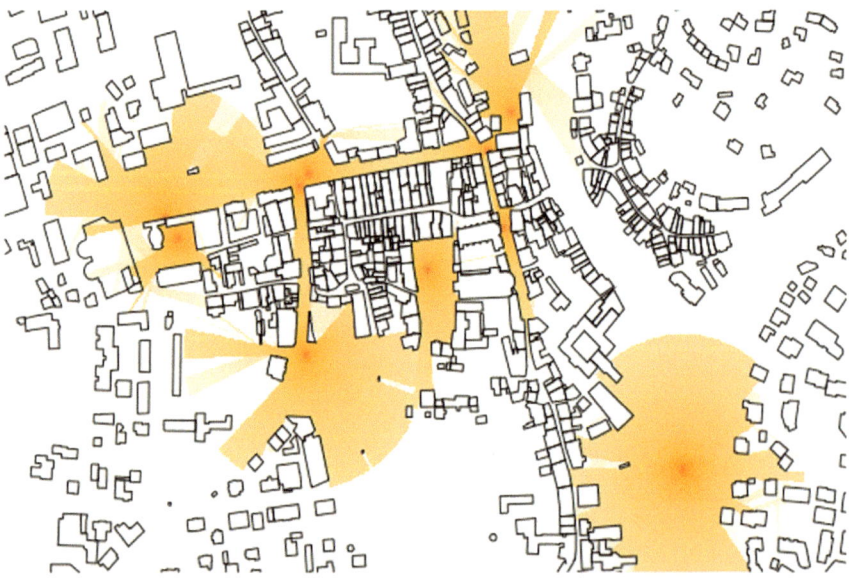

Fig. 5.11 Coverage map estimation with the method of images and multipath propagation model

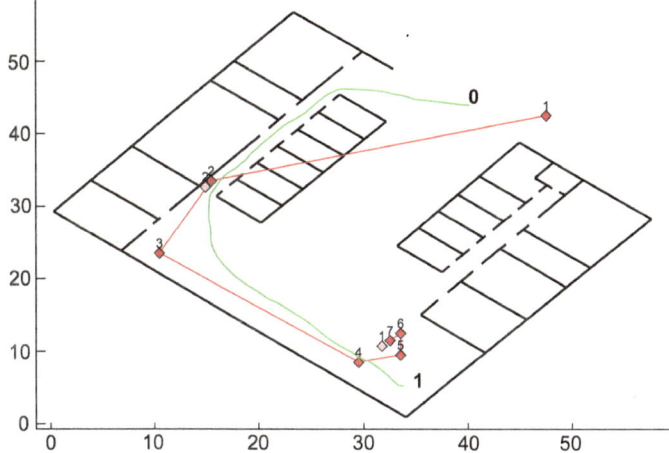

Fig. 5.12 Indoor environment

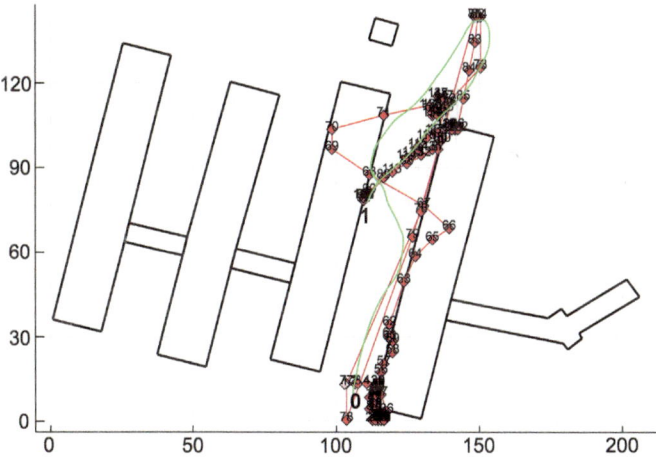

Fig. 5.13 Outdoor environment

is closer to radial due to open space. This makes the positioning with few APs harder due to monotonic gradient patterns. Two APs are used to cover a larger area. The device automatically switches between Aps. (Refer to Fig. 5.13 where two APs are installed; an open source project OpenStreetMap was used to obtain the building outlines).

The proper testing procedure should be performed by collecting the ground truth trajectory data and the RSSI measurements along the path. Then, the pointwise positioning error is measured. It makes sense when the locations are estimated independently from each other, and it is sufficient to walk long enough trajectory to have a reasonable estimate of positioning precision. However, in our case, there is a high correlation between points in the trajectory, and variation on intermediate steps may result in significant error propagations. For this reason, it is much more considerate to perform testing on more than one trajectory. Unfortunately, the data collection process is too long, and we resort to estimation with simulation. Then, we test the positioning system with a real path, and we are going to test whether the empirical precision is much different from the simulated one.

In order to perform simulation, one needs to have or generate the test paths. We use generative model defined by the Eq. (5.2) to generate a set of trajectories. Then, the paths are reconstructed with the same model they were generated. We set the standard deviation of the acceleration $\sigma_a = 1$. The simulation was performed for the indoor environment depicted in the Fig. 5.12. The length of the simulated trajectories was varied and set to 10, 20 or 40 steps. The average error in all cases was 0.39 m.

We did an empirical evaluation for indoor and outdoor environments. For the indoor area, we used the RSSI from the associated AP to reconstruct the path. Both, the mobility and the signal model diverged from the actual. For the outdoor environment, we employed two APs and ensured automatic switching between these APs. The reconstructed paths show the trajectory similar to the ground truth path.

Besides conducting the test on real measurements, we performed a computer simulation. The path of a person was generated according to the mobility model, and the RSSI signal was sampled from the analytical coverage map with noise. Thus, the divergence of the generation and estimation models is minimal, which implies that the error performance of this test is biased upwards. Additionally, we considered how the number of access points affects the positioning performance. Out finding is shown in Fig. 5.14.

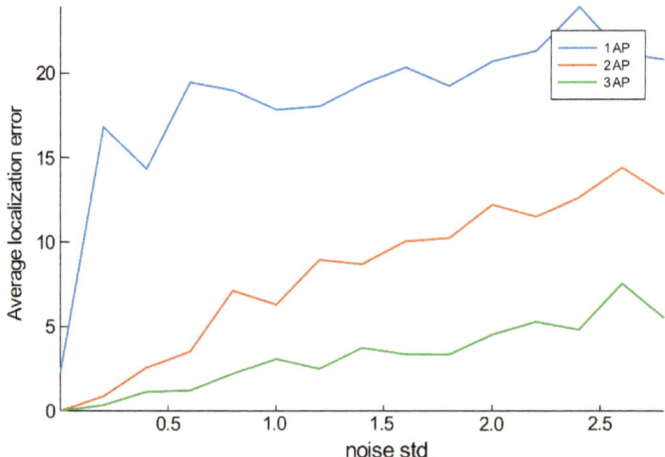

Fig. 5.14 Performance analysis of positioning systems concerning signal noise for different number of available APs

5.8 Conclusion and Future Research

In this chapter, we covered basics of positioning with wireless sensor networks. We have discussed many different techniques; the most viable one being the RSSI based positioning. Wireless WiFi AP plays a suitable role as a sensor in the network and often allows to precisely identify a location of a wearable mobile device, which is often equivalent to detecting a person. Device-free positional methods also exist, but they usually rely on a large number of sensors or are computationally intractable. Modern techniques that rely on modeling of people's movement through the environment can achieve a decent positioning accuracy even with only one AP in operation.

It appears that the future research will be heavily dependent on the future of the IEEE 802.11 standard. So far, RSSI based positioning has held an upper hand due to the ease of operation but with some hardware limitations. This might change when the network devices can operate at GHz clock frequency. These will then be able to measure small time periods of few nanoseconds. This would allow switching the attention to more precise time-based technique. If this occurs, this transformation will come with the price, and before then RSSI will be the primary information source about people's location for cheaper sensor networks.

Our analysis has shown that the positioning systems are highly sensitive to noise and the number of available APs. Positioning can be performed reliably when the signal noise is low, but the performance deteriorates rapidly as the noise increases. Thus, the evident way to reduce the error is to increase the number of available APs.

References

1. Vo QD, De P (2016) A survey of fingerprint-based outdoor localization. IEEE Commun Surv Tutor 18(1):491–506
2. Li H et al (2014) Achieving privacy preservation in WiFi fingerprint-based localization
3. Youssef MA, Agrawala A (2007) Analysis of the optimal strategy for WLAN location determination systems. Int J Model Simul 27(1):53–59
4. Figuera C et al. (2009) Nonparametric model comparison and uncertainty evaluation for signal strength indoor location. IEEE Trans Mob Comput 8(9):1250–1264
5. Pourhomayoun M, Fowler M (2012) Improving WLAN-based indoor mobile positioning using sparsity. In: Conference record—asilomar conference on signals, systems and computers, pp 1393–1396
6. Sen S et al. (2013) Avoiding multipath to revive inbuilding WiFi localization. In: Proceeding of the 11th annual international conference on mobile systems, applications, and services—MobiSys'13
7. Dardari D et al. (2015) Indoor tracking: theory, methods, and technologies. IEEE Trans Veh Technol 64(4):1263–1278
8. Sapiezynski P et al. (2016) Inferring person-to-person proximity using WiFi signals
9. Chilipirea C et al. (2016) Presumably simple: monitoring crowds using WiFi, pp 220–225
10. Bellavista P et al. (2017) Human dynamics of mobile crowd sensing experimental datasets. In: IEEE international conference on communications, pp 0–5
11. Li Z et al. (2017) Discovering mass activities using anomalies in individual mobility motifs. In: Proceedings—13th IEEE international conference on ubiquitous intelligence and computing, 13th IEEE international conference on advanced and trusted computing, 16th IEEE

international conference on scalable computing and communications, IEEE international, pp 321–326

12. Traunmueller M et al. (n.d.) Digital traces: modeling urban mobility using WiFi probe data
13. Basalamah A (2016) Crowd mobility analysis using WiFi sniffers. Int J Adv Comput Sci Appl 7:374–378
14. Scheuner J et al. (2016) Probr—a generic and passive WiFi tracking system. In: Proceedings—conference on local computer networks, LCN, pp 495–502
15. Vanderhulst G et al. (2015) Detecting human encounters from WiFi radio signals. In: Proceedings of the 14th international conference on mobile and ubiquitous multimedia—MUM'15, pp 97–108
16. Ma W et al. (2015) Detecting pedestrians behavior in building based on Wi-Fi signals. Proceedings of 2015 IEEE international conference on smart city, SmartCity 2015, Held jointly with 8th IEEE international conference on social computing and networking, SocialCom 2015, 5th IEEE international conference on sustainable computing and communication, pp 1–8
17. Acer UG et al. (2016) Capturing personal and crowd behavior with Wi-Fi analytics. In: Proceedings of the 3rd international on workshop on physical analytics—WPA'16, pp 43–48
18. Wu FJ, Solmaz G (2017) We hear your activities through Wi-Fi signals. In: 2016 IEEE 3rd world forum on internet of things, WF-IoT 2016, pp 251–256
19. Seifeldin M et al (2013) Nuzzer: a large-scale device-free passive localization system for wireless environments. IEEE Trans Mob Comput 12:1321–1334
20. Zhou Z et al. (2013) Towards omnidirectional passive human detection
21. Wang Y et al. (2014) E-eyes: device-free location-oriented activity identification using fine-grained WiFi signatures. In: Proceedings of the 20th annual international conference on Mobile computing and networking—MobiCom'14, pp 617–628
22. Wu C et al. (2015) Non-invasive detection of moving and stationary human with WiFi. IEEE J Sel Areas Commun
23. Abdelnasser H et al. (2015) WiGest: {A} Ubiquitous WiFi-based gesture recognition system. CoRR, Volume abs/1501.0, pp 1472–1480
24. Mrindoko NR, Minga LM (2016) A comparison review of indoor positioning techniques. Int J Comput (IJC) 21:42–49
25. Liu H et al. (2007) Survey of wireless indoor positioning techniques and systems. IEEE Trans Syst Man Cybern Part C (Appl Rev) 37(6):1067–1080
26. Deak G et al. (2012) A survey of active and passive indoor localisation systems. Comput Commun 35(16):1939–1954
27. Xiao J et al. (2016) A survey on wireless indoor localization from the device perspective. ACM Comput Surv 49(2):25
28. Pirzada N et al (2013) Comparative analysis of active and passive indoor localization systems. Elsevier B.V, New York
29. Hossain AKMM, Soh WS (2015) A survey of calibration-free indoor positioning systems. Comput Commun 66:1–3
30. Farid Z et al (2013) Recent advances in wireless indoor localization techniques and system. J Comput Netw Commun 2013:1–12
31. Makki A et al. (2015) Survey of WiFi positioning using time-based techniques. Comput Netw 88:218–233
32. Honkavirta V et al. (2009) A comparative survey of WLAN location fingerprinting methods, pp 243–251
33. He S, Chan SHG (2016) Wi-Fi fingerprint-based indoor positioning: recent advances and comparisons. IEEE Commun Surv Tutor 18(1):466–490
34. Yiu S et al. (2017) Wireless RSSI fingerprinting localization. Sig Process 131:235–244
35. Saxena M et al. (2008) Experimental analysis of RSSI-based location estimation in wireless sensor networks. In: 2008 3rd international conference on communication systems software and middleware and workshops (COMSWARE'08), pp 503–510
36. Mariakakis A et al. (2014) SAIL: single access point-based indoor localization In: MobiSys'14

37. Xiao Z et al. (2015) Robust indoor positioning with lifelong learning. IEEE J Sel Areas Commun
38. Wu C et al. (2013) DorFin: WiFi fingerprint-based localization revisited. arXiv preprint arXiv: 1308.6663
39. Wu C et al. (2013) Smartphones based crowdsourcing for indoor localization
40. Liu H et al. (2012) Push the limit of WiFi based localization for smartphones. In: Proceedings of the 18th annual international conference on mobile computing and networking (Mobicom'12), p 305
41. Jing H et al. (2016) Wi-Fi fingerprinting based on collaborative confidence level training. Pervasive Mob Comput 30:32–44
42. Zàruba GV et al. (2007) Indoor location tracking using RSSI readings from a single Wi-Fi access point. Wirel Netw 13(2):221–235
43. Ni Y et al. (2016) An indoor pedestrian positioning method using HMM with a fuzzy pattern recognition algorithm in a WLAN fingerprint system. Sensors (Switzerland) 16(9):1447
44. He S et al. (2015) Fusing noisy fingerprints with distance bounds for indoor localization
45. Xie Y et al. (2015) Precise power delay profiling with commodity WiFi. In: Proceedings of the 21st annual international
46. Chapre Y et al. (2014) CSI-MIMO: indoor Wi-Fi fingerprinting system. In: Proceedings—conference on local computer networks, LCN, pp 202–209
47. Jiang ZP et al. (2014) Communicating is crowdsourcing: Wi-Fi indoor localization with CSI-based speed estimation. J Comput Sci Technol 29(4):589–604
48. Kumar CP et al. (2014) Single access point based indoor localization technique for augmented reality gaming for children. s.l., s.n
49. Yang Z et al (2013) From RSSI to CSI: indoor localization via channel response. ACM Comput Surv (CSUR) 46:25
50. Wu C et al. (2012) WILL: wireless indoor localization without site survey
51. Jin Y et al. (2010) Indoor localization with channel impulse response based fingerprint and nonparametric regression. IEEE Trans Wirel Commun 9(3):1120–1127
52. Zhou Z et al. (2014) LiFi: Line-Of-Sight identification with WiFi. s.l., s.n
53. Phillips C et al (2013) A survey of wireless path loss prediction and coverage mapping methods. IEEE Commun Surv Tutor 15:255–270
54. Bose A, Chuan HF (2007) A practical path loss model for indoor WiFi positioning enhancement. s.l., s.n
55. Ji Y et al. (2006) ARIADNE: a dynamic indoor signal map construction and localization system. In: Proceedings of the 4th international conference on mobile systems, applications and services—MobiSys 2006, p 151
56. Kraxberger S et al. (2010) WLAN location determination without active client collaboration. In: Proceedings of the 2010 international conference on wireless communications and mobile computing, IWCMC 10, pp 1188–1192
57. Ji Z et al (2001) Efficient ray-tracing methods for propagation prediction for indoor wireless communications. IEEE Antennas Propag Mag 43:41–49
58. El-Kafrawy K et al. (2010) Propagation modeling for accurate indoor WLAN RSS-based localization. In: IEEE vehicular technology conference, pp 1–5
59. Cocheril Y, Vauzelle R (2007) A new ray-tracing based wave propagation model including rough surfaces scattering. Prog Electromagn Res 75:357–381
60. Kusaka M (2015) Efficient ray tracing algorithm with the avoidance of duplicate image generation, pp 152–157
61. Viol N et al. (2012) Hidden Markov model-based 3D path-matching using raytracing-generated Wi-Fi models. In: 2012 international conference on indoor positioning and indoor navigation, IPIN 2012—conference proceedings, pp 13–15
62. Kausar ASMZ et al. (2013) Efficient radio propagation prediction algorithm including rough surface scattering with improved time complexity. Prog Electromagn Res 53:127–145
63. Klepal M (2003) Novel approach to indoor electromagnetic wave propagation modelling. Czech Technical University in Prague

64. Pahlavan K (1998) Site-specific wideband and narrowband modeling of indoor radio channel using ray-tracing, vol 6, pp 65–68
65. Schmitz A, Kobbelt L (2011) Efficient and accurate urban outdoor radio wave propagation. In: Proceedings—2011 international conference on electromagnetics in advanced applications, ICEAA'11, pp 323–326
66. Schmitz A et al. (2011) Efficient rasterization for outdoor radio wave propagation. IEEE Trans Vis Comput Graph 17:159–170
67. Salem M et al (2011) Validation of three-dimensional ray-tracing algorithm for indoor wireless propagations. ISRN Commun Netw 2011:1–5
68. Rothe D (2012) Indoor Localization of mobile devices based on Wi-Fi signals using raytracing supported
69. Khodayari S et al (2010) A RSS-based fingerprinting method for positioning based on historical data. In: 2010 international symposium on performance evaluation of computer and telecommunication systems, pp 306–310
70. Gu Z et al. (2016) Reducing fingerprint collection for indoor localization. s.l., s.n
71. Jiang Y et al. (2012) ARIEL: automatic Wi-Fi based room fingerprinting for indoor localization. In: Proceedings of the 2012 ACM conference on ubiquitous computing, pp 441–450
72. Jiang Y et al. (2013) Hallway based automatic indoor floorplan construction using room fingerprints. In: Proceedings of international conference on ubiquitous computing
73. Goswami A et al. (2011) WiGEM: a learning-based approach for indoor localization. CoNEXT
74. Ferris B D et al. (2007) WiFi-SLAM using Gaussian process latent variable models. Science 2480–2485
75. Wang B et al. (2015) Indoor positioning via subarea fingerprinting and surface fitting with received signal strength. Pervasive Mob Comput 23:43–58
76. Kaji K, Kawaguchi N (2012) Design and implementation of WiFi indoor localization based on Gaussian mixture model and particle filter. In: 2012 international conference on indoor positioning and indoor navigation
77. Xiao Z et al. (2014) Lightweight map matching for indoor localisation using conditional random fields. s.l., s.n., pp 131–142
78. Chintalapudi K et al. (2010) Indoor localization without the pain. In: 16th Annual international conference on mobile computing and networking—(MobiCom'10), p 173
79. Yang Z et al. (2012) Locating in fingerprint space: wireless indoor localization with little human intervention. In: Proceedings of the annual international conference on mobile computing and networking, pp 269–280
80. Wang H et al. (2012) No need to war-drive: unsupervised indoor localization. In: Proceedings of the 10th international conference on mobile systems, applications, and services (MobiSys'12), pp 197–210
81. Shen G et al. (2013) Walkie-Markie: indoor pathway mapping made easy. In: Proceedings of the 10th USENIX conference on networked systems design and implementation, pp 85–98
82. Xiong H, Tao D (2017) A Diversified generative latent variable model for WiFi-SLAM. In: Proceedings of the 31th Conference on Artificial Intelligence (AAAI 2017), pp 3841–3847
83. Herranz F et al (2016) WiFi SLAM algorithms: an experimental comparison. Robotica 34:837–858
84. Faragher RM et al. (2012) Opportunistic radio SLAM for indoor navigation using smartphone sensors. In: Record—IEEE PLANS, position location and navigation symposium, pp 120–128
85. Bruno L, Robertson P (2011) WiSLAM: improving FootSLAM with WiFi. s.l., s.n
86. Huang J et al. (2011) Efficient, generalized indoor WiFi GraphSLAM. s.l., s.n
87. Kivimäki T et al. (2014) A review on device-free passive indoor positioning methods. Int J Smart Home 8(1):71–94
88. Zhao Y, Patwari N (2011) Noise reduction for variance-based device-free localization and tracking. s.l., s.n., pp 179–187
89. Wilson J, Patwari N (2011) See-through walls: motion tracking using variance-based radio tomography networks. IEEE Trans Mob Comput 10:612–621

90. Wang J et al (2013) Robust device-free wireless localization based on differential RSS measurements. IEEE Trans Industr Electron 60:5943–5952
91. Nannuru S et al (2013) Radio-frequency tomography for passive indoor multitarget tracking. IEEE Trans Mob Comput 12:2322–2333
92. Guo Y et al. (2015) An exponential-rayleigh model for RSS-based device-free localization and tracking. IEEE Trans Mob Comput
93. Hong J, Ohtsuki T (2015) Signal eigenvector-based device-free passive localization using array sensor. IEEE Trans Veh Technol 64:1354–1363
94. Xu C et al (2016) The case for efficient and robust RF-based device-free localization. IEEE Trans Mob Comput 15:2362–2375
95. Lui G et al. (2011) Differences in RSSI readings made by different Wi-Fi chipsets: a limitation of WLAN localization, pp 53–57
96. Jun J et al. (2013) Social-Loc: improving indoor localization with social sensing. s.l., s.n., p 14
97. Geng X et al. (2013) Hybrid radio-map for noise tolerant wireless indoor localization
98. Kaemarungsi K, Krishnamurthy P (2012) Analysis of WLAN's received signal strength indication for indoor location fingerprinting. Pervasive Mob Comput 8:292–316
99. Farshad A et al. (2013) A microscopic look at WiFi fingerprinting for indoor mobile phone localization in diverse environments. s.l., s.n
100. Wen Y et al. (2015) Fundamental limits of RSS fingerprinting based indoor localization. s.l., s.n
101. Jiang P et al. (2015) Indoor mobile localization based on Wi-Fi fingerprint's important access point. Int J Distrib Sens Netw
102. Valenzuela R (1993) A ray tracing approach to predicting indoor wireless transmission. In: IEEE 43rd vehicular technology conference, pp 214–218

Chapter 6
Need of Ambient Intelligence for Next-Generation Connected and Autonomous Vehicles

Adnan Mahmood, Bernard Butler, Quan Z. Sheng, Wei Emma Zhang and Brendan Jennings

Abstract The automotive industry is shifting its focus from performance and features to safety, entertainment, and driver comfort. In this regard, driver assistance and autonomous driving technology are gaining more attention. Such technology has the potential to reduce road accidents, traffic congestion, and fuel usage. However, vehicles cannot become fully autonomous, until they are able to sense their context efficiently (context sensing), and to use ambient learning to respond appropriately and within short timescales to the data they have sensed. Context sharing will also become essential, because a single vehicle will not be able to gain a holistic view of its context without cooperation from other nearby vehicles and from the roadside infrastructure. Indeed, there are further advantages when a group of vehicles make intelligent decisions based on a common understanding of their context. This chapter highlights the significance of ambient intelligence for next-generation connected and autonomous vehicles, describes its current state of the art, and also shows how its potential might be achieved. One of the main challenges refers to how to provision and coordinate cloud-based services to meet the needs of real-time (low latency) data-intensive (high data rate) ambient intelligence, particularly for safety-critical vehicular safety applications. It indicates how autonomous or semi-autonomous vehicles are likely to make seamless use of any available wireless networking technologies to improve both coverage and reliability and, where feasible, to cache critical content near the network edge so as to minimize the number of network hops and hence service latencies. Both of these approaches should improve the network quality of service afforded to driving applications.

Keywords Ambient intelligence · Context sensing · Context sharing
Heterogeneous vehicular networks · Edge computing
Software-Defined networks · VANETs

A. Mahmood (✉) · Q. Z. Sheng · W. E. Zhang
Department of Computing, Macquarie University, Sydney, NSW 2109, Australia
e-mail: adnan.mahmood@mq.edu.au; adnanqureshi@ieee.org

A. Mahmood · B. Butler · B. Jennings
Telecommunications Software and Systems Group (Science Foundation Ireland—CONNECT),
Waterford Institute of Technology, Waterford, Republic of Ireland

© Springer Nature Switzerland AG 2019
Z. Mahmood (ed.), *Guide to Ambient Intelligence in the IoT Environment*, Computer
Communications and Networks, https://doi.org/10.1007/978-3-030-04173-1_6

6.1 Introduction and Preamble

Vehicular Ad hoc Networks (VANETs) have gained significant interest from both academia and industry over the past few decades. One proposed application of VANETs is to reduce the number of road traffic accidents. According to World Health Organization estimates, more than 1.25 million road fatalities are reported each year with many occurring in low- and middle-income countries, due to inadequate/unsafe road transport infrastructure. VANETs facilitate the placement of distributed sensors on vehicles and the roadside infrastructure, allowing each vehicle to become aware of its local context while the networks they offer enable the wider context to be derived by means of context sharing [1]. Vehicle-to-Vehicle communication (V2V), Vehicle-to-Infrastructure (V2I) communication, and Vehicle-to-Pedestrian (V2P) communication form the foundations for the Vehicle-to-Everything (V2X) communication networks of the future [2]. Vehicles released in 2015 are typically equipped with more than 100 onboard sensors. By 2020, that number is expected to grow to 200 sensors. Consequently, the data to be analyzed will grow in both quantity and variety. Understanding this data is needed to derive the current state (context) of a given vehicle, and hence to decide what actions the driver, or the vehicle's control system, should take. This data needs to be accumulated, stored, computed, and analyzed within specified latency thresholds. Failure to respond in time could lead to human fatalities. Work is also under way to migrate some of the processing to edge nodes, rather than requiring of long signal paths to nodes in the cloud. VANETs will play their part in ensuring that the network infrastructure can meet the demands placed on it by applications offering ambient intelligence for vehicular safety purposes.

However, ambient intelligence in next-generation Intelligent Transportation Systems (ITS) is still in its infancy. Driving is a complex activity requiring interactions among three key components: *the drivers, the vehicles, and the surrounding environments*. With semi-autonomous vehicles, the control system needs to provide alarms to warn drivers of the need to transition from automated to manual driving. The alerting strategy needs to take account of the cognitive ability of the (human) driver to understand the situation and respond to it accordingly [3]. In practice, this means that the alerting system should not just issue alarms, it should also provide extra context and prompts to help the driver to make the right decision(s) quickly. Typically, vehicular safety applications have delay thresholds ranging from 10 ms to 100 ms. Fully autonomous vehicles lack the option of delegating actions to human drivers. Hence, their ambient intelligence needs to be extremely reliable. For example, Tesla's self-driving car accident in May 2006 occurred when the vehicle's control system failed to distinguish between a white tractor-trailer and a bright sky background, resulting in the death of its occupant [4, 5]. Thus, just as humans sense their environment and learn how to react safely when driving, vehicle control systems should draw upon ambient intelligence to deal with unfamiliar and/or complex situations. Of course, driving involves decision-making and not just pattern recognition. Given the enormous space of possible scenarios, it is impossible to program autonomous vehicles with every fine-grained road traffic scenario that it is likely to encounter. For instance, humans know that buses can discharge large numbers of pedestrians at or near the bus stops, some of which might need to cross the road at short notice. Indeed, we as humans benefit from a lifetime of experience and can often make some good decisions in new situations. We suggest that this is part of the common sense we have accumulated since childhood. The question therefore arises: *How to empower semi- or fully- autonomous vehicles with ambient intelligence, equivalent to (human) common sense?*

Therefore, the aim of this chapter is (a) to discuss the principles and current state of the art of ambient intelligence in the automotive sector and (b) to identify the requirements for realizing the safe deployment of connected vehicles within a smart city. In this case, two critical but related aspects are needed: *context sensing* and *context sharing*. The former (context sensing) relates to measurement, aggregation, filtering, and subsequent decision-making. The latter (context sharing) requires seamless communication using any available wireless networking technologies. When context sensing and sharing are combined, the result is ambient intelligence, where the whole is greater than the sum of its parts.

This chapter also considers the issues and challenges pertinent to Big Data in the realm of vehicular networks/Internet of Vehicles (IoVs) and of how it can be harnessed to provide better data for decision-making while achieving the lowest possible delay between sensing and making control decisions. A use-case involving Augmented Reality is also presented. Finally, conclusions are drawn.

6.2 Context Sensing for Connected and Autonomous Vehicles

As mentioned in the introduction, the number of onboard sensors is increasing rapidly [6]. Cameras and automotive radars are the most commonly found safety sensors installed on modern vehicles. Cameras and video processing can reduce the risk of collision at known danger areas such as blind intersections. Increasingly, the cameras are being enhanced with infrared (i.e., night vision) capabilities to enhance night vision. An automotive radar, if fitted, augments visible light sensing to measure the position and velocity of vehicles in the vicinity. Furthermore, Light Detection and Ranging Systems (LIDARs) utilize lasers to generate high resolution, dynamically updated depth, and range maps. These three main sensing modes can work together to reduce collision risks, and facilitates adaptive cruise control, lane change assistance, parking assistance, and vehicle platooning.

With all this instrumentation, context sharing and ambient decision-making, a massive amount of vehicular data is generated, especially in the case of dense and/or slow-moving vehicles. Indeed, it appears that, over the same time period, a single self-driving car would generate and process more data than a single human being. Intel estimates that cameras and LIDARs produce around 20–40 Mbps and radars around

Table 6.1 Levels of automation for autonomous vehicles

Level 0	No automation	Vehicles do not possess any automated features and drivers primarily perform all critical driving tasks of steering, accelerating, slowing down, and breaking.
Level 1	Driver assistance	At this level, the drivers still control most of the driving functions, but some specific tasks pertinent to Advanced Driver Assistance Systems (lane departure warning, blind spot warning, automatic braking, etc.) enhances and augments the drivers' sense of control.
Level 2	Partial automation	Some of the driver's load (adaptive cruise control, automated parking, automatic lane changing, etc.) has been taken over by the vehicles.
Level 3	Conditional automation	A driver might cede full operation to the vehicle under certain environment and traffic conditions. Control would revert to the driver once the control system is unable to ensure safe operations.
Level 4	High automation	Humans are treated primarily as passengers. Humans still tell vehicles where to go and are still needed to take over the control in case of any unusual environments outside the Operational Design Domain of the vehicle (i.e., those scenarios that have been used to train the control system).
Level 5	Full automation	This level includes full level autonomy, wherein all critical controls are performed by the vehicle's control system. No human control is required nor there is a fallback mode that allows for it.

10–100 Kbps of data [7], so a single autonomous vehicle is expected to produce 40 terabytes of data for every 8 hours of driving. For comparison, the average person's Internet, video, and chat use currently stands at 650 MB per day and is expected to increase to 1.5 GB per day by the end of 2020. Also, the maps being downloaded by (semi-)autonomous vehicles would need to be updated continuously to take account of changing conditions [8].

As well as increasing in number, sensors are generating richer data streams, e.g., as camera resolution and frame rates increase. Table 6.1 provides a summary of the different levels of automation recommended for the autonomous vehicles [9].

Table 6.2 classifies automotive systems in terms of safety, comfort, and drivetrain along with their associated sensors [10]. More sensors obviously support greater autonomy; however, they may also lead to massively increased data flows requiring much more network capacity.

The real challenge though is to assimilate the data generated by each vehicle and integrate it with data from its peers. The hardware in the onboard units in modern vehicles can often be extended as needed for new services. For instance, Solid-State Drives (SSDs) can be deployed in NVIDIA's self-driving data collection system for higher performance, and such drives can be expanded to several terabytes as price/performance improves. Smartphones belonging to drivers and passengers have their own sensors (cameras, accelerometers, etc.) and potentially gigabytes of disk

Table 6.2 Classification of automotive systems and their sensors

Data acquisition (*In-vehicle sensors*)	Safety	Ranging radar (automatic cruise control and precrash)
		Tilt sensor (headlight aiming)
		High-pressure sensor (Electronic Stability Program—ESP)
		Torque sensor (power-assisted steering)
		Steering wheel angle sensor (ESP)
		Video camera (lane keeping, traffic sign recognition, etc.)
		Acceleration sensor (airbag)
		Seat occupation sensor (airbag)
		Acceleration sensor (antilock braking system)
		Yaw rate sensor (ESP and rollover sensing)
		Tilt sensor (safeguarding and anti-theft system)
		Speed sensor (antilock braking system)
	Comfort	Yaw rate sensor (navigation)
		Air quality sensor (Air Conditioning—AC)
		Humidity/Temperature sensor (heating and AC)
		Distance sensor, Ultrasound (backup monitoring)
		Pressure sensor (central locking system)
		Rain sensor (windshield wiper control)
	Drivetrain	Pressure sensor (transmission control)
		Boost pressure sensor (electronic diesel control)
		Air mass meter (motronic)
		Knock sensor (motronic)
		Atmospheric pressure sensor (motronic)
		High-pressure sensor (common rail)
		Speed sensor (transmission control)
		Tank pressure sensor (onboard diagnostic)
		Accelerator pedal sensor (electronic throttle control and electrohydraulic failure)
		Angle of rotation and position sensor (motronic)

available for data storage and caching. Roadside units (RSUs) might also provide tens of gigabytes [11] of supplementary storage per vehicle, particularly when upgraded with external USB storage. Furthermore, drop boxes could be deployed along the roadside to store content such as maps and entertainment and made available to passing vehicles. Also, depending on data access frequency, some data could be offloaded to storage services in the cloud, such as Google Drive, Microsoft's One Drive, or Dropbox, for later access when required. This feature begs the question: *which data needs to be stored locally and which can be offloaded to cloud services?* Depending on access, we can distinguish between fast storage, medium storage, and slow storage. Questions relating to storage in vehicular networks will be discussed later in this chapter.

6.3 Context Sharing for Connected and Autonomous Vehicles: Ensuring Seamless Ubiquitous Communication

So far, we have considered sensing and storage of data obtained by vehicles, RSUs, and their sensors. It is also imperative to consider how the data from these sensors needs to be transmitted between entities in the network: vehicles, pedestrians, nodes at the network edge, and those in the cloud. Given a variety of Radio Acess Technologies (RATs) and a vertical handover system that makes efficient use of these RATs, we seek to ensure that the entities operate in an *"Always Best Connected"* mode, i.e., with the highest bandwidth and lowest delay that is possible between two endpoints.

Currently, IEEE 802.11p/Dedicated Short-Range Communication (DSRC) is regarded as the de facto standard for V2V communication with its particular focus on low latency communication. However, DSRC suffers from relatively short communication range and low data rates. Other wireless communication technologies, i.e., Wi-Fi, WiMAX, and LTE/LTE Advanced have been recently employed for the vehicular communication purposes. The Third-Generation Partnership Project (3GPP) promotes the use of cellular communication technologies for (V2X) communication, often referred to as C-V2X [12, 13]. Cellular communication is able to provide higher data rates than DSRC but is unable to meet the low latency requirements of safety-critical vehicular applications [14, 15]. Table 6.3 highlights the potential RATs that could be used for vehicular communication.

Since there is no single RAT that can offer higher data rates, low latency, long range, low cost, and plentiful access points *simultaneously*, heterogeneous networking has been proposed for vehicular networks [16]. With heterogeneous networking, the network is free to choose the most appropriate RAT for each network hop. The Fifth-Generation Public–Private Partnership Group (5G PPP) promotes heterogeneous networking in their "5G Vision" [17], stating that 5G wireless will support a heterogeneous set of integrated air interfaces. Candidate RATs would include the evolution of current RATs and completely new RATs. The goal of such heterogeneous networks is seamless connectivity, hence maximizing the reliability of vehicular communications.

Note that heterogeneous networking often needs to satisfy conflicting criteria. For reasons of reliability and/or higher data rates, it might be better to use more RATs and hence to have more handovers between these RATs on the data path. However, each handover introduces extra delay, so it is advisable to minimize the number of such handovers. Indeed, vertical handovers (i.e., handovers between two different networking technologies) need to be managed carefully to minimize the wastage of precious network resources [18]. Thus, algorithms for the following tasks need to operate well together:

- Handover Necessity Estimation—determine if a handover from a given RAT is needed at any given time and location;
- Handover Target Selection—if a handover is needed, choose the best candidate RAT from those available; and

Table 6.3 Potential radio access technologies for vehicular communication

Characteristics	802.11 Wi-Fi	802.11p DSRC	LTE	LTE Advanced	802.11ad mmWave
Maximum range	100 m	1 km	100 km	100 km	10 m
Maximum bandwidth	20 MHz	10 MHz	1.4, 3, 5, 10, 15, 20 MHz	100 MHz	2.16 GHz
Connectivity	Intermittent	Intermittent	Pervasive	Pervasive	Intermittent
Capacity	Average	Average	Very high	Very high	Extremely high
Frequency band	2.4 GHz, 5.2 GHz	5.9 GHz	700 MHz–2690 MHz	450 MHz–4.99 GHz	30 GHz–300 GHz
Peak data rates	54 Mbps	6–27 Mbps	300 Mbps	1 Gbps	7 Gbps
Support for mobility	Low	Moderate	Very high (350 km/h)	Very high (350 km/h)	Low
V2V connectivity	Ad hoc	Ad hoc	No	D2D	Ad hoc
V2I connectivity	Yes	Yes	Yes	Yes	Yes
Market penetration	Very high	Low	Very high	Potentially high	Low

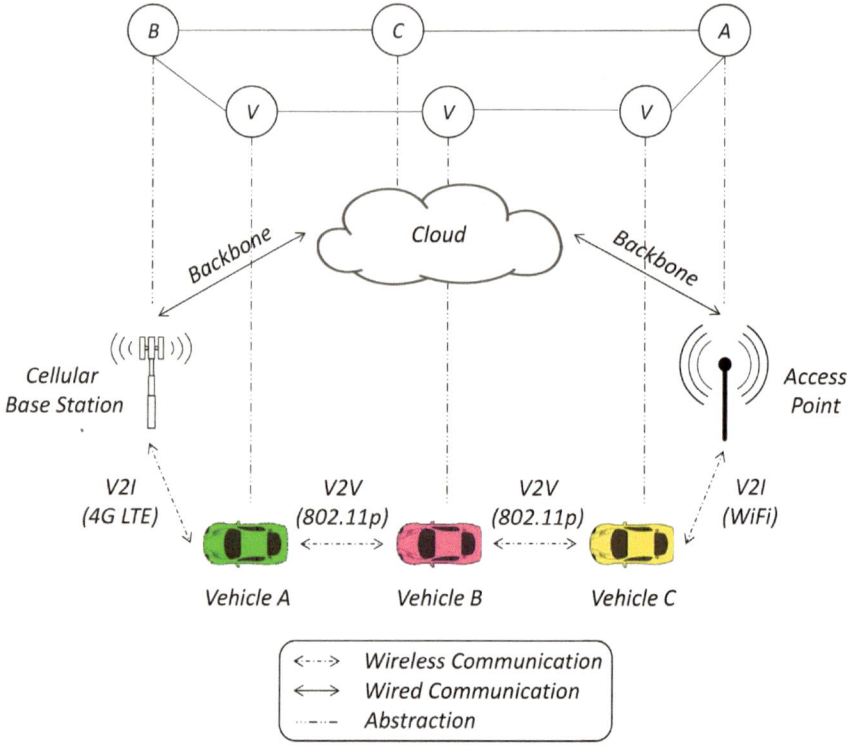

Fig. 6.1 An overview of heterogeneous vehicular communication

- Handover Triggering Condition Estimation—determines the best time to trigger handover from the present RAT to the selected target RAT.

An overview of heterogeneous vehicular communication is depicted in Fig. 6.1. Work on sophisticated next-generation wireless networking technologies is also in progress. For instance, millimeter wave communication (mmWave) has been explored recently for vehicular communication purposes and is already available to consumers in the form of IEEE 802.11ad [6]. Similarly, terahertz (THz) communication is being explored for vehicular communication too [19], based on its potential for extremely high data rates. Both mmWave and THz communication are expected to play a role in the future, possibly beyond 5G wireless networks.

The purpose of these heterogeneous networks is to achieve reliable communication with sufficient bandwidth and low latency for safety-critical vehicular applications. Ambient intelligence plays its part by ensuring that such applications are able to make the best use of data collected to support such applications.

6.4   Edge-based Computing for Guaranteeing Low Latency Ambient Intelligence in Vehicular Networks

The availability of several radio access technologies in parallel could help achieve extended communication range and higher throughput for V2X applications and services. However, if data needs to be fetched from the (remote) cloud rather than from nearby nodes, the additional delay could breach the latency conditions of the service level agreement for any safety-critical application deployed on a vehicle. To improve the Quality of Service (QoS), data needs to be cached on or near the network edge so that the required data is local to the vehicular network [20].

For edge caching to be effective, network intelligence needs to be pushed near to the edge, in the form of edge-based caching that has been a feature of Content-Distribution Networks (CDNs) and similar. However, edge-based caches have limited size and computing resources relative to what is available in the cloud [21]. Thus, *cache insertion strategies* have been proposed in the research literature, such as, leave copy down, randomly copy one, cache less for more, probabilistic caching, latency-aware caching, etc. Conversely, *cache eviction strategies* need to be considered,

Table 6.4   Key performance indicators for V2X use-cases

Use-cases	Key performance indicators			
	End-to-End Latency[1] (ms)	Reliability[2] (10^{-x})	Communication Range[3] (cm)	Data Rate[4] (Mbps)
Automated overtake	10	10^{-5}	30	–
Cooperative collision avoidance[a]	10^a 100^b	10^{-3a} 10^{-5b}	30	–
High density platooning	10	10^{-5}	30	–
See-through	50	–	–	10
Vulnerable road user discovery	–	–	10	–
Bird's eye view	50	–	–	40

[a]*For Trajectory Handshake*
[b]*For Status Updates*
[1]*End-to-End Latency* specifies tolerable elapsed time from the moment a data packet is generated at the source to the moment it is received at the destination
[2]*Reliability* refers to tolerable packet loss at the application layer. If a packet is not received at the destination within the tolerable elapsed time, it is considered to be lost
[3]*Communciation Range* is a distance between the source and destination within which the specified reliability could be achieved
[4]*Data Rate* refers to bit rate required for any vehicular application to function in an appropriate manner

and the most widely used ones include First-in, First-out (FIFO), RANDOM, Least Frequently Used (LFU), Time to Live (TTL) based caching, etc. [22].

However, vehicular networks are highly dynamic and distributed, and so edge-based caching faces specific challenges in this domain. For instance, unlike static networks, content popularity in dynamic networks changes much more frequently, so the update strategy needs to be more effective at tracking popular content. The cache update strategy itself requires significant computational resources, and thus trade-offs need to be made given the resource constraints near the network edge. Furthermore, due to vehicle mobility, the connectivity (in the form of the network topology and the properties of each link) between any two nodes in the network is constantly changing. Table 6.4 outlines the key performance indicators for several V2X use-cases described in the European Commission's "5G Automotive Vision" [23].

Edge-based caches and distributing the computation to the network edge, therefore, enables both context sensing (by providing local endpoints for sensed data) and context sharing (by providing local sources of sensed data and processed knowledge of the context). Given the cost in terms of data transport delays of sending the data to the cloud and back, edge-based caching is one of the most effective strategies for keeping end-to-end communication times within the specified latency thresholds.

6.5 Software-Defined Networking for Intelligent Network Resource Management and Orchestration

Software-Defined Networking (SDN) was originally developed for data center networks because SDN offered re-programmability, scalability, agility, flexibility, etc. More recently, it has been proposed for vehicular networks [20, 24]. SDN decouples the data plane from the control plane and the network intelligence is passed to a logically centralized controller to store link information and make routing decisions. However, VANETs are distributed in nature than traditional networks and so it is extremely difficult to govern the entire network from a *single* logical point of control. Edge-based computing (as discussed in the previous section) has been proposed to operate with SDN-based vehicular networks [25] to make the task of the SDN controller a little easier. A distributed network control (localized view) is maintained in parallel with a logically centralized control (globalized view). The architecture of such a hybrid software-defined heterogeneous vehicular network is depicted in Fig. 6.2.

Such a hybrid approach serves to minimize the network management overhead. In the case of centralized control and of a dense vehicular environment, each SDN-enabled switch needs to report its status to a centralized controller, with the potential for network congestion in the control plane. Hence, if localized control were introduced, link state updates could be managed locally, with less network overhead. Also, instead of each SDN-enabled switch (e.g., vehicle) communicating with the central controller, vehicles can group together (in vehicular clouds, clusters, or platoons).

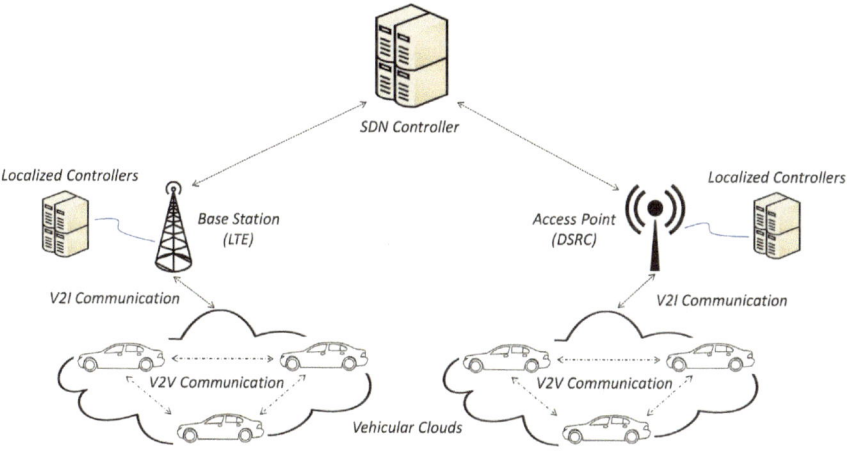

Fig. 6.2 Hybrid approach for software-defined heterogeneous vehicular networks

Grouping in this way ensures that it is sufficient that each cluster head reports the relevant status of the cluster as a whole to the centralized controller instead of every member of the cluster reporting its status to the controller.

With its globalized view of the entire underlying network topology, SDN is better aware of the trajectory of each vehicle and can use this information to predict where and when to cache content in vehicular networks. Such trajectory-based prediction is necessary because vehicles move from one network or geographical coverage region to another in very little time [26, 27]. Vehicle state could be encapsulated in a Virtual Machine (VM) that is hosted in the supporting infrastructure (in the cloud or near the network edge). Migrating a vehicular VM from one serving cloud to another requires bulk data transport in real time, and so accurate prediction of trajectories is necessary. Hence, a service with access to data available from (a network of) SDN controllers can anticipate where vehicle VMs should be placed. Knowledge of location can be derived from the onboard GPS of modern vehicles, possibly supplemented with in-network vehicle location tracking.

6.6   Big Data Management and Analytics for Connected and Autonomous Vehicles

6.6.1   Big Data Accumulation

Data in vehicular networks may be accumulated from the vehicles, their users, supporting roadside infrastructure, and the network itself. Data generated by the vehicles can be classified into two types, i.e., onboard data and onroad data. Onboard data generally arises from in-vehicle sensors used to measure a vehicle's velocity, driving parameters, engine status, and brake status, etc. By contrast, onroad data encompasses data derived from events external to the vehicle, including intervehicular distances, RADARs positions, LIDARs distances, and video from cameras, or data shared by other nodes in the neighborhood including traffic light status, road maps, and crowd-sourced applications. Furthermore, consumer smart devices in vehicles also generate a massive amount of data. GPS chips and gyro sensors on smartphones or tablets could provide supplementary sensing data such as vehicle's trajectory and acceleration. Several research studies [28–30] have used smartphone sensors as probes for monitoring road traffic. Furthermore, it is possible that transient social networks of vehicles and their passengers will be formed and generate their own data. Finally, data might also be acquired from the supporting roadside units as well as space and aerial platforms. High Altitude Platforms (HAPs), satellites, and drones are promising sources of vehicular data by providing enhanced coverage in areas where connectivity is poor. HAPs with a working altitude of 20 km could potentially support citywide data accumulation and communication services. Indeed, V2X technology and the masses of data it generates could be utilized to predict the traffic conditions and to calculate the optimal navigation routes dynamically [31]. By providing substantial connectivity to the internet, the fundamental software on the vehicles (such as that for the engine management) can be updated frequently and transparently to vehicle owners. Tesla offers such an auto-update feature for its models.

6.6.2 Big Data Transmission

In vehicular networks, the size and complexity of the data flows pose challenges for the underlying infrastructure. Applications have been classified as (a) safety applications; and (b) non-safety applications. A safety application is typically smaller in size, i.e., 800 bytes and accordingly demands a low data rate and requires extremely low latency. Safety applications tend to report their status information continuously, including their velocity, geographical position, traveling direction, and turn signal status, with the recipients being other nearby vehicles. By contrast, non-safety applications demand higher data rates and often do not possess stringent latency requirements (in fact, some of them could be delay-tolerant applications). Transmission strategies also need to be devised intelligently to enhance transmission reliability.

For ensuring reliability, two models are generally used for transmission purposes, i.e., push model and pull model. Safety-critical applications usually employ a push-based approach, wherein data is periodically broadcast regardless of whether it has been requested by another node, in order to notify any recipient of any imminent threat and for updating the shared state of neighboring nodes. Delay-tolerant applications, on the other hand, adopt a pull-based model in a bid to avoid wastage of communication resources. However, often both strategies are combined in the form of a hybrid model to facilitate diverse vehicular applications that have been deployed simultaneously on the same network [32].

Due to the dynamic nature of vehicular networks, it is challenging to devise intelligent routing mechanisms. Vehicular routing protocols can be classified as (a) topology-based and (b) position-based routing protocols. Topology-based protocols utilize network topology information to select the relay nodes for an optimal route. Since it is difficult (and often expensive in terms of network resource usage) to update the network topology frequently in vehicular networks, proactive, reactive, and hybrid approaches have been proposed [33]. In the case of proactive routing schemes, each node employs one or more routing tables to store the topology of the entire network and the optimal path is then selected by guaranteeing the shortest possible path and/or least transmission delay. This information is captured and updated regularly regardless of whether a specific node is involved in the route discovery process or not. Nevertheless, such schemes can lead to significant routing management overhead in dense traffic. By contrast, reactive routing protocols are bandwidth efficient on-demand ad hoc routing protocols and the route discovery process is initiated only when a route to a destination is required. In addition, hybrid protocols are an amalgamation of proactive and reactive protocols, which intends to minimize both the management overhead of proactive routing protocols and the additional delay caused by the route discovery process of reactive routing protocols. These routing protocols were developed for mobile ad hoc networks and have also been adopted for VANETs but are not yet suitable due to the highly dynamic behavior of vehicles. Finally, position-based routing protocols rely on each vehicle's geographical position to make routing decisions.

6.6.3   Big Data Storage

Vehicular big data storage can be classified as (a) onboard storage, (b) roadside storage, and (c) Internet storage. In the case of onboard storage, the onboard units could store much of the data themselves as modern vehicles increasingly possess extensible storage such as the NVIDIA platform mentioned earlier [34]. For this to work well, edge caches need to use onboard storage effectively, and the cache strategy needs to be aligned with the routing strategy. Also, roadside units typically comprise cellular Base Stations (BSs) and more general Access Points (APs) for networking purposes, but increasingly they contain excess and extensible storage capabilities. For reference, each Codha Wireless roadside unit possess 10 GB onboard storage [35]. Similarly, roadside Wi-Fi hotspots could also be installed with massive storage. Lastly, with seamless connection to the internet, accessing remote storage becomes conceptually easy, though possibly at the expense of greater latency. Thus, vehicles could access cloud services and provisionally upload content there in a bid to save precious onboard resources and release space for more time-critical operations. The contents stored on the remote clouds could be later accessed on demand by delay-tolerant services.

In addition to this, there are various storage mechanisms which could be classified as (a) fast, (b) medium, and (c) slow storage [36]. Fast storage typically refers to storage either on the vehicle itself or on a passenger's smart device (smartphones and tablets), so the data can be accessed with a guaranteed minimum delay. Generally, delay-sensitive applications and services should utilize fast storage in the first instance and with higher priority than other applications. Medium storage refers to storage external to the vehicle that can be reached via a reliable network connection but with a slightly larger delay than fast storage. Medium storage includes opportunistic data sharing among the vehicles using V2V and V2I communication with the roadside units. Finally, slow storage includes external storage with a reliable Internet connection and normally refers to storage on centralized clouds which trades off storage quantity and access delays. Vehicular non-safety (e.g., infotainment) applications are normally expected to be stored in the clouds. Apart from the (centralized) cloud, both fog and edge clouds work in tandem with centralized clouds in order to bring computing resources for intelligence nearer to the edge of the network.

6.6.4 Big Data Computing

Onboard computing has been in vehicles for decades and can be traced to when Volkswagen became one of the first automotive companies to introduce an Electronic Fuel Injection (EFI) system for vehicles [37]. EFI system has now become commonplace. Before V2X communication become the norm, onboard computing primarily focuses on processing the sensor data related to engine and vehicular controls. If the market for connected and autonomous vehicles is to grow, it requires a quantum leap in technology and will result in massively increased data flows between vehicles and infrastructure, analogous to how smart devices have revolutionized personal communications, computing, and social interaction. Vehicles could become networked computing centers [38, 39] and, in time, computing from multiple vehicles could be employed for cooperative operations, i.e., capturing and sharing road conditions [40], calculating the next hop or routing directions for big data relaying in a vehicular environment [41], and self-organized vehicle platooning and clustering [42]. This is also evident from the fact that the NVIDIA Drive PX platform (powered by real-time vision processing, sensor fusion, and deep learning functionalities) can support 24 trillion operations per second (Tflops) [43]. Furthermore, enhanced roadside units employing multicore CPUs and modern operating systems could be utilized for vehicular big data computing. Finally, if local computational capabilities are unable to meet the demand at a given time and location, remote resources can be used, albeit with a higher delay.

6.7 Augmented Reality for Enhancing Ambient Intelligence in Connected and Autonomous Vehicles

To assist understanding of how such ambient intelligence could play a part in the IoV, we discuss a scenario in which Augmented Reality (AR) in ITS assists drivers. The word "augmented" means to add or enhance something. Hence, AR enhances a human's perception by superimposing (useful) contextual information in real time. When human drivers are in control, certain maneuvers, such as overtaking a slow-moving lorry on expressways as depicted in Fig. 6.3, have inherent risks and AR might help to make them safer.

From this scenario, it can be observed that occlusions or obstacles in a driver's line of sight could lead to fatal accidents with vehicles approaching from the opposite side of the road. It can also be observed that a vehicle (either fully or partially controlled by its driver) needs to be aware of its ambience (context) before undertaking any critical maneuverer. In case of overtaking slow-moving lorries, AR could help by projecting a *"see-through view"* from the leading vehicle (the lorry) to the vehicles following it, especially that vehicle which is directly behind. Even if the see-through feature is not available for whatever reason, the fallback might be that vehicles could still be informed with alerts regarding various vulnerabilities. However, this all becomes possible only if ambient intelligence is available that is built upon context sensing

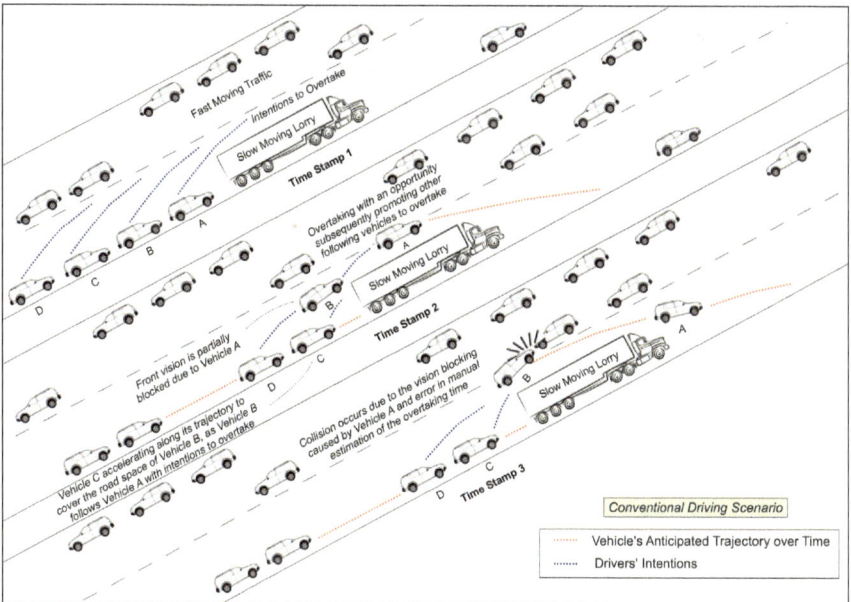

Fig. 6.3 AR visualization for intelligent transportation systems (*Depicting Human Behavior in Overtaking Scenarios*)

(sensor data accumulation), context analytics and context sharing (via low latency high-speed communication links). In principle, such links can be realized using SDN (for flexible control of the network), heterogeneous networking (for increased reliability and resilience of communications in a highly mobile transport scenario), and edge-based caching (to minimize delay by ensuring that frequently accessed content is available with minimum delay).

6.8 Conclusion

The notion of Internet of things has spread everywhere and today the *"things"* ranging from smartphones to automobiles are being connected to the Internet at an unprecedented rate. According to estimates by Gartner, by the year 2020, more than 20 billion things will be connected to the Internet [44]. These devices will generate a slew of data at the network edge. Most of this data is better left for processing at the network edge, and where remote servers are needed, heterogeneous networking needs to be used to achieve the conflicting demands when connectivity and/or bandwidth are limited. Also, retrieving the computed insights from this data could become impossible within the permittable latency thresholds. Therefore, heterogeneity of diverse radio access technologies along with intelligent network resource management remains the most promising way to guarantee extended communication range, high reliability, extremely high data rates, and low latency as needed for vehicular applications.

Empowering modern vehicles with ambient intelligence is necessary to achieve fully connected and autonomous vehicles. Instead of simply relying on certain hard-coded policies for safe driving, the rich domain of cognitive sciences offers the promise of endowing vehicles with a kind of common sense so that they can quickly adapt to new situations. Deep learning is one possible way of learning from previous experience, however, it is difficult to determine when this learner has learnt enough. Therefore, to validate such models, researchers need to understand how humans make sense of the real world and then to mimic it safely for connected and autonomous vehicles.

This chapter aims to bring forth the significance and need of ambient intelligence in next-generation connected and autonomous vehicles. If the ambient intelligence algorithms for autonomous vehicles could be devised and validated soon, autonomous vehicles would undoubtedly become an example of the latest generation of intelligent, highly sophisticated robots interacting with people in the real world.

References

1. He Z, Zhang D, Liang J (2016) Cost-efficient sensory data transmission in heterogeneous software-defined vehicular networks. IEEE Sens J 16(20):7342–7354
2. Fontes RDR, Campolo C, Rothenberg CE, Molinaro A (2017) From theory to experimental evaluation: resource management in software-defined vehicular networks, vol 5. IEEE Access, pp 1–8
3. Zheng K, Zheng Q, Yang H, Zhao L, Hou L, Chatzimisios P (2015) Reliable and efficient autonomous driving: the need for heterogeneous vehicular networks. IEEE Commun Mag 53(12):72–79
4. 3GPP (2015) 3rd Generation partnership project; technical specification group services and system aspects; study on LTE support for vehicle to everything (V2X) services (3GPP TR 22.885 V14.0.0—Technical Report). https://portal.3gpp.org/desktopmodules/Specifications/SpecificationDetails.aspx?specificationId=2898
5. Thompson C (2010) New details about the fatal Tesla Autopilot crash reveal the driver's last minutes. http://www.businessinsider.com/details-about-the-fatal-tesla-autopilot-accident-released-2017-6. Accessed 06 May 2018
6. Choi J, Gonzalez-Prelcic N, Daniels R, Bhat CR, Heath RW (2016) Millimeter wave vehicular communication to support massive automotive sensing. IEEE Commun Mag 54(12):160–167
7. Nelson P (2016) Just one autonomous car will use 4000 GB of data/day. https://www.networkworld.com/article/3147892/internet/one-autonomous-car-will-use-4000-gb-of-dataday.html. Accessed 06 May 2018
8. Zheng K, Zheng Q, Chatzimisios P, Xiang W, Zhou Y (2015) Heterogeneous vehicular networking: a survey on architecture, challenges, and solutions. IEEE Commun Surv Tutor 17(4):2377–2396
9. Bagloee SA, Tavana M, Asadi M et al (2016) Autonomous vehicles: challenges, opportunities, and future implications for transportation policies. J Mod Transp 24(4):284–303
10. Xu W, Zhou H, Cheng N, Lyu F, Shi W, Chen J, Shen X (2018) Internet of vehicles in big data era. IEEE/CAA J Autom Sin 5(1):19–35
11. Sun S-H, Hu J-J, Peng Y, Pan X-M, Zhao L, Fang J-Y (2016) Support for vehicle-to-everything services based on LTE. IEEE Wirel Commun 23(6):4–8
12. Amadeo M, Campolo C, Molinaro A (2016) Information-centric networking for connected vehicles: a survey and future perspectives. IEEE Commun Mag 54(2):98–104
13. Mach P, Becvar Z (2017) Mobile edge computing: a survey on architecture and computation offloading. IEEE Commun Surv Tutor 19(3):1628–1656
14. Lu N, Cheng N, Zhang N, Shen X, Mark JW (2014) Connected vehicles: solutions and challenges. IEEE Internet Thing J 1(4):289–299
15. He Z, Cao J, Liu X (2016) SDVN: enabling rapid network innovation for heterogeneous vehicular communication. IEEE Netw 30(4):10–15
16. 5G PPP (2015) 5G Vision—5G infrastructure public private partnership: the next generation of communication networks and services. https://5g-ppp.eu/wp-content/uploads/2015/02/5G-Vision-Brochure-v1.pdf. Accessed 06 May 2018
17. Marquez-Barja JM, Ahmadi H, Tornell SM, Calafate CT, Cano JC, Manzoni P, DaSilva LA (2015) Breaking the vehicular wireless communications barriers: vertical handover techniques for heterogeneous networks. IEEE Trans Veh Technol 64(12):5878–5890
18. Mumtaz S, Jornet JM, Aulin J, Gerstacker WH, Dong X, Ai B (2017) Terahertz communication for vehicular networks. IEEE Trans Veh Technol 66(7):5617–5625
19. Malandrino F, Chiasserini CF, Kirkpatrick S (2016) The impact of vehicular traffic demand on 5G caching architectures: a data-driven study. Veh Commun 8:13–20
20. Deng DJ, Lien SY, Lin CC, Hung SC, Chen WB (2017) Latency control in software-defined mobile-edge vehicular networking. IEEE Commun Mag 55(8):87–93
21. Liu J, Wan J, Zeng B, Wang Q, Song H, Qiu M (2017) A scalable and quick-response software defined vehicular network assisted by mobile edge computing. IEEE Commun Mag 55(7):94–100

22. Modesto FM, Boukerche A (2017) An analysis of caching in information-centric vehicular networks. In: 2017 IEEE international conference on communications (ICC). Paris, pp 1–6
23. 5G-PPP (2018) 5G automotive vision. https://5g-ppp.eu/wp-content/uploads/2014/02/5G-PPP-White-Paper-on-Automotive-Vertical-Sectors.pdf. Accessed 06 May 2018
24. Yaqoob I, Ahmad I, Ahmed E, Gani A, Imran M, Guizani N (2017) Overcoming the key challenges to establishing vehicular communication: Is SDN the answer? IEEE Commun Mag 55(7):128–135
25. Azizian M, Cherkaoui S, Hafid AS (2017) Vehicle software updates distribution with SDN and cloud computing. IEEE Commun Mag 55(8):74–79
26. Yao H, Bai C, Zeng D, Liang Q, Fan Y (2015) Migrate or not? Exploring virtual machine migration in roadside cloudlet-based vehicular cloud. Concurr Comput Pract Exp 27(18):5780–5792
27. Joerer S, Segata M, Bloessl B, Lo Cigno R, Sommer C, Dressler F (2014) A vehicular networking perspective on estimating vehicle collision probability at intersections. IEEE Trans Veh Technol 63(4):1802–1812
28. Händel P, Ohlsson J, Ohlsson M, Skog I, Nygren E (2014) Smartphone-based measurement systems for road vehicle traffic monitoring and usage-based insurance. IEEE Syst J 8(4):1238–1248
29. Ghose A, Biswas P, Bhaumik C, Sharma M, Pal A, Jha A (2012) Road condition monitoring and alert application: Using in-vehicle Smartphone as Internet-connected sensor. In: 2012 IEEE international conference on pervasive computing and communications workshops. Lugano, pp 489–491
30. Liang X, Li X, Luan TH, Lu R, Lin X, Shen X (2012) Morality-driven data forwarding with privacy preservation in mobile social networks. IEEE Trans Veh Technol 61(7):3209–3222
31. Mueck M, Karls I (2018) Networking vehicles to everything (Evolving automotive solutions). DelG Press, Berlin
32. Ahmed SH, Bouk SH, Kim D, Rawat DB, Song H (2017) Named data networking for software defined vehicular networks. IEEE Commun Mag 55(8):60–66
33. Sanaei Z, Abolfazli S, Gani A, Buyya R (2014) Heterogeneity in mobile cloud computing: taxonomy and open challenges. IEEE Commun Surv Tutor 16(1):369–392
34. Bojarski M, Testa D, Dworakowski D, Firner B, Flepp B, Goyal P, Jackel LD (2016) End to end learning for self-driving cars. arXiv:1604.07316
35. Cohda Wireless (2018). http://www.cohdawireless.com/sectors/v2x/. Accessed 06 May 2018
36. Darwish TSJ, Abu Bakar K (2018) Fog based intelligent transportation big data analytics in the internet of vehicles environment: motivations, architecture, challenges, and critical issues. IEEE Access 6:15679–15701
37. Hou X, Li Y, Chen M, Wu D, Jin D, Chen S (2016) Vehicular fog computing: a viewpoint of vehicles as the infrastructures. IEEE Trans Veh Technol 65(6):3860–3873
38. Baccarelli E, Naranjo PGV, Scarpiniti M, Shojafar M, Abawajy JH (2017) Fog of everything: energy-efficient networked computing architectures, research challenges, and a case study. IEEE Access 5:9882–9910
39. Vigneri L, Spyropoulos T, Barakat C (2016) Storage on wheels: offloading popular contents through a vehicular cloud. In: IEEE 17th International symposium on a world of wireless, mobile and multimedia networks (WoWMoM). Coimbra, pp 1–9
40. Zhou Z, Yu H, Xu C, Zhang Y, Mumtaz S, Rodriguez J (2018) Dependable content distribution in D2D-based cooperative vehicular networks: a big data-integrated coalition game approach, pp 1–12
41. Cui L, Yu FR, Yan Q (2016) When big data meets software-defined networking: SDN for big data and big data for SDN. IEEE Netw 30(1):58–65
42. Su D, Ahn S (2017) In-vehicle sensor-assisted platoon formation by utilizing vehicular communications. Int J Distrib Sens Netw 13(7):1–12

43. Nvidia Drive (2018) Scalable AI platform for autonomous driving—World's first functionally safe AI self-driving platform. https://www.nvidia.com/en-us/self-driving-cars/drive-platform/. Accessed 06 May 2018
44. Gartner Inc. (2017) Gartner says 8.4 billion connected "things" will be in use in 2017, Up 31 percent from 2016. https://www.gartner.com/newsroom/id/3598917. Accessed 06 May 2018

Chapter 7
Intelligent Control Systems for Carbon Monoxide Detection in IoT Environments

Champa Nandi, Richa Debnath and Pragnaleena Debroy

Abstract Carbon Monoxide (CO) is a ubiquitous product of partial burning of materials containing carbon. It is a poisonous gas, inhalation of which causes headache, nausea, dizziness which may sometimes lead to death. Thus, safety and security of human being from CO is of paramount significance. Designing a carbon monoxide detection system has, therefore, become very much essential to prevent such serious incidents. This chapter discusses the design and implementation of a secure and cost-effective real-time carbon monoxide detection and control system for living environments (e.g., air-conditioned rooms, factory spaces, and automobiles) by using embedded intelligent control mechanisms. The chapter provides a basic overview of the carbon monoxide gas, its sources and effects, related carbon monoxide gas sensors, and embedded intelligent controllers. The chapter also illustrates a review of various accidental cases due to exposure to CO and discusses a mathematical model for the embedded intelligent controllers. Lastly, a brief description of the software and hardware implementation of the embedded intelligence in an IoT platform has also been discussed. The chapter concludes with the significant contribution of this system by suggesting future research opportunities in this field.

Keywords Carbon monoxide · CO · Sensor · Embedded intelligent controller
CO gas sensor · Electrochemical sensors · Infrared sensors · Optical sensors
IoT · GPS · GSM

7.1 Introduction

With much advancement of and innovation in current technologies, a great deal of effort has been made for the improvement of human safety and comfort. But with this development, taking care of the surroundings has been forgotten in which humans

C. Nandi (✉) · R. Debnath · P. Debroy
Department of Electrical Engineering, Tripura University,
Suryamaninagar, Tripura, India
e-mail: cnandi@tripurauniv.in; chmpnandi@gmail.com

© Springer Nature Switzerland AG 2019
Z. Mahmood (ed.), *Guide to Ambient Intelligence in the IoT Environment*, Computer
Communications and Networks, https://doi.org/10.1007/978-3-030-04173-1_7

live and spend most of their time. Thus, the environment gets polluted and the quality of air reduces. One example of these innovations is a motor car. These vehicles release approximately 25% of hazardous gases in the form of carbon monoxide (CO) into the air [1]. With time, vehicles are being designed with the improvement of engine and combustion technologies. However, due to the irregular maintenance of the vehicles, engines get deteriorated that then results in incomplete combustion of fuels causing increased production of CO. This incomplete combustion of hydrocarbons is caused due to lack of enough oxygen present in the system, or insufficient time to complete the combustion in the chamber. If the combustion temperature of hydrocarbons drops in the combustion chamber, CO is then produced [2].

Industries are one of the major sources of carbon monoxide emission. CO gets emitted from various industrial plants through the ubiquitous combustion of coal, natural gas, and coke. Many industrial processes including carbon black manufacturing, metal manufacturing, chemicals production, oil and gas extraction from land or sea, food manufacturing, petroleum refining, plaster and concrete manufacturing, electricity supply, metal ore mining, etc. can cause a huge amount of CO production and ejection in the air [3].

As half of the global civilization lives in urban areas, utilization of new inventions is widely increased. Homes having fuel-burning appliances like water heaters, geysers, stoves, ovens, clothes dryers, charcoal grills, generators, power tools, gas and wood fire burners, wood stoves, furnaces, etc. may all have CO problems [4]. During winter, emissions of CO from vehicles increase dramatically. In low temperatures, more fuel is required to start engines and so the emission control devices such as catalytic converters do not operate efficiently. In this case, the CO released from car engines stays in the closed space, e.g., garages, for a longer period of time and the air containing CO flows into the living spaces, e.g., houses. In cold weather, houses are mostly closed up and maximum air enters the house through the attached car garages [5, 30].

The traffic congestion plays a crucial role in increasing CO level in urban areas. According to one analysis [6], CO emission from the traffic is approximately 90% during rush hours. As the number of vehicles is increasing day by day, it is not surprising that air quality of car cabins is becoming worse especially when a large number of exhaust pipes are only a short distance away from adjacent vehicles. All this increases the possibility of CO to get into the car cabins [7].

In urban and industrialized areas, the concentration of CO is even higher. So, there is a good possibility of CO to enter the buildings, rooms or car cabins from the external environment via air conditioning system, window openings, gaps, and imperfect seals. CO may also be produced inside the rooms by the household appliances [7].

CO is a fragrance-free, uncolored, toxic gas. Exposure to CO frequently goes unrecognized until it is too late; exposure can also be fatal [8]. As the area inside a vehicle cabin or room is small and there is often lack of proper ventilation, the effect of CO becomes more dangerous in air-conditioned rooms, car cabins, and such closed spaces. It can cause serious health issues to the occupants [1]. The inhalation of this poisonous gas causes headache, nausea, dizziness and sometime s may lead to death. In the last few years, many death cases have been recorded due to CO poisoning in an air-conditioned car [10–12]. To prevent such fatal cases related to CO poisoning, there should be a vigilant alert system to inform people inside car cabins or rooms when the excessive concentration of CO is detected.

In the literature, there have been several studies that looked at the reasons for CO production. There is some research on Carbon Monoxide Detection Systems for Motor Vehicles [13–19]. A pilot study has been carried out in which the urban air pollution has been mapped using mobile CO sensor and Global Positioning System (GPS) receiver for tracking the sensors [20].

In this context, the Internet of Things (IoT) based embedded intelligent controllers integrated with alert systems for CO detection can be designed to prevent CO accidental incidents. To increase the security of the occupants of a vehicle, a Global Positioning System (GPS) module can be introduced for tracking the location of the vehicle [20, 21] and a Global System for Mobile communication (GSM) module can also be integrated to send alert messages to the authorized users.

So, much work does exist on CO detection, monitoring, and control, but the related challenges have sometimes been left unresolved. This chapter reviews various issues and challenges, presents a number of accidental case studies due to CO exposure and provides review on CO sensors. This chapter also discusses intelligent control design in the area of CO detection and control in GPS-enabled vehicles and rooms. The embedded intelligent control system methodology and design are also explored in some detail. The chapter concludes with future research directions in this field.

7.2 Carbon Monoxide (CO) Gas Sensors

Carbonous oxide, commonly known as carbon monoxide (CO) is a colorless, practically odorless, tasteless, and non-irritant gas as it is 3% lighter than air [22]. It burns with a violet flame, slightly soluble in water and considerably soluble in alcohol and benzene [23]. In the modern world, CO intoxication is the most common type of poisoning that leads to death. Neither people nor animals can realize when they breathe the CO contaminated air. That is why, it is often considered as *Silent Killer* [8].

Carbon monoxide (CO) gas is an industrial hazard produced by incomplete combustion of fuels, natural gas, and other substances containing carbon. In normal conditions, the combustion process takes place when carbon gets combined with oxygen (O_2) and produces carbon dioxide (CO_2). But if there is a scarcity of oxygen (O_2) in the combustion process or heating of carbon is faulty, there is a chance of production of CO [24]. CO may be caused through different other sources like gas fires, vehicle engines, furnaces, the improper working of air conditioning systems, etc. [9, 12].

7.2.1 Sources of Carbon Monoxide

In recent models of vehicles, CO emission has reduced due to the improvement of engine design and by the chemical treatment of exhaust gases. But incomplete combustion of fuel can cause the production of CO due to incorrect air–fuel ratio. This problem is caused due to lack of enough oxygen present or insufficient time to complete the combustion in the chamber [25]. CO can also be produced when the temperature falls below the combustion temperature of hydrocarbons [2]. Cigarette smoking also causes CO emission [26].

The vehicle exhaust system produces CO along with other noxious gases such as nitrous oxide and carbon dioxide [27, 28]. An amount of CO gas may enter the vehicle through a hole or a gap in car body during the warming up of the vehicle engine in a garage [29]. There is also a possibility of generating CO if the air conditioning system of the vehicle is not working properly [7, 12]. In cold weather, emissions of CO from vehicles increase dramatically. In low temperature, more fuel is required to start engine and the emission control devices as catalytic converters do not operate efficiently [30].

CO may enter the room or car cabin through the air conditioning (AC) system when outside air room or car become more polluted and consists of a higher amount of CO concentration [7]. CO may also be produced inside a living room through the use of water heaters, geysers, stoves, ovens, clothes dryers, charcoal grills, fireplaces, etc. [31].

7.2.2 Effects of Carbon Monoxide on Human Body

For humans, when breathing in the surrounding air, oxygen gets combined with red blood cells (RBC) in the body and then carried through the human body. On the other hand, during breathing out, the generated carbon dioxide gets released and frees up the RBC to pick up oxygen for the next inhale. The most harmful effect of carbon monoxide is that it also gets bounded with RBC-like oxygen and forms carboxyhemoglobin (COHb). In fact, the hemoglobin possesses a characteristic of being attracted towards carbon monoxide 200 times more than oxygen [32, 33]. So, the ability of hemoglobin to carry oxygen gets decreased and thus causing tissue hypoxia [33].

If there is an excessive inhalation of CO, the symptoms like headache, nausea, dizziness, fatigue, etc. can occur. Sometimes extreme CO inhalation can lead to death [32]. Table 7.1 summarizes some health effects due to prolonged exposure to various concentrations of CO, as well as some government recommended limits and Pocket CO alarm levels. It has been compiled from various data sources, including the National Fire Protection Association (NFPA) [34].

Table 7.1 CO Concentration with toxic symptoms

PPM of CO	Time	Symptoms
0–5	8 h	None or decrease exercise tolerance
10–20	8 h	Tightness across the forehead and slight headache
35	8 h	Headache and dizziness
100	2–3 h	Mild headache
200	2–3 h	Mild headache, fatigue, nausea and dizziness
400	1–2 h	Serious headache and life threatening after 3 h
800	45 min	Convulsion, dizziness and nausea
1600	20 min	Headache, dizziness and nausea. Death within 1 h
3200	5–10 min	Headache, dizziness and nausea. Death within 1 h
6400	1–2 min	Headache, dizziness and nausea. Death within 25–30 min
12800	1–3 min	Death

Note PPM refers to Parts per Million; it can be expressed as milligrams per liter (mg/L)

7.2.3 Case Studies of Carbon Monoxide Accidents

The poisonous CO is a by-product of partial ignition of hydrocarbons. It binds rapidly with hemoglobin (Hb) in human body leading to the formation of Carboxyhemoglobin (COHb). As a result, the ability of blood to carry oxygen decreases and leads to death [8, 32, 33]. Some such cases have been surveyed from different case reports, newspapers, and articles.

A case report [10] was published by Chand Meena in 2014 that investigates a death case of a 35-year-old-male mechanics due to CO poisoning in a car garage. He was found unconscious in that automobile garage and was dead after one hour in the hospital. On external examination by doctors, Rigor Mortis had occurred throughout the whole body and Cherry red postmortem lividity was found present over the back (refer to Fig. 7.1). The chemical analysis report of blood confirmed that the death was due to carbon monoxide toxicity.

In 2010, Kumar and Rautji [11] published a case report in which an 18-years-old young female was found unresponsive in a car with the engine and air conditioner in the switched-on mode. The car was parked inside a closed garage. Refer to Fig. 7.2. That girl was feeling giddiness earlier and her boyfriend who was also inside the car was experiencing drowsy. Her back had cherry red stains as reported on postmortem and fingernails beds were cyanosed.

The Daily News and Analysis [35] published news of a suicide case by sniffing CO, on September 22, 2014. Suicide was committed by inhaling CO. One of the deceased's colleagues, who was staying in the same apartment also suffered severe injuries due to partial inhalation of CO; he was later admitted in ICU.

The Hindustan Times, on July 22, 2014, reported that three men died in south Delhi in a suspected case of CO gas poisoning. Police said that they were found dead sitting in a car with engine and AC in running condition [36].

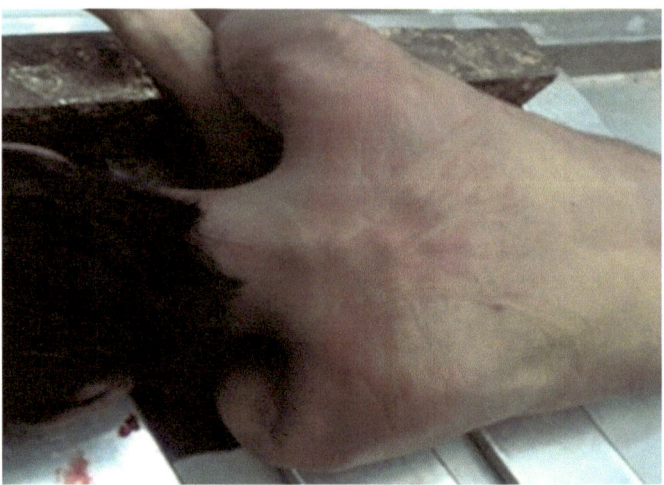

Fig. 7.1 Cherry red lividity was present on the back

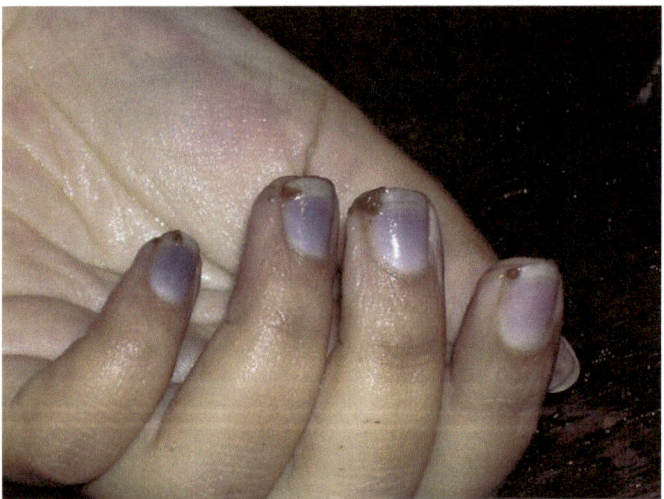

Fig. 7.2 Cyanosed fingernails

On December 9, 2011, a renowned fire accident took place in AMRI Hospital Kolkata in India. The fire began at about 3 am. Seventy patients and three hospital staffs were among those died at this private hospital. Most of them died due to suffocation caused by carbon monoxide accumulation in the building [37].

In 1975, Seah Han Cheow and Chao Tzee Chang [12] published a report based on 47 death cases due to Carbon monoxide poisoning at the Government Department of Pathology, Singapore. The cause of deaths was mainly due to CO saturation in the blood. Among these 47 cases, 91% of deaths were accidental and 9% were suicides.

7.2.4   Review of Carbon Monoxide Sensors

A gas sensor is a device that senses and measures the concentration of gas in its surroundings [38]. For practical applications, the sensing performance like sensitivity, selectivity, and the rate of response of gas sensors depends on location, purpose and condition of sensor operations. Reliability, stability, and interfering gases are also considered as important aspects of material design for gas sensors.

As the simplicity, cost, and size are critical aspects of sensor design, only three types of sensors may be considered for indoor air quality monitoring [7]:

- Semiconductor Metal Oxide (SMO) Sensors
- Electrochemical (EC) Sensors
- Infrared/Optical (IR) Sensors.

7.2.4.1   Semiconductor Metal Oxide (SMO) Sensors

Semiconductor Metal Oxide (SMO) Sensors, also known as *chemiresistors* are well suited for air quality monitoring application of motor vehicle cabin as they are small in size, reliable, long-lasting, and cheaper. The detection concept of SMO sensors is based on the change of the surface resistance of the sensor upon adsorption of the gas molecules on the surface of a semiconductor. This type of sensor uses tin dioxide (SnO_2) as the sensing element. This sensor is heated to allow the free electrons to flow through the grain boundary of tin dioxide crystals.

In clean air, i.e., in the presence of around 21% of oxygen (O_2), oxygen gets absorbed on the surface of the metal oxide. As oxygen has high electron affinity, it attracts the free electrons of the metal oxide and forms a potential barrier at the grain boundaries. Thus, it prevents the electron flow through the metal surface causing the high surface resistance of the sensor in clean air. When a sensor is exposed to a pollutant gas, the oxidation reaction with that gas takes place at the metal surface. Hence, the density of absorbed oxygen on the surface of tin dioxide decreases which reduces the height of the potential barrier. It results in the electrons to flow easily through the metal surface and the sensor resistance decreases as the sensor resistance is inversely proportional to the concentration of target gas [7]. Some examples of SMO sensors include MQ-7, MQ-2, TGS 2442, TGS 3870, and diamond diode CO sensors [39–42, 47].

7.2.4.2   Electrochemical (EC) Sensors

Electrochemical (EC) Sensors consist of chemical reactance as electrolytes or gel and two terminals; an anode and a cathode. The anode is used for oxidization process and the cathode for the reduction process. In EC sensors, electrodes are placed in contact with the electrolyte to create potential. Negative ions flow towards the anode and positive ions flow towards the cathode. The reducible gases may be oxygen, nitrogen oxides and chlorine at the cathode and oxidizable gases may be carbon monoxide, nitrogen dioxide, and hydrogen sulfide at the anode. As the pollutant gases diffuse, these react with the electrodes and change the potential which is proportional to the gas concentration [7]. EC4 500 CO is an example of an electrochemical sensor used for carbon monoxide sensing [45].

7.2.4.3   Infrared/Optical (IR) Sensors

IR based sensors are convenient for air quality monitoring applications because of their small size, low power consumption, good resolution, high selectivity, and reduced cost. These sensors have a lifetime of over 6 years depending upon IR source degradation. Based on the analysis of unique IR absorption spectra, IR sensors identify the pollutant gases. When these sensors are exposed to the target gas, their optical sensing elements undergoes light transmission changes proportionally to the target gas concentration. These comprise an IR source like an incandescent lamp, an IR detector like thermopile, pyroelectric detectors, photodiode, etc. and an optical filter to select appropriate wavelengths of the transmitted light and a sample cell. The IR source is placed at one end and the IR detector at the other. The optical bandpass filter passes only the light of the appropriate wavelength of the target gas being measured. When the concentration of target gas increases, the gas molecule absorbs the IR ray. As a result, the sensor output reduces. This phenomenon can be expressed by the Beers Law of Absorption as:

$$I = I_0 e^{(-kcl)} \tag{7.1}$$

Here, I is the intensity of the IR radiation at the IR detector, I_0 is the Irradiation emitted from the IR source, k is the absorption coefficient, c is the gas concentration and l is the optical path length.

For indoor air quality monitoring, semiconductor metal oxide (SMO) sensors are preferable to IR and electrochemical sensors because of their small size and low cost. That is why SMO sensors may be used for the embedded systems that are being discussed in this chapter.

7.2.5   Comparison of Carbon Monoxide Gas Sensors

There are different types of CO sensors available in the market. Based on the market survey, characteristics of some CO sensors are listed in Table 7.2 [43].

Table 7.2 compares various CO sensing technologies against seven key criteria. Both metal oxide and optical sensors are good candidates for automotive applications. IR-based sensors are costly compared to metal oxide (SMO) and electrochemical sensors. Electrochemical sensors, on the other hand, fall short as they have a maximum lifetime of approximately 2–5 years, which renders them unacceptable for automotive applications. Today, most air quality monitoring

Table 7.2 Comparison of CO sensing technologies

Criteria	Metal oxide	Electrochemical	Infrared
Typical sensor cost	$10	$20	>$50
Lifetime (years)	>6	2–5	>6
Selectivity	Poor	Very Good	Excellent
Sensitivity	Very Good	Very Good	Very Good
Response time	Seconds	Seconds	Seconds
Size	Small	Medium	Medium
Ease of use	Excellent	Excellent	Good

Table 7.3 Comparison of different types of SMO CO sensors

Criteria	MQ-7	MQ-9	TGS 2442	TGS 3870
Target gas	CO	CO and combustible gas	CO	Methane and CO
Maximum detection range (ppm)	2000	10000	1000	1000
Conditioning period before the test	Less than 2 days	2 days	2 days or more	More than 5 days
Price (Approx)	Rs 150	Rs 200	Rs 1200	Rs 1300
Operating temperature	20±2 °C	20±2 °C	20±2 °C	20±2 °C

sensors employ SMO sensors. However, as the cost of IR-based sensors rapidly decreases, they are more willingly found in integrated and aftermarket solutions. For indoor air quality monitoring, semiconductor metal oxide (SMO) sensors are preferable to IR-based and electrochemical sensors because of their small size and low cost. That is the main reason that SMO sensors may be usefully employed for the embedded systems [7]. This is further described in this chapter.

Table 7.3 compares various SMO sensors against five key criteria. All are good candidates for CO detection. However, MQ-7 is the cheapest among all these and has greater sensitivity towards CO, high stability, and long life. It can be used as gas leakage detector for domestic and industrial purposes, as well. Sensors which are listed in Table 7.3 have some limitations and these are given below:

- As per the datasheet, the operating temperature range of these sensors is 20±2 °C. So, these cannot withstand extreme temperature fluctuation inside a closed vehicle or room.
- For coherent measurement of CO, around 21% of oxygen is needed in the environment [44].

There are some sensors whose operating temperature is high, e.g., EC4-500-CO (-20 °C to $+50$ °C) [45], 4CM CO sensor (-40 °C to $+55$ °C) [46], and Diamond Diode CO sensor ($+50$ °C to $+500$ °C) [47]. Hence, different types of CO sensors with different operating temperature ranges may be used according to the temperature requirements.

7.3   Design of Intelligent Controllers

An intelligent controller may be designed to detect the presence of CO and indicate accordingly. This controller may consist of a CO sensor, microcontroller, window controller unit, AC controller unit, alert system, Global System for Mobile (GSM) module, and a Global Positioning System (GPS) module, as illustrated in Fig. 7.3.

For a complete picture, components of Embedded Intelligent Controller as shown in Fig. 7.3 include the following:

- CO Sensor: It may consist of two sensors: one for measuring the CO concentration of the inside environment and the other for outside environment.
- Microcontroller: This is the main unit to control the entire system.
- AC controller unit: This may be used for controlling the air conditioning system of the vehicle or room according to Microcontroller logic.
- Power window controller unit: In today's world, power windows or electric windows have become a quality feature of almost all smart vehicles and smart buildings. Power window controller may be used for monitoring the window of room or car cabin according to Microcontroller logic.
- Alert System: It may comprise Buzzer, LED and LCD for indicating purpose.
- IoT: This is the network for device connectivity and exchange of data.
- GPS Module: This is useful for tracking the movement of a vehicle.
- GSM Module: This is required for sending messages to authorized users in case of unhealthy situations.
- Power Supply: An absolute necessity for delivering power to the system.

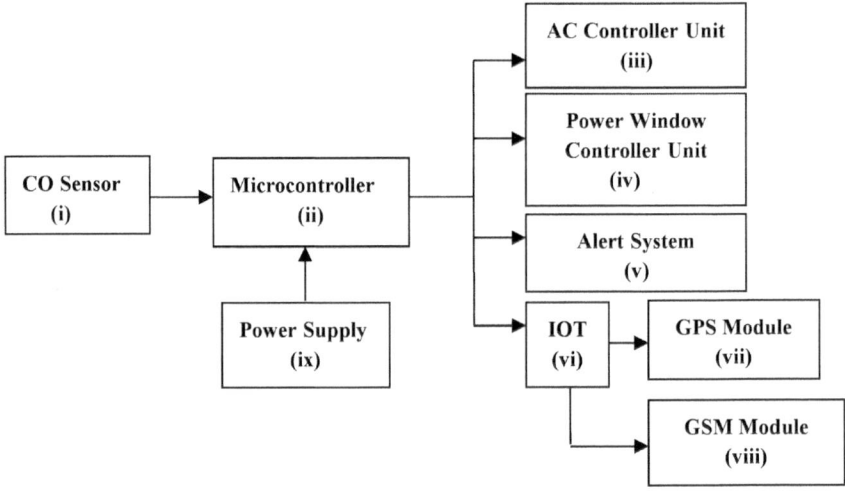

Fig. 7.3 Block diagram of the embedded intelligent controller

In this intelligent controller, one sensor (sensor-1) may be used for measuring the CO concentration of inside environment and another sensor (sensor-2) for outside environment. The signals from both sensors will be received in the microcontroller. The received signal will be compared with the predefined threshold value. If the received value from both the sensors exceeds the safety limit then microcontroller will send a signal to AC controller unit of the smart room or smart car to turning off or adjust the AC (air conditioner) or automatically turn it on from the off position when air quality becomes acceptable. When, or if, CO concentration exceeds safety limit inside the smart room or smart car cabin, sensor-1 sends a signal to microcontroller unit for opening the power window of the room or car cabin. In all cases, the microcontroller will buzz an alarm to warn the occupants. LED and LCD may also be used for indication purposes. Here IoT is the Wi-Fi network of vehicles or rooms to connect and exchange data through GPS and GSM modules. GPS and GSM modules may be included for tracking the vehicles and to send SMS to authorized users so they can rescue people during accidental or dangerous situations.

7.3.1   Flow Chart of the Embedded Intelligent Controller

Figure 7.4 shows the flowchart of the embedded intelligent controller. It is self-explanatory.

7.3.2   System Design for CO Detection Model

The CO detection system needs several mathematical models for the analysis of system dynamics and the design and evaluation of the control system. A closed-loop structure of CO detection system may be designed as shown in Fig. 7.5. It may consist of some basic elements such as a comparator, controller, plant, CO sensor as feedback element and some disturbances may also be considered in this system.

The controlled variable will be first measured and an electrical signal will be created to activate the closed-loop controller to control the process variable. The measured value will then be compared with the setpoint value of the process. According to the comparator, output controller will take necessary action to control the process. The task of this system will be to maintain CO concentration within the safety level range.

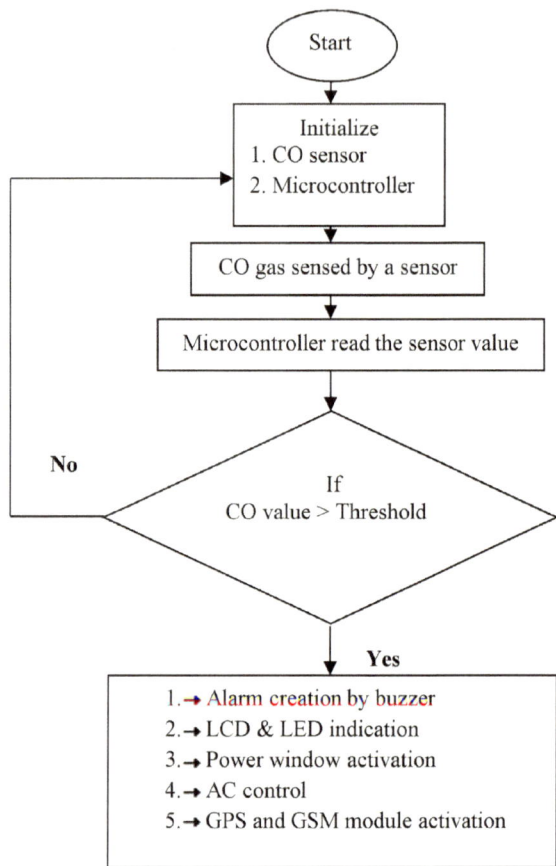

Fig. 7.4 Flowchart of the embedded intelligent controller

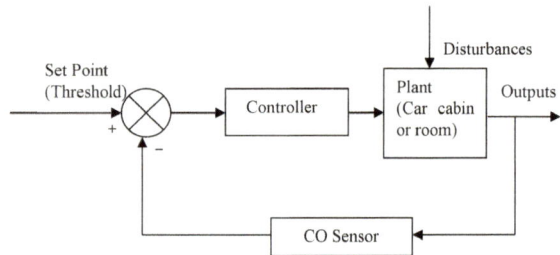

Fig. 7.5 Closed-loop control of the embedded intelligent controller

7.3.3 Mathematical Modeling of the Embedded Intelligent Controller

To implement any system, a mathematical model of the system must first be designed. The mathematical model of an intelligent controller is given in Fig. 7.6 which shows the schematic diagram of the plant where system parameters are given as below:

Fig. 7.6 Schematic diagram of the plant air system

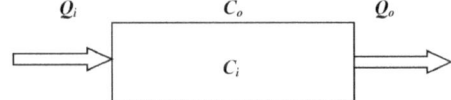

Q_i is the inlet flow rate of the plant.

Q_o is the outlet flow rate of the plant.

C_i is the CO concentration inside the plant.

C_o is the CO concentration of outside air

    In this case, the mass balance equation of CO gas for the plant can be written as:

$$\frac{dm}{dt} = C_0 Q_o - C_i Q_i \tag{7.2}$$

    If the pressure within the plant is constant, the flow rates of incoming and outgoing air by ventilation are the same. So,

$$Q_i = Q_o = Q$$

Hence Eq. (7.2) can be written as

$$\frac{dm}{dt} = C_o Q - C_i Q$$

$$\text{or, } \frac{dm}{dt} = (C_o - C_i)Q \tag{7.3}$$

    Now, the relationship between CO mass, CO concentration and plant air volume is

$$C_i = \frac{m}{V} \tag{7.4}$$

Where V is the volume of the plant air and m is the CO mass of the plant. The derivative of Eq. (7.4) is

$$dC_i = \frac{dm}{V}$$

$$\text{or, } dm = V dC_i \tag{7.5}$$

    Now, putting the value of dm in Eq. (7.3) [9], we get

Fig. 7.7 The model block
diagram for plant

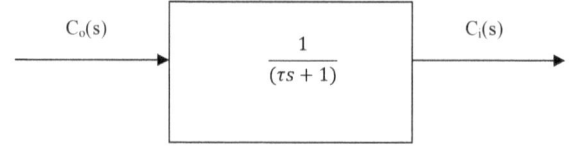

$$V\frac{dC_i}{dt} = (C_o - C_i)Q$$

$$\text{or, } \frac{V}{Q}\frac{dC_i}{dt} = (C_o - C_i)$$

$$\text{or, } \tau\frac{dC_i}{dt} + C_i = C_o \tag{7.6}$$

Taking Laplace Transformation of Eq. (7.6), we get

$$\tau s C_i(s) + C_i(s) = C_o(s)$$

$$\text{or, } (\tau s + 1)C_i(s) = C_o(s)$$

$$\therefore \frac{C_i(s)}{C_o(s)} = \frac{1}{(\tau s + 1)}, \tag{7.7}$$

where $\tau = $ Time constant mathematically written as:

$$\tau = \frac{1}{Air\ flow\ rate} = \frac{V}{Q}$$

Hence, Eq. (7.7) is the transfer function of the plant, which is shown in Fig. 7.7.

7.3.4 Mathematical Model of Semiconductor Metal Oxide CO Sensor

A standard circuit diagram of the sensor is presented in Fig. 7.8. The surface resistance of the sensor, R_s is obtained by affected voltage signal output of the load resistance, and R_o that is series connected, are shown in Fig. 7.8. Then the relationship between surface resistance, load resistance, and voltage output can be written as [1]:

$$\frac{R_s}{R_o} = \frac{V_c}{V_o} - 1, \tag{7.8}$$

where

R_s = Surface resistance of the sensor.
R_o = Load resistance, where electrical resistance of the sensor is zero ppm.
V_c = Source voltage (5 V).

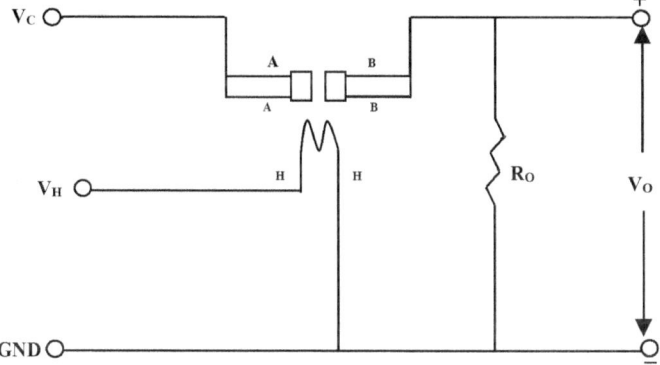

Fig. 7.8 Standard circuit diagram of SMO Sensor

V_o = The voltage analog output of the sensor.

The electrical resistance of the sensor is

$$R = \frac{R_s}{R_o} \tag{7.9}$$

The relationship between sensor resistance and concentration of target gas usually follows the Power Law and can be written as follows:

$$R = K \times (C)^{-n}, \tag{7.10}$$

where

K = Constant of the sensor material.
C = Concentration of CO in ppm.
n = Sensitivity according to change of gas concentration with value 0.7.

Equation (7.10) can be modified by taking the logarithm on both sides

$$\log(\frac{R_s}{R_o}) = \log(K \times (C)^{-n})$$

$$\text{or, } \log(\frac{R_s}{R_o}) = \log K + \log(C)^{-n}$$

$$\text{or, } \log(\frac{R_s}{R_o}) = \log K - [n \times log(C)] \tag{7.11}$$

Considering the normal temperature and pressure (NTP), Eq. (7.11) can be written as:

$$-1.4 + \log(\frac{R_s}{R_o}) = -0.7 \times log(C) \tag{7.12}$$

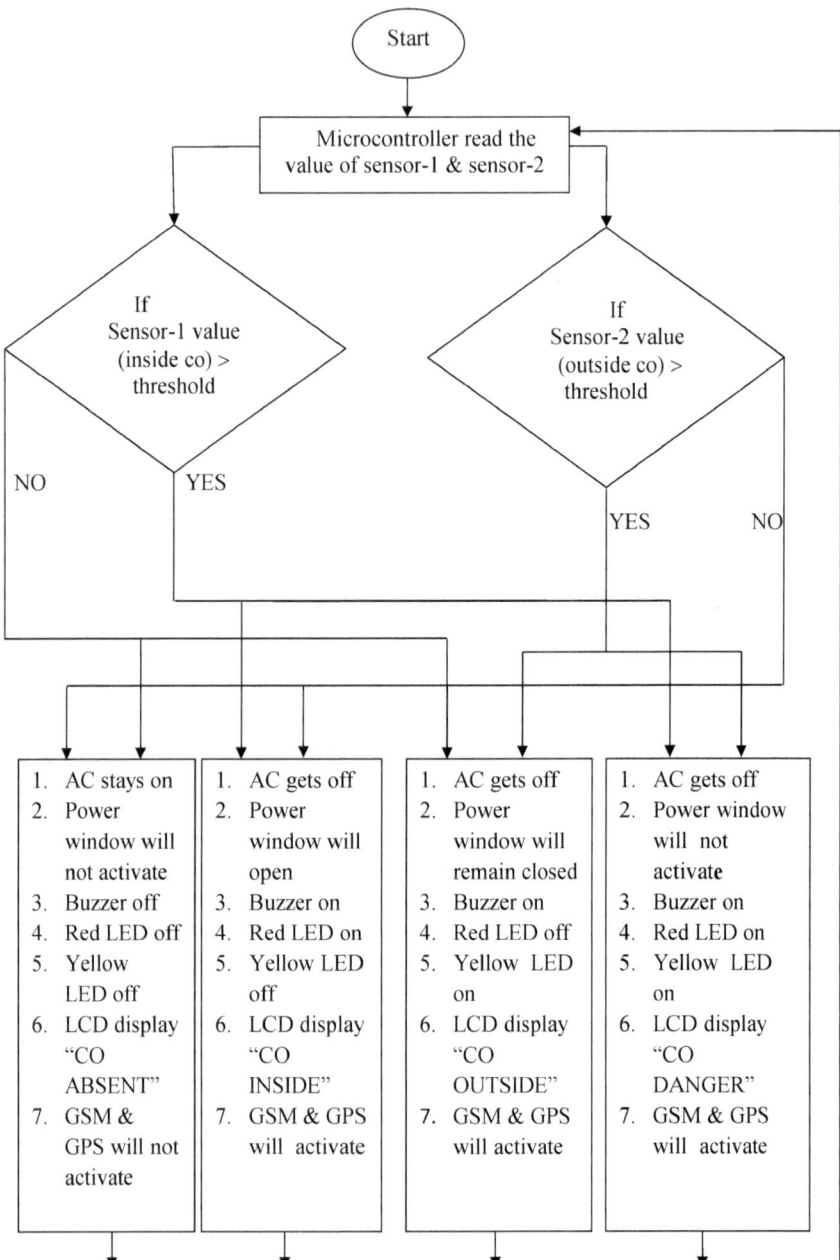

Fig. 7.9 Flowchart of microcontroller logic design

Comparing Eqs. (7.11) and (7.12) at NTP, we note that

$$n = 0.7$$

$$\log K = 1.4$$
$$\text{or, } K = 10^{1.4}$$
$$\text{or, } K = 25.12$$

From the power law (Eq. 7.10),

$$R = K \times (C)^{-n}$$
$$\text{or, } \frac{V_c}{V_o} - 1 = K \times (C)^{-n}$$
$$\text{or, } (\frac{5}{V_o}) - 1 = 25.12 \times (C)^{-0.7}$$
$$\text{or, } (C)^{-0.7} = [\frac{(\frac{5}{V_o}) - 1}{25.12}]$$
$$\text{or, } C = [\frac{(\frac{5}{V_o}) - 1}{25.12}]^{-1.43}$$
$$\therefore C = 100.468 \times [(\frac{5}{V_o}) - 1]^{-1.43} \tag{7.13}$$

Hence, Eq. (7.13) is the expression of the carbon monoxide concentration which is the output of the sensor.

7.3.5 Logic Design of Microcontroller

Figure 7.9 presents a logic design diagram for a microcontroller. It is, hopefully, self-explanatory.

7.4 Proposal for an Embedded Intelligent Control System

In this chapter, an embedded intelligent controller is presented and discussed for CO detection. It is a compact system which may be assembled and installed in air-conditioned motor vehicles or living rooms to restrict the CO poisoning. This embedded system can be placed within the Internet of Things (IoT) for online measurement and monitoring of CO presence. By coordinating with other relevant devices in the IoT, the system may also be used for sending messages on mobile devices to authorized users if the CO level in air-conditioned motor vehicles or rooms exceeds the predefined values.

This intelligent controller may use semiconductor gas sensor for sensing the presence of CO. An alert system may be integrated with this whole system to alert the occupants when CO concentration reaches the margin point. This alert system may comprise an LCD display, buzzer and LED indicator. For providing proper ventilation, a windows controller unit may be used to control automatically the power windows of air-conditioned vehicle or room. The air conditioning system can be controlled to restrict CO entering the room or car cabin through AC at the desired condition. GPS and GSM module may also be introduced in this system for better security to track the location of the vehicle and send a message to the authorized user respectively. One microcontroller must be incorporated into the system for controlling the whole system operation.

7.4.1 Software Implementation

The proposed embedded intelligent controller may be implemented as a software system. There are numerous software tools available for electrical and electronic circuit design which can be used to execute this controller to validate its working. The software design of the controller may contain many components for different activities. Two sensors may be used: one for sensing the presence of CO for inside environment and another for the outside environment. One microcontroller must be included in this system to process the sensor outputs and accordingly take actions. For displaying the presence of CO according to the given situations, one LCD display may also be incorporated. A potentiometer may be used with LCD for adjusting the resistance to control the electric current. Three LED indicators of different colors may be utilized; one for showing the existence of CO of inside environment, the other for showing the existence of CO of outside environment, and the third can represent the air conditioning system of the room or car cabin. One buzzer may be included along with a transistor and a battery to create an alarm when the sensor values go high. One motor could be added for representing the power window. To run the motor in both directions, a driver circuit may be required. A GSM module may be used in this model for sending messages. A GPS module can be included for tracking purposes. A virtual terminal can be used to display the messages sent by GSM and locations tracked by GPS. The whole system will be designed for simulation and to design software tool by compacting all these components. The coding of the microcontroller will be executed ideally in an Integrated Development Environment (IDE). After compilation, the program code will be fetched for uploading code of the Microcontroller. The system operation may depend upon four circumstances, as follows:

- The first condition is when CO concentration is within safety level in both inside and outside environments (normal the healthy condition). In this condition, both sensors values will be in the LOW state. The microcontroller will read sensor outputs and take actions accordingly. The LEDs used for indicating the presence of CO of inside and outside environment will remain OFF. LCD will display "CO ABSENT" message. The motor will not run and buzzer will not sound alarm. The LED representing air conditioner will remain ON as the condition is normal. No signal will be sent to GPS and GSM and the virtual terminal will show nothing.

- The second condition is when CO concentration of inside environment exceeds the threshold but CO concentration of outside environment is within safety levels. Here, the sensor of inside environment will be at the HIGH indication and the other one showing a LOW state. For this condition, Microcontroller will send a signal to the motor to open the power window so that CO can be expelled out. LCD will display "CO INSIDE" message. The LED for the inside environment will glow but the LED for outside environment will remain OFF. The buzzer will sound the alarm. GPS will track the location with its longitude and latitude. GSM will send an alert message with the link to Google map of this tracked location to the virtual terminal. The LED representing air conditioner will be OFF.
- The third condition is when CO concentration of inside environment does not exceed threshold but CO concentration of outside environment is above safety levels. Here, the sensor of inside environment will be at the LOW state and the other one showing the HIGH state. Accordingly, the motor will not be activated and power window will remain closed (as normally the power windows of air-conditioned car or room remain closed) so that CO entering the room or car cabin from outside can be prevented. The LED for the outside environment will glow but the LED for inside environment will remain OFF. The buzzer will be ON. LCD will display "CO OUTSIDE" message. GSM will send an alert message with the link to Google map of the tracked location which will be tracked by the GPS to the virtual terminal. The LED representing air conditioner will be OFF.
- The fourth condition is when CO concentration exceeds the safety levels in both inside and outside environments. This may be the most dangerous condition of these four. Here, both the sensors will be at HIGH state representing high CO concentration in both environments. In this condition, the motor will be OFF and LCD will display "CO DANGER." The LEDs used for indicating the presence of CO of inside and outside environment will be ON. The buzzer will create sound. GPS and GSM will activate through Wi-Fi network and the virtual terminal will show the outputs of them. The LED representing air conditioner will be OFF.

7.4.2 Hardware Design

For implementing the proposed system in practice, a hardware model of this with embedded intelligent controller can also be designed. The CO gas sensors will be interfaced with a microcontroller which will be further connected with LEDs, LCD display, buzzer, and a motor. The microcontroller will be connected to the laptop or computer via a USB cable to upload the programming code which will be written using an Integrated Development Environment (IDE). The whole system may be connected with the Internet of Things (IoT) by adding GPS and GSM modules.

The hardware implementation model may be composed of two sensors for inside and outside environments, three LED indicators along with three resistors, one buzzer, one LCD display, one motor, a GPS module and a GSM module. One potentiometer may be required for maintaining the contrast of LCD display. To run the motor in both directions, one motor driver may be used. GSM module may demand an adopter for getting the power supply. One power window model can also be designed to show its activation with the rotation of the motor.

This intelligent controller may work for four possible scenarios as software implementation, as follows:

- When both the inside and outside environments have CO concentration within the acceptable limits, i.e., both the sensor values are within safety limit, only the LED representing the air conditioning system of car cabin or room will remain ON and other components, i.e., other two LEDs and buzzer will remain OFF. LCD will display "CO ABSENT" and motor will remain OFF so that the power window of the air-conditioned room or car cabin will remain closed. GPS and GSM modules will not be activated here as the condition is normal.
- If the CO level of inside environment crossed the predefined value but the CO level remained acceptable in outside environment, the sensor for inside environment will have the output value greater than the threshold and the value of the other sensor for outside environment will be within safety limits. Here, the LED for inside environment will glow but the LED for outside environment will be OFF. The buzzer will alarm to warn the occupants. LCD will display "CO INSIDE" and the motor will start rotating to open the power window. GPS module will start tracking the location as soon as it receives the trigger plus from microcontroller and GSM module will send the alert message along with the link to Google map of that particular location to the authorized user. The LED which represents the air conditioner will remain OFF.
- Another situation may arise when CO concentration of inside environment does not exceed the threshold but CO concentration of outside environment is above the safety level. For this situation, the output of the sensor for inside environment will be within the threshold and the other sensor will have an output greater than safety limit. The microcontroller will process the sensor values and take actions accordingly. The LED that acts for indicating the CO existence of outside environment will be ON and the LED for inside environment will remain OFF. LCD will display "CO OUTSIDE." The motor will not be activated and power window will remain closed (as normally the power windows of air-conditioned car or room remain closed) so it can restrict the incoming of CO. GPS and GSM will activate here for tracking the location and sending the message along with the Google map link of that location to the authorized numbers respectively. The LED that is used as air conditioner will remain OFF.

- Lastly, when CO concentration becomes excessively high in both inside and outside environments, both the sensors values exceed the safety limit. For this, both the LEDs representing the indicators for inside and outside environment will be ON indicating the dangerous condition. LCD will display "CO DANGER." The buzzer will warn the occupants. In this condition, the motor will remain OFF. GSM will send an alert message with the link to Google map of the tracked location which will be tracked by the GPS. The LED representing air conditioner will be OFF in this case.

7.5 Conclusion

As carbon monoxide gas is highly toxic that has no taste or smell, it remains unrecognized when people inhale it but this inhalation can make people unwell. Deaths can also be attributed to CO poisoning in a badly ventilated room or motor vehicles. So, the detection of the excessive amount of CO has become a burning need in today's world to prevent such deaths. This chapter has proposed an embedded intelligent controller for carbon monoxide (CO) sensing of the indoor environment based on IoT architecture. This system is able to detect the carbon monoxide concentration inside an air-conditioned vehicle or room integrated with an alarm system to notify the occupants or nearby people so as to avoid possible accidents caused by carbon monoxide poisoning. If this system is appropriately implemented, it can help in reducing the number of deaths caused by carbon monoxide poisoning. Thus, it may have a great impact on the safety of occupants of air-conditioned automobiles and rooms by alert them through GSM and use of GPS; that in turn will hopefully be very helpful for human safety and comfort.

7.6 Open Research Directions

In this chapter, several dynamic CO sensors have been discussed. Each of these has some unique characteristics and addresses some aspect of operating range. Although this area has received a lot of research attention, there are still open research issues and directions that need investigation in the future, such as those briefly discussed below.

7.6.1 Sensor Temperature Range

Temperature varies throughout the day and night so an appropriate CO sensor should be used which can work in different temperature environments. In different locations of the world, temperature varies so choosing the right CO sensor is very important.

Temperature constraints should be considered as an important factor for the design of practically implementable CO detection schemes.

7.6.2   Sensor Detection Range

Appropriate design of CO sensors is a field that is open to further exploration in many respects including the sensor detection range. This factor is very important and it is also most complex in nature. However, the sensor detection range constraints should be considered for the design of the practically implementable hardware scheme.

7.6.3   Sensor Price

A two-side pricing mechanism needs to be developed: one that ensures that consumer can buy the sensor and the other that ensure that the company also makes some profit. Pricing mechanism needs to be built with new computational intelligence techniques, including novel algorithms.

7.6.4   Other Generic Research Directions

The smart intelligent controller must ensure the safety of the consumer and public. Web and mobile applications also need to be developed to assist the public in the decision-making with the smart intelligent technologies and with the higher diffusion of programs.

The innate advantages of SMO-based CO sensors (such as high sensitivity and stability, potential for miniaturization, low power consumption, low price for consumers, and potential for integration into arrays) make them very strong nominee for providing appropriate safety solutions. Moreover, the unparalleled spread of mobile devices offers the opportunity for sensor integration with the Internet of Things. To be able to meet the needs of all or part of those applications, we need to further improve the performance of sensors and that requires advances in both science and technology. The public should be aware of their environmental conditions and thus manage their safety accordingly. The future safety should be designed in such a way that it can protect the interest of both the parties: consumer as well as the utility.

The data collected from the sensors can be huge. It is a challenging task to handle this huge amount of data. But with proper analysis, the safety forecast can become easier and more accurate. There is also a need to design models for safety forecasting by using these data.

References

1. Patil SS, Singh J (2015) Monitoring and controlling of hazardous gases inside vehicle and alerting using GSM technology. Int J Adv Res Comput Sci Softw Eng
2. Greiner TH (1997) Carbon monoxide poisoning: dangers, detection, response, and poisoning (AEN-193). Department of Agricultural and Biosystems Engineering, Iowa State University. https://www.abe.iastate.edu/extension-and-outreach/carbon-monoxide-poisoning-dangers-detection-response-and-poisoning-aen-193. Accessed Apr 1997
3. Indiana Department of Environmental Management (2014) Criteria pollutants: carbon monoxide (CO), fact sheet. https://www.in.gov/idem/files/factsheet_air_quality_co.pdf. Accessed Apr 2014
4. Minnesota Department of Health (2018) Carbon monoxide (CO) poisoning in your home. http://www.health.state.mn.us/divs/eh/indoorair/co/. Accessed 7 Apr 2018
5. Committee on Carbon Monoxide Episodes in Meteorological and Topographical Problem Areas, National Research Council (2002) The ongoing challenge of managing carbon monoxide pollution in fairbanks, alaska, sources and effects of carbon monoxide emission. National Academic Press. https://www.nap.edu/read/10378/chapter/3
6. Sathitkunarat S, Wongwises P, Aram RP, Meigen Z (2006) Carbon monoxide emission and concentration models for Chiang Mai urban area. Springer
7. Galatsis K, Wlodarski W (2006) Car cabin air quality sensors and systems. American Scientific Publishers
8. Nordqvist C (2017) Carbon monoxide, the Silent Killer. Medical News Today. https://www.medicalnewstoday.com/articles/171876.php. Accessed 11 Dec 2017
9. Universal Security Instrument, INC. (2014) What are some common sources of carbon monoxide (CO)? http://www.universalsecurity.com/faqs/carbon-monoxide-alarm/what-are-some-common-sources-of-carbon-monoxide-co
10. Meena MC (2014) Accidental death due to carbon monoxide: case report. Int J Med Toxicol Forensic Med
11. Kumar A, Rautji R (2010) Fatal unintentional carbon monoxide poisoning inside a frances garage a case report. J Indian Acad Forensic Med
12. Cheow SH, Chang CT (1975) Carbon monoxide poisoning in Singapore. Singapore Med J, 175
13. Fleming R, Fleming AA (2001) Automobile carbon monoxide detection and control device. US Patent 6,208, 256 B1, Mar 2001
14. Phillips F (2000) Automotive carbon monoxide detection system. US Patent 6,057,755, May 2000
15. Brooks EC et al (1994) Carbon monoxide concentration indicator and alarm. US Patent 5,276,434, Jan 1994
16. James RC, Cherney DM (1994) Carbon monoxide sensor and control for motor vehicles. US Patent 5,333,703, Aug 1994
17. Murphy RF (1996) Carbon monoxide safety system. US Patent 5,576,739, Nov 1996
18. Marhulies S (1998) Carbon monoxide detection system for motor vehicles. US Patent 5,739,756, Apr 1998
19. Stern DA (2000) System for detecting and purging carbon monoxide, US Patent 6,110,038, Aug 2000
20. Milton R, Steed A (2007) Mapping carbon monoxide using GPS tracked sensors. Springer
21. Milton R, Steed A (2005) Correcting GPS readings from a tracked mobile sensor. Springer
22. Jay Markanich (2018) Where should i put a carbon monoxide detector? Jay Markanich Real Estate Inspections, LLC. http://www.jaymarinspect.com/carbon-monoxide-detector.html
23. Copland & Son (2018) An introduction to carbon monoxide (CO). Air Conditioning And Heating Service, INC. https://www.copelandandson.com/webapp/p/248/introduction-to-carbon-monoxide. Accessed Apr 2018
24. Gas Networks Ireland (2018) What is carbon monoxide (CO), carbon monoxide the facts. http://www.carbonmonoxide.ie/htm/whatis.htm. Accessed Apr 2018

25. CarbonMonoxideKills.com   (2016)   Carbon   monoxide   poising.   http://www. carbonmonoxidekills.com/are-you-at-risk/carbon-monoxide-poisoning. Accessed Apr 2018
26. CarbonMonoxideKills.com   (2016)   Carbon   monoxide   in   cigarettes.   http://www. carbonmonoxidekills.com/are-you-at-risk/carbon-monoxide-in-cigarettes/.   Accessed   Apr 2018
27. Chemistry LibreTexts (2018) Catalytic converter. https://chem.libretexts.org/Core/Physical_ and_Theoretical_Chemistry/Kinetics/Case_Studies%3A_Kinetics/Catalytic_Converters. Accessed 18 Mar 2018
28. Woodford C (2017) How do catalytic converters work? Explain That Stuff! http://www. explainthatstuff.com/catalyticconverters.html. Accessed 15 June 2017
29. Greiner TH (1998) Carbon monoxide poisoning: vehicles (AEN-208). Department of Agricultural and Biosystems Engineering, Iowa State University. https://www.abe.iastate.edu/ extension-and-outreach/carbon-monoxide-poisoning-vehicles-aen-208/. Accessed Sept 1998
30. Science Daily (2017) Warmed up and raring to go: What happens when a car starts in the cold? https://www.sciencedaily.com/releases/2017/12/171205091937.htm. Accessed 5 Dec 2017
31. Lowa State University (1999) Carbon monoxide. https://www.extension.iastate.edu/Pages/ communications/CO/co1.html
32. Laukkonen J (2018) How to avoid carbon monoxide poisoning in your car, lifewire. https:// www.lifewire.com/avoid-carbon-monoxide-poisoning-in-car-4134877. Accessed 19 Feb 2018
33. Blumenthal I (2001) Carbon monoxide poisoning. J R Soc Med
34. CO Knowledge Centre (2018) CO health risk. https://www.detectcarbonmonoxide.com/co-health-risks/. Accessed Apr 2018
35. Bhuyan PJ (2014) Techie commits suicide by sniffing carbon monoxide, DNA. http://www. dnaindia.com/india/report-techie-commits-suicide-by-sniffing-carbon-monoxide-2020511. Accessed 22 Sept 2014
36. Hindustan Times (2014) Delhi car deaths: Here's what you should know about carbon monoxide poisoning. https://www.hindustantimes.com/health-and-fitness/delhi-car-deaths-here-s-what-you-should-know-about-carbon-monoxide-poisoning/story-IX9jznVOb4sTBlEDdD2DXL. html. Accessed 22 July 2014
37. Pandey JM (2011) AMRI hospital fire: 73 killed, several injured. The Times of India City. https://timesofindia.indiatimes.com/city/kolkata/AMRI-hospital-fire-73-killed-several-injured/articleshow/11044875.cms. Accessed 9 Dec 2011
38. Wikipedia (2018) Gas detector. https://en.wikipedia.org/wiki/Gas_detector
39. Hanwel Electronics (2019) Technical Data MQ-7 gas sensor. www.sparkfun.com/datasheets/ Sensors/Biometric/MQ-7.pdf. Accessed Mar 2018
40. GitHub Inc. (2018) MQ-9 gas sensor data sheet. https://raw.githubusercontent.com/ SeeedDocument/Grove-Gas_Sensor/.../MQ-9.pdf. Accessed Mar 2018
41. FIGARO (2007) TGS 2442 gas sensor data sheet. https://cdn.sos.sk/productdata/af/2e/ 9901fb15/tgs-2442.pdf. Accessed July 2007
42. FIGARO (2005) TGS 3870 gas sensor data sheet. www.diltronic.com/wordpress/wp-content/ uploads/TGS_3870.pdf
43. Veluri SS, Prabhudesai S (2017) Cypress semiconductor, in-cabin carbon monoxide sensing for automotive applications
44. Martian Rover Project (2018) Passive sensors. http://users.polytech.unice.fr/~pmasson/rover-sensors.php
45. SGX (2009) sensortech performance, EC4-500-CO Carbon Monoxide Electrochemical Sensor data sheet. https://www.sgxsensortech.com/content/uploads/2014/07/EC4-500-CO1.pdf
46. CiTiceL (2016) 4CM Carbon monoxide data sheet. https://www.shawcity.co.uk/documents/ SensorPDFs2017/CO-4CM.pdf
47. Gurbuz Y, Kang WP, Davidson JL, Kerns DV (1999) A new diode-based carbon monoxide gas sensor utilizing $Pt–SnO_x$/diamond. Elsevier

Chapter 8
IoT-Based Ambient Intelligence Microcontroller for Remote Temperature Monitoring

Balwinder Raj, Jeetendra Singh, Santosh Kumar Vishvakarma
and Shailesh Singh Chouhan

Abstract The aim of this book chapter is to provide a comprehensive assessment of the ambient intelligence (AmI) microcontrollers suitable for low-power Internet of things (IoT) applications. The current challenges and trends in the evolution of low-power and high-performance microcontroller are also explored. The key focus is on the performance analysis of such devices as they facilitate the IoT vision with increased reliability. A detailed discussion of various microcontrollers, their architectures, low-power modes, and available temperature monitoring systems is also provided. In this context, design and architecture of a low-powered microcontroller is proposed and TCAD simulations are carried out for a better understanding of the suggested system. The intended audience is expected to be research and scientific community working in the field of IoT-based smart and intelligent microcontrollers for environmental study applications. The book chapter could be used for a course of higher education and for researchers in the fields of computer science, microelectronics, nanotechnology, and VLSI design. The microcontroller features and content related to IoT, as presented in this contribution, will hopefully be most valuable to the readers to understand the underlying concepts and to develop advanced high-performance circuits and systems. Illustrations, tables, and figures are also provided to supplement the text.

Keywords Microcontrollers · Microprocessor · Ambient intelligence · AmI
Internet of things · IoT · Temperature monitoring · RISC architecture · AVR
PIC

B. Raj (✉) · J. Singh
Nanoelectronics Research Lab, Department of Electronics and Communication Engineering,
National Institute of Technology (NIT), Jalandhar 144011, India
e-mail: balwinderraj@gmail.com

S. K. Vishvakarma
VLSI Circuit and System Design Lab, Discipline of Electrical Engineering,
Indian Institute of Technology (IIT), Indore, MP, India

S. S. Chouhan
Embedded Internet Systems Lab, Department of Computer Science,
Electrical and Space Engineering, Luleå University of Technology, Luleå, Sweden

© Springer Nature Switzerland AG 2019 177
Z. Mahmood (ed.), *Guide to Ambient Intelligence in the IoT Environment*, Computer
Communications and Networks, https://doi.org/10.1007/978-3-030-04173-1_8

8.1 Introduction

Senior citizens and people who are unwell are usually left alone at home during the day as grownups go to work and children go to schools. So, there may be a constant need during the day to get an update on the health and well-being of such individuals. There are many ways in which a check can be made remotely, e.g., through the use of CCTV cameras but it is difficult to always be able to see the real-time videos from the location of such elderly and unwell persons. Sensors can also be used to detect the activity or inactivity of the persons at home, which can then send the necessary data to a processing device or to the cloud, as we intend to demonstrate later in this book chapter.

In an industry scenario, there are often situations in which the temperature of a place needs to be controlled: raised or lowered as the environment dictates. In this case, there is a need to have or develop a system that monitors the temperature and executes an appropriate action: to sound the alarm, for example, or send a message to another smart device or a human operator for further actions. Now, there are requirements for temperature sensors along with Wi-Fi and GSM facilities together with connectivity and communication systems as well as satisfactory human–computer interface (HCI). Ambient intelligence (AmI) thus becomes a core requirement when developing future application-specific devices such as microcontrollers [1, 2].

Internet of things (IoT) is a useful paradigm for the advanced development of AmI environments [3, 4]. AmI in microcontrollers for remote monitoring of temperature within the IoT environment is currently a desired need for various application areas including industrial machinery, household appliances, consumer products, healthcare devices, and environmental calamities monitoring [5, 6].

Microcontrollers, as their name suggests, were derived from microprocessors. These consist of a processor to execute instructions, memory, and mechanisms to store data, input–output periphery, clock, and all other necessary systems. Microcontrollers are employed in control systems of embedded electronics. Each and everything are inbuilt into a single package, which is the reason why a microcontroller is also known as a computer on a chip.

Microcontrollers control and guide the operation of various gadgets and other smart devices, according to the program or instructions encrypted in the memory [7]. The Firmware, i.e., the permanent program needed for the appropriate functioning of microcontrollers is fed into the read-only memory (ROM) [8] that is a nonvolatile memory meaning that its contents remain available even when the power is switched off. Some latest ROMs have appeared in the market which can be reprogrammed, so with the help of ROM, any hexadecimal format file can be generated and loaded for any given explicit purpose.

Use of microcontrollers is increasing rapidly and proving especially useful for the IoT environments [9]. Currently, these are used in innumerable devices especially in electronic appliances such as washing machines, electronic garbage collection bins, microwave ovens, refrigerators and other electronic gadgets including mobile phones and tablets [10–14]. Microcontrollers can be implanted into any device due to their low-price tags and small size. The practical outcomes of any program can be easily seen and verified by using appropriate simulators on Personal Computers (PCs). Thus, the working of an embedded project can be visualized virtually without assembling real components and chips.

8.1.1 Role of AmI Microcontrollers in IoT Applications

Ambient Intelligence (AmI), in relation to electronic network systems, refers to context awareness and interaction and communication with other connected smart devices within the IoT environment. The development of AmI is useful for efficient utilization of the Internet culture within the mainstream society of prevalent smart communication systems [15, 16] with devices embedded with microcontrollers. Potential benefits of AmI are profound and highly applicable to smart devices and systems as part of the IoT. The network sensors embedded with AmI give us smart connectivity, high performance, and low-cost. These are considered as the key features of these advanced technologies especially in the case of real-time-distributed applications [17, 18]. Thus, microcontrollers designed for the IoT systems are a crucial element toward providing AmI, in the sense that a majority of the smart devices rely on mobile and wireless technologies for communication with each other and for sharing information. In this context, in the rest of this contribution, we explore the roles and principles of AmI that can be applied in futuristic low-power microcontrollers for various IoT-based applications.

In this chapter, the design specifications for AmI-IoT microcontrollers are discussed including various power modes, e.g., idle and sleep modes, etc., through VHDL coding in Xilinx. One objective in this chapter is to develop a low-power AmI-based microcontroller for remote temperature monitoring system suitable for devices in the IoT environments. Hopefully, the proposed system will be suitable for research as well as for practical applications in industrial settings where temperature is the main factor to be controlled during production phases, as many products require right temperatures for their proper functioning.

8.1.2 Chapter Organization

This chapter consists of six sections. Section 8.1 introduces the role of AmI in the evolution of low-power microcontroller's technology with IoT related applications. In Sect. 8.2, current developments in microcontroller technology, as per the current market requirements with the arrival and use of novel technologies, are discussed. Section 8.3 explains primitive applications of microcontroller with IoT such as the smart home systems and monitoring of environmental parameters. Section 8.4 comprises brief discussion on microcontroller varieties and their characteristics in terms of bit, memory, and instruction set. Furthermore, the families of microcontrollers with their detailed architecture and applications are described. The proposed architecture of a microcontroller for remote temperature monitoring system is outlined in Sect. 8.5. The implementation of xBoard miniv2.1, AVR programmer, GSM/GPRS modules, and modem are also discussed in terms of temperature monitoring system design suitable for devices in the IoT environments. The adopted low-power modes and power optimizing techniques are also presented with an aim to achieve low-power consumption. Finally, Sect. 8.6 summarizes the entire chapter and provides hints on future directions.

8.2 Demands and Trends that Shape Microcontrollers

Microcontrollers are devices that have the widest applications in the field of electronic gadgetry [10–14]. From an alarm clock to the washing machine, almost every electronic item is embedded with microcontrollers. They are used in various applications on the basis of their power consumption, performance, and speed. Today's market demands vary generally with respect to two main factors: low-power performance of device and integration with other advanced technological paradigms such as the IoT and AmI. In the IoT implementation, it is also required to choose the right kind of microcontrollers that are suitable to drive the systems in the most economical manner.

In this context, some significant methods along with emerging techniques and underlying technologies are discussed below mainly to help to select efficient, energy-saving, and cost-effective microcontrollers.

8.2.1 Low-Power Techniques

Numerous varied devices are beginning to appear in the market that has minimal use of power. Thus, to keep in pace with them, the development of low-power microcontrollers is the need of the hour. There are various ways through which the power consumption of microcontrollers can be reduced, e.g., by reducing gate delays, reducing switching, and introducing different modes of operation, e.g., idle, sleep, shut down ALU, or power mode. Within these modes, the microcontroller will selectively wakeup its different components as required and thus results in reduced power consumption. As all microcontrollers are application specific to different operating conditions, different modes are incorporated into the microcontroller architecture. Some of these are mentioned below, which are helpful in reducing the power consumption [19, 20].

- Peripherals exhibiting current consumption during the off state when retention of state and RAM are disabled.
- Reducing the power consumption in running RTC (Real-Time Clock) with RAM retention by enabling the running of RTC.
- Introducing the wakeup time.
- Reducing the range of supply voltage.

The current consumption, as well as the system-level optimization, should be known to determine the MCU (Microcontroller Unit) that has the least power consumption. Certain vendors, e.g., Texas Instruments have produced comparative study reports for different microcontrollers with a view to designing ultra-low-power microcontrollers using MSP432 architecture. Some innovative techniques like switching currents, program execution speed, etc., can also be utilized for extremely low-power consumption in 8-bit microcontrollers. Apart from this, remedies like power gating, clock gating, and usage of interrupts are also useful techniques for reducing the power loss in microcontrollers used for a wider variety of applications [21].

8.2.2 High-Performance Capabilities

Developing high-performance microcontrollers is one of the well-known challenges. For the design of CPUs, several techniques are available including independent multiplier, multi clocks configuration, divider module, and hardwired control unit, to achieve high speeds. A high-speed AVR consumes less power and can be extensively used for high-performance processing. Also, it was the initial MCU that utilized the flash chip for the first time for memory storage [22]. TELOS is an ultra-low-power module developed by UC Berkeley, which uses wireless sensor application to achieve high performance [23]. Also, an 8-bit microcontroller with a combination of different modules is a useful way to enhance the performance [24].

8.2.3 Dense Storage Ability

Although, a program can be executed using any types of memory (whether Flash, ROM, EPROM, EEPROM, or NvRAM), DRAM and Flash remain dominant, while challenges relating to Dual In-Line Memory Module (DIMM) DRAM [25] and memristor-based memories [26] still remain. With the development of newer technologies, the embedded memories of industrial and automotive electronics microcontroller need to be updated with the consistent growth of semiconductor nonvolatile memories for program code, storing configuration settings, and application parameters. High bandwidth memory (HBM) with a wider bus also offers significantly higher on-chip storage than off-chip storage. There are three types of memory that exist in microcontroller architectures, viz.:

- Program memory that stores program instructions
- Data memory that has general purpose registers (GPR) and Special function registers (SFR). GPR stores the transient data when the programs are interrupted, whereas SFR controls many special functions such as program counter, input–output, stack pointer, peripheral controls, timer, and processor status
- Data EEPROM that is a nonvolatile type of memory.

8.2.4   Internet of Things (IoT).

The IoT is a network of interconnected processing devices, computers, smart items (e.g., smart watches), mobile phones, and other electronic gadgets that are fitted with intelligence and communication protocols. It is *an open and exhaustive system of smart objects that have the ability to auto-arrange, share data, responding and acting according to varying particular situations, circumstances, and changes in the environment* [27, 28]. IoT empowers the ability of gathering and trading information among system administration of all the connected physical objects and structures. Empowerment is through the internet with the help of embedded hardware, programming, and sensor networks in an AmI environment. Global Standards Initiative on the Internet of Things (GSI-IoT) has identified the IoT as a worldwide framework for the information sharing and processing. In this regard, the physical domain is referred to as physical devices; and information domain referred to as virtual things that are interrelated and interconnected in light of existing and developing information and communication technologies.

The IoT enables the objects to be discovered, accessed, and managed; these objects being in some cases geographically long way away. The established interconnection of the physical world with personal computers and smart phones brings the expected proficiency in distributed computing, and financial advantages without human intervention or engagement. The advancements result in advanced physical digital systems that comprise innovated systems like virtual power plants, smart homes and cities, astute transportation, and keen urban areas.

When the internet infrastructure is connected to a diverse variety of smart objects and devices through the embedded intelligence systems, then each has its unique identity for discovery and access purposes. Every smart object is exceptionally perceived through its installed processing framework. It is estimated that IoT will include around 30 billion devices constantly connected by 2020. The IoT paradigm needs advanced and smart connectivity for linking of various gadgets, frameworks, and services. Within the IoT domain, things can allude to a wide assortment of devices like biochip transponders for the domestic and other animals, DNA examination equipment for food/environmental/pathogen, heart monitoring systems, electric clams within the coastal waters, automobiles with built-in sensors, and battlefield equipment needed for search and rescue operation in case of emergency situations [29].

In the IoT environment, devices are inbuilt with relevant advanced technologies like Wi-Fi, GSM, GPRS modules and sensors to gather and autonomously transfer all important information to other relevant devices [30]. An example in case of a smart home includes automatic control of electrically operated equipment, e.g., room lights, fans, washing machines, refrigerators, geysers with smart thermostats, ovens, air conditioners, air purifiers, and ventilators as well as many more devices equipped with AmI and Wi-Fi for remote monitoring. More detail on AmI is provided in the next subsection.

8.2.5 Ambient intelligence (AmI)

AmI is an evolving technological concept that adds intelligence to the existing distributed computing environment (e.g., IoT) to make it more sensitive to environmental conditions, adaptive in case of changing ambient conditions, and responsive to human presence and their requirements. It combines the advanced technologies like artificial intelligence, machine learning, neural networks, pervasive computing, and sensors-based networks [16]. The AmI enables automatic reasoned decision making according to the nature of the surrounding environment. The information needed for decision making is gathered by the use of sensors-based network consisting of actuators, thermostats, measurement devices, etc. Thus, the algorithms of AmI rely heavily upon the data obtained from the said sensor-based networks. AmI mechanism has the following characteristics.

- Sensing: This is required so that system can sense various features and quantify variables such as temperature, pressure, velocity, position, light, and radiation.
- Reasoning: This is needed in order to make the system useful, adaptive, and responsive.
- Decision making: This is required for making changes to the environment. It depends upon the reasoning results obtained after recognizing behavior or perceiving the activity within the environment.
- Activity: This provides the capability to act according to the decision obtained in the previous step.

8.2.6 Wireless Sensor Networks (WSN)

The information required for implementation of a smart environment is collected by wireless sensor-based networks (WSN). These form part of pervasive computing and are effectively used to create smart digital environments. For the collection of relevant information, the WSN performs functionalities including sensing the environmental conditions, accessing the measurable data, performing analysis, and taking adaptive actions [31]. A wireless sensor-based network works more effectively if it has appropriate network topology and relevant communication protocols. It needs to be programmable, scalable, and secure. In addition to this, various methods for reducing power consumption need to be employed. The choice of the correct processor is also necessary that has the capability of minimizing power utilization [32].

8.3 Typical AmI Microcontrollers for IoT Applications

Two examples of home automation system and monitoring of environmental parameters are being discussed in this section. This is in the background of among several applications of AmI microcontrollers in the context of the IoT vision.

8.3.1 Home Automation Systems

Home Automation Systems (HASs) provide real-time communication between the smart homes and their occupants. The underlying idea is that a smart home is embedded with numerous automated electronic appliances embedded with AmI, ideally with minimum power consumption. Intel Galileo-based HAS, that integrates the combined usage of cloud networking and wireless communication, has been found highly suitable [33]. This facilitates the users to observe and control the home appliances, lights, home security, baby and pet care within their home while also keeping the data safe and private. It is mainly the sensor data that helps to change the AmI-based system automatically. AmI-based microcontrollers developed for home automation systems within the IoT environment make them more efficient and powerful tool to monitor and examine the ambient activity inside the home environment without human intervention [6, 34].

8.3.2 Remote Monitoring of Environmental Parameters

The monitoring of various environmental parameters like temperature, pressure, humidity, wind direction and velocity, air quality, and soil quality is imperative for many applications, especially in the industrial settings. Temperature monitoring has become probably the most important need for our daily lives. Many of daily use devices work at specific temperature conditions and if there are significant changes in the ambient temperature, devices may well stop to work satisfactorily. Often, we need to keep an eye on various environmental parameters (e.g., temperature) for many applications, e.g., those relating to food warehouse environments, server rooms, hospital wards, and greenhouses. With the use of normal temperature sensors, we need a human person to go around frequently to check and adjust the sensor readings. Whereas, sensor information details are normally sent to a phone or PC in the form of text via the use of Wi-Fi modules and GSM networks [35], automatic environmental parameters monitoring offers a diverse platform for AmI microcontrollers by means low-power consumption Internet gateways such as Wi-Fi and GSM [36].

8.4 Microcontrollers Versus Microprocessors

Microcontrollers and microprocessors are processors designed to run computing devices. Microprocessors, also called central processor units (CPU) are general purpose processors found in all computers whereas microcontrollers, sometimes known as System on a Chip (SOC) are specialized form of processors that once programmed, will execute stored set of instructions as and when required. A microcontroller can be said to be a small microprocessor that has a CPU, RAM, ROM and the I/O ports all on a single chip. Figure 8.1 illustrates the various components of microcontroller and microprocessors.

A comparative study of microcontrollers and microprocessors on the basis of their cost, speed, purpose, dependency, and resources is presented in Table 8.1.

Microcontrollers can be categorized depending upon the number of bits, memory size, instruction set, and architecture. The available microcontrollers are either 8, 16, 32, or 64 bits devices:

- An 8 bits microcontroller performs arithmetic and logic operations. Intel 8031 and 8051 are examples of these 8 bits microcontrollers [37].
- In contrast to 8 bits, a 16 bits microcontroller provides more accuracy and efficiency while performing the arithmetic and logic operations. Intel 8096 is an example of 16-bit microcontroller [38].

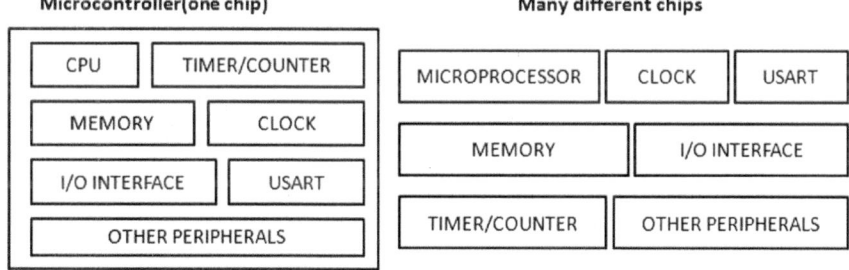

Fig. 8.1 Microcontroller versus Microprocessor

Table 8.1 Comparative study of microcontrollers and microprocessors

Parameters	Micro-controller	Micro-processor
Area	All components on a single chip	External components are needed to interact
Resources	Limited	Unlimited
Speed	Slow (in MHz Range)	Fast (in GHz range)
Function	Specific	General
Cost	Low-cost	Costly

- A 32 bits microcontroller is used mostly in automatically controlled appliances such as implantable medical appliances, office machines, etc. The arithmetic and logic functions of such microcontrollers are carried out by 32 bits instruction set [39].
- 64 bits microcontrollers also need 32-bit instructions to perform logic and arithmetic operations. These are currently being developed and further researched. These will be available to be used for large and complex computations.

In terms of storage, microcontrollers are of two varieties as follows:

- In the first group, there are microcontrollers with external memory. In this case, if an embedded structure is assembled with a microcontroller that does not contain the entire required functioning block on the chip especially the memory, then this type of microcontroller is called external memory microcontroller. The 8031 microcontroller is an example of this variety, which does not have program memory on the chip [40].
- In the case of the second group, if an embedded system is made with a microcontroller comprising all the fundamental functioning blocks along with memory located on the same chip, then this class of microcontrollers is known as embedded memory microcontrollers. The 8051 microcontroller is an example of embedded memory microcontrollers, which have counters, timers, interrupts, I/O ports along with all programs, and data memory [41].

Moreover, Harvard [42] and Princeton [43] are the two principle architectures for microcontrollers. The Harvard architecture uses the same memory and data path for storage and instructions whereas Princeton architecture uses different memory and data paths for storage and instructions.

8.4.1 The 8051 Microcontroller

In 1981, Intel Corporation launched the 8051, an 8-bit microcontroller, which is now available with 40-pin dual inline package (DIP). It is built with 128 bytes of RAM capacity and 4 kilobytes of ROM programmable capacity on the chip [39, 44]. Also, depending upon the necessity, external memory of up to 64 KB can be added to the microcontroller. It is facilitated with four easily programmable and addressable parallel 8 bits ports. An on-chip crystal oscillator is also availed with 120 MHz of crystal frequency. A simplified internal structure of XX51 is shown in Fig. 8.2a and its schematic with input and output ports is shown in Fig. 8.2b. The architecture consists of two timers each of 16 bits. These timers can be deployed for internal and external functioning of the microcontroller. The microcontroller has five interrupt signals ; Timer Interrupt 1, Timer Interrupt 0, Serial Port Interrupt,

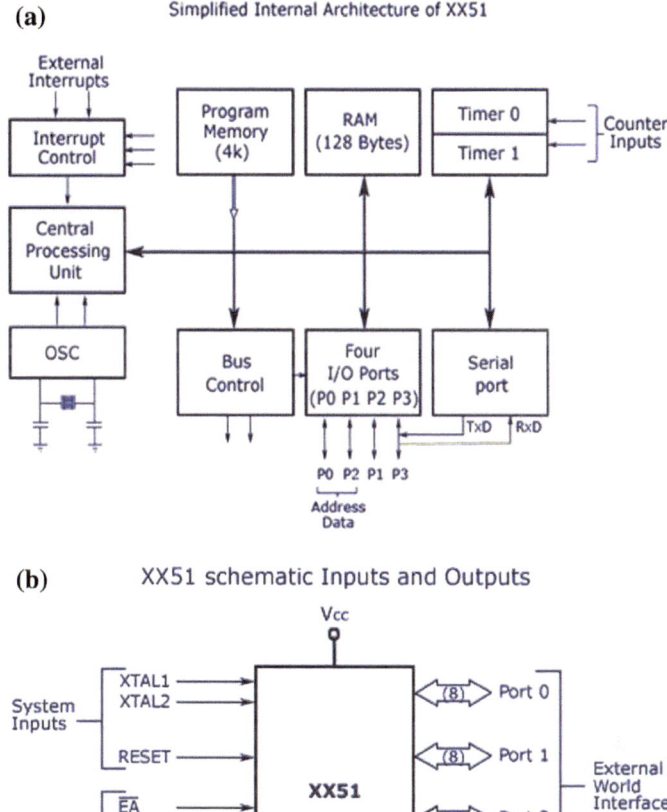

Fig. 8.2 **a** Simplified internal architecture of XX51. **b** XX51 schematic inputs and outputs

External Interrupt 0, and External Interrupt 1. Also, it is incorporated with SFRs (Special Function Registers), General Purpose Registers (GPR), and Special Purpose Registers (SPR) programming modes. The 8051 is used in a wide variety of application, e.g., in automobile, aeronautics, rail transport, industrial processing, mobile communication, remote sensing, radio and networking, robotics, safety and security (smart cards, e-commerce), and medical electronics facilities (e.g., hospital equipment and mobile monitoring).

8.4.2 Peripheral Interface Controller (PIC)

General Instruments' Microelectronics Division has been highly instrumental in the development of this family of microcontrollers as part of microchip technology. The PIC is one of the smallest microcontrollers with high performance and low-cost. The basic structure of PIC consists of a subset of MIPS ISA, one accumulator and 30–35 instruction sets. It is based upon the Harvard architecture as shown in Fig. 8.3. It reflects the following features.

- Field-programmable EPROM or ROM was used as program storage in the early models of PIC, although some of them have the provision of memory erasing. The recent models use flash memory for program storage that allows the PIC to reprogram them. These models also have separate program memory and data memory.
- In current models, the data memory is built with lengths of 8, 16, and 32 bits wide, where the program instructions may be 12, 14, 16, or 24 bits long. The instruction set varies with different models and bit counts. These also have chips adding instructions which empower them for digital signal processing functions.

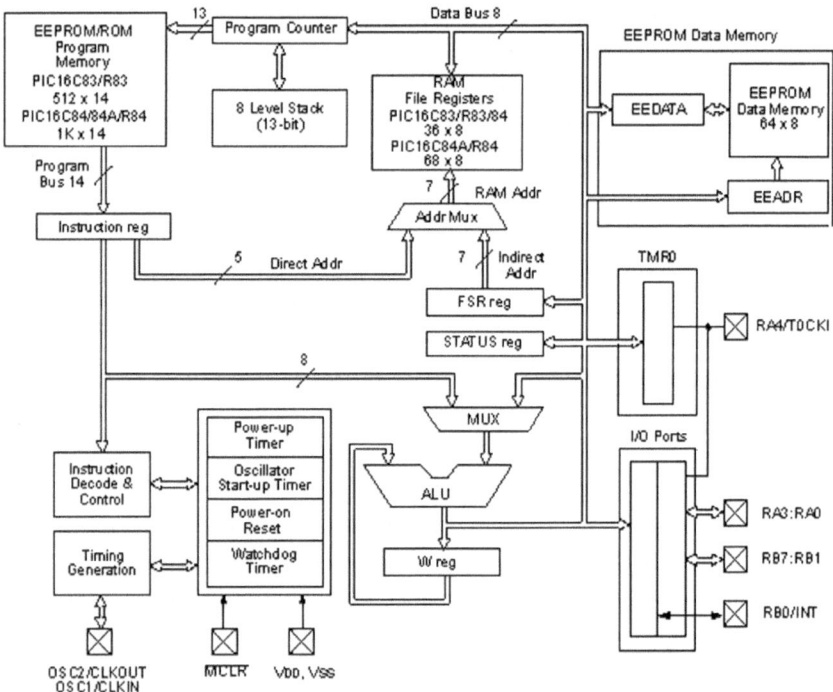

Fig. 8.3 Basic PIC architecture

- There are several hardware viabilities because of different ADC and DAC modules, 6-pin SMD, 8-pin DIP chips, up to 144-pin SMD chips, and communications ports including CAN, I2C, UART, and even USB, with discrete I/O pins of PIC devices. High-speed and low-power variations also exist in these types of microcontrollers [45].
- As for the computing software, C/C++ compilers, MPLAB, and assemblers are available from various manufacturers for development along with programmer/debugger hardware within the MPLAB and PICKit series.

The PICs are found in numerous everyday applications. Automotive sectors and Do it Yourself (DIY) projects are just two examples. Other common applications include computing systems, mobile phones, handheld smart devices, automobile airbags, mini robots, industrial control systems, as well as smart devices as part of the IoT and AmI vision.

8.4.3 Alf Vegardwollen RISC Processor (AVR)

This family of microcontrollers was developed by Atmel [46]. These single-chip microcontrollers are based on modified 8-bit reduced instruction set computing (RISC) Harvard architecture. Among the microcontroller families, AVR was the first to use flash memory for storage of programs, unlike other microcontrollers that are embedded with single-time programmable ROM, EPROM, or EEPROM. These have wide applications in the electronics field mostly in the embedded systems. These are extensively used in the Arduino line of open source board designs. The program memory of AVR exists from 4 KB to 256 KB with entire code compatibility and flexible pin configuration ranging from 8 to 100.

The AVR is an improved Harvard architecture microcontroller. As shown in Fig. 8.4, program and data have different paths and are stored in the individual physical memory structure. The data and program appear at different addresses and the data items are read from program memory on the special instructions of AVR. There are 32 GPRs (General Purpose Registers, R0–R31) which are directly connected to ALU so that all arithmetic and logical operations can be performed on these registers. RAM is accessed to load and store instructions where only a few instructions are performed on 16-bit register pairs. The processor provides high level of integration with strong instruction set for C and assembly languages, with very efficient core throughput of 20 MIPS at 20 MHz. It comprises various applications in different domains including navigation systems, car radio automatic controls, DES encryption/decryption, Reed–Solomon (error correction) encoder/decoders, refrigerators' low-power controls, safe battery chargers, embedded strong web servers, label laser/recite printers, etc.

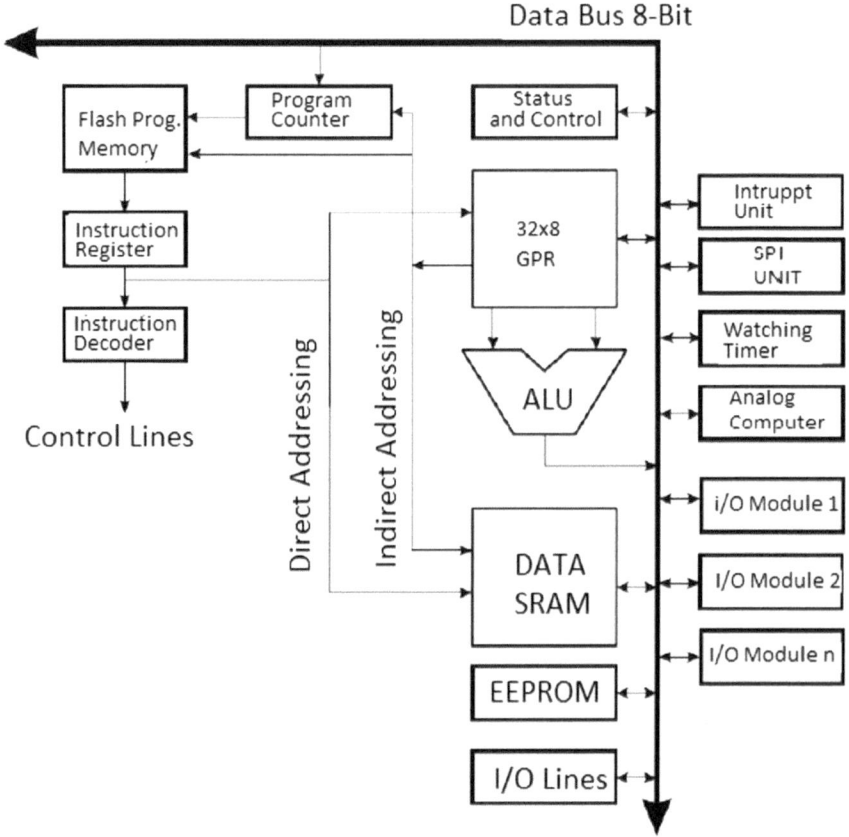

Fig. 8.4 Basic AVR architecture

8.4.4 Advanced RISC Machine (ARM)

The ARM is a 32-bit advanced Reduced Instruction Set Computing (RISC) family of microcontrollers. It was first developed by Acron Computers in 1987 and is now a strong contender in the market for low-power and cost-effective embedded applications. The ARM is developed by a collaboration of manufacturers including ST Microelectronics and Motorola [47]. It requires less silicon area, low power, and yields much higher performance. There are different versions of ARM including ARMv1, ARMv2, etc., with slightly different characteristics. It has an extended architecture along with the typical structure. The generic primitive structure of ARM is shown in Fig. 8.5. The extended architecture has provision for automatic increment/decrement of addressing modes and multiple store/load registers. It

Fig. 8.5 Basic architecture of ARM

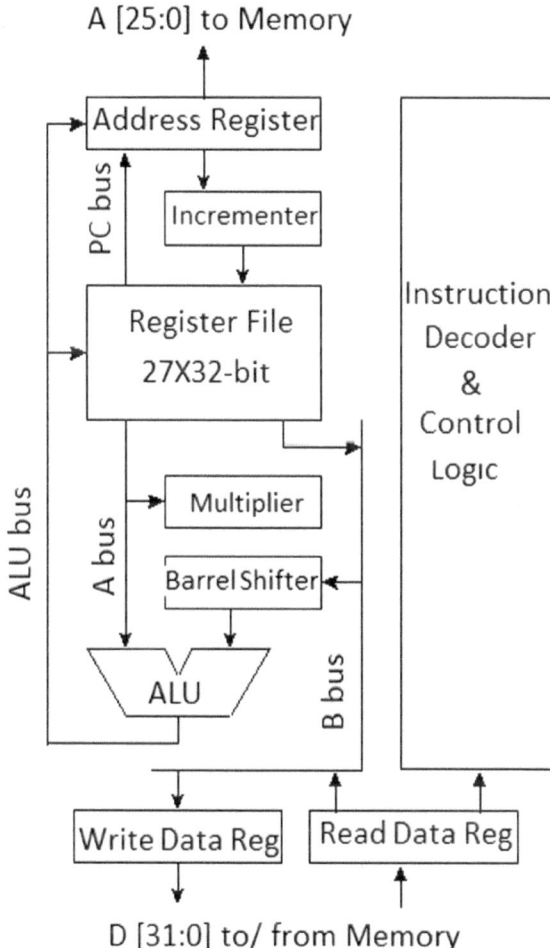

A [25:0] to Memory

D [31:0] to/ from Memory

has conditional execution of programs and each instruction individually controls the ALU and the shifter. Typical architecture contains uniform and fixed-length instruction fields, large and uniform register file, and store/load structure in which data is processed by registers and stored by addresses in a simple addressing mode. ARM is deployed in a vast variety of application fields in electrical and electronics, e.g., in automatically operated electronic airbags, smart sensor devices, smart printers and networks, mixed-signal devices, automatic banking gateways, smartphones, televisions, and many more.

A comparison of different microcontrollers on the basis of various features like memory architecture, power consumption, families, speed, and communication protocols is summarized in Table 8.2.

Table 8.2 Concluding comparison of various microcontrollers considering different features

	8051	PIC	AVR	ARM	MSP430
Bus width	8-bit	8/16/32-bit	8/32-bit	32/64-bit	16-bit
Memory architecture	von Neumann	Harvard architecture	Modified Harvard architecture	Modified Harvard architecture	von Neumann
ISA	CISC	RISC	RISC	RISC	RISC
Power consumption	High	Low	Low	Low	Ultra-low
Memory	ROM, SRAM, Flash	SRAM, Flash	Flash, SRAM, EEPROM	Flash, SDRAM, EEPROM	Flash, SDRAM, EEPROM
Communication protocols	USART, SPI	UART, USART, LIN, Ethernet	UART, USART, SPI,12C, etc.	UART, USART, SPI,12C, etc.	UART, SPI, BLUETOOTH,12C, etc.
Speed	12 clock/instruction cycle	4 clock/instruction cycle	1a clock/instruction cycle	1 clock/instruction cycle	1 clock/instruction cycle
Families	8051 variants	PIC16, PIC17, PIC18, PIC24	Tiny, Atmega, Xmega	Tiny, Atmega, Xmega	MSP430F5x/6x MSP430F2x/4x
Popular Microcontroller	AT89C51, P89v51, etc.	AT89C51, P89v51, etc.	Atmega8, 16, 32	LPC2148, ARM Cortex-M0 to M7, etc.	MSP432

8.5 Implementation of Remote Temperature Monitoring System

This section discusses the implementation of microcontrollers for remote Temperature Monitoring system (RTMS) within the IoT environment. Atemga8 AVR microcontroller, GSM module, and AVR program burner are used for RTMS. The temperature monitoring systems used for plants (creamery and winery) [48], is extended and applied to industries and remote conditions. For this purpose, instead of Arduino, the current implementation is with AVR microcontroller. The wireless sensors used in remote monitoring of temperature and humidity [49] give the idea of humidity and temperature monitoring in industries, where temperature plays an important role in storage and manufacturing. Another IoT application which incorporates this strategy is the remote inactivity detection [50] and this will provide the idea of application process of an accelerometer Wi-Fi module.

In order to design remote temperature monitoring system, a microcontroller architecture has been developed along with the following hardware components: microcontroller Atmega8, GSM/GPRS modules for transferring the temperature recorded on mobile phone and PCs, AVR programmer, burner to successfully burn the hex file over the AVR so as to execute the program, and extreme burner. AVR C and USBasp software are also used, and some of the architectural and design aspects are discussed below providing more details.

8.5.1 Proposed Architecture

The proposed architecture of microcontroller is developed after examining the merits and demerits of previously reported architectures that utilize the low and efficient power techniques, choosing an appropriate processor, and best interfacing schemes. Figure 8.6 presents the proposed architecture of a microcontroller in which arithmetic logic unit (ALU) is directly accessible by the input/output of processor control unit (CTR), RAM, ROM, and input/output devices; it performs all the arithmetic and logic operation on the operands. The input of ALU carries instruction set or operation code; also called opcode and operands. The output contains the results stored in registers after successful operation. Decoder shares the data from the CTR and is connected to ALU and DEBUG circuit to obtain and decode data in the required format. The CTR processor directs and shares the information with all parts of the device.

8.5.2 xBoard MINI v2.1

xBoard MINI v2.1 is the latest development available with 28 pins for ATmega8, and ATmega168, which are derived from Atmel AVR microcontrollers. It has a facility for in-circuit programming with USB AVR programmer v2.1. This board as shown in Fig. 8.7, comprises 16 MHz inbuilt crystal, on/off switches, regulated supply of

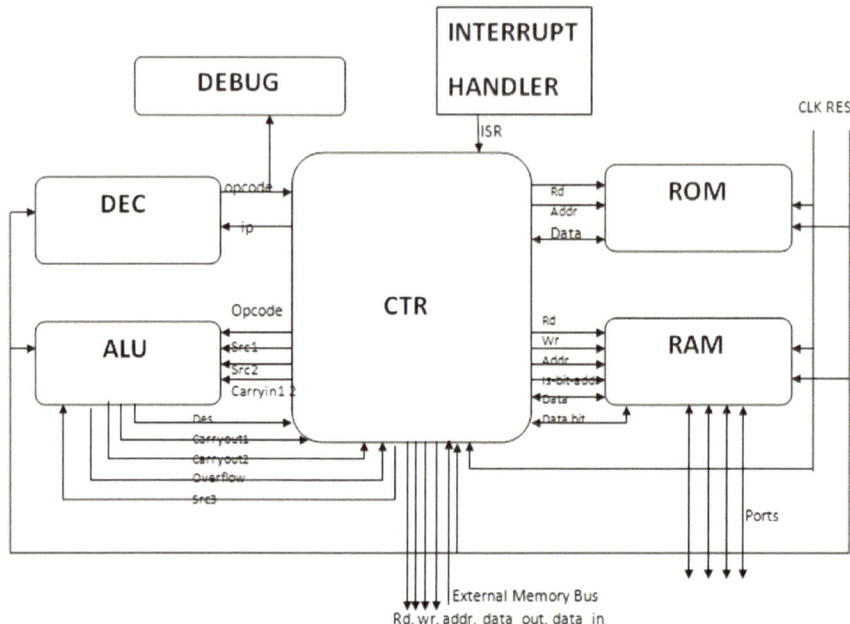

Fig. 8.6 Architecture of the proposed microcontroller

Fig. 8.7 10x board mini v2.1

5 V, and a reset switch. It was initially used for microcontroller programming. At the advanced stage, it is used as a base for building complex projects. The main benefit of this is that it is a completely verified and well-equipped board with advanced levels of peripherals.

It has everything embedded on the board to diminish outside wiring amid the primary learning period. It has four user push buttons, one user LED, serial port with TTL and RS232 output (MAX232 Circuit Inbuilt), 16 × 2 alphanumeric LCD modules, NEC Format remote control receiver (TSOP1738) and a decoder (software library), LM35 Temperature sensor, Philips RC5 decoder software xAPI library for the C language. To perform general tasks, DS1307 based on real-time clock with coin cell backup, and huge amounts of fully debugged and quality C sample code serve as a base for the design. This makes tasks much easier to accomplish, even the toughest tasks of programming and interconnecting, which initially seemed impossible.

8.5.3 AVR Programmer

An AVR (or Alf and Vegard's RISC) processor programmer is used to program an AVR microcontroller. The AVR microcontroller board diagram, as shown in Fig. 8.8, is based on an 8-bit RISC microcontroller. These microcontrollers are often used in hardware development and robotics.

An AVR programmer allows the user to place operating instructions on the microcontroller that tell it to perform a specific task. This programmer is usually plugged into the serial port of the computer, and a special program is required to get the code

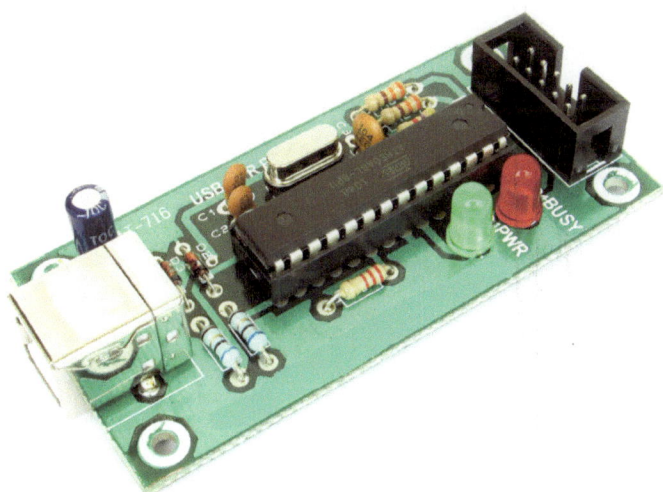

Fig. 8.8 AVR programmer

from the computer to the microcontroller. This code is typically saved as a HEX file and then uploaded onto the flash memory of the microcontroller. The code is then compiled into a binary file. The process of transferring the program from the computer to the microcontroller is called flashing.

The programmer is connected to certain pins on the microcontroller that allows it to be programmed. Most AVR microcontrollers have only a few kilobytes of memory, so programs put on the microcontroller must be fairly small.

In case of AVR microcontrollers, input and outcomes are taken care of through the various pins on the microcontroller. It is conceivable to snare up sensors, switches, lights, engines, and numerous other things to the pins, which permit it to an incredible extent of uses. AVR microcontrollers are frequently utilized for prototyping and automated projects. The AVR programmer is often more expensive as compared to the price of a typical microcontroller and can be purchased separately or as a kit.

8.5.4 GSM and GPRS Modules

GSM global system for mobiles is used to set up communication through mobile phones using voice and data while the Global Packet Radio Service (GPRS) module is used to share information via the internet between mobile phones or between mobile

Fig. 8.9 GSM module

phones and computers in the form of voice, data, images, and videos. A typical GSM/GPRS module is shown in Fig. 8.9. It consists of a GSM/GPRS modem and an interfacing port such as USB or RS-232 along with power supply.

The GSM/GPRS modem, which is the core part of the modules, enables the application of IoT and AmI in monitoring and adjusting the temperature and other ambient parameters. It is designed to establish connection between computers or other devices with the help of mobile devices or Internet and comes under the class of wireless modem devices. In order to actuate the connection between two devices, it needs a Subscriber Identity Module (SIM) card as in the mobile phones. Moreover, this is also associated with an International Mobile Equipment Identity (IMEI) number for their recognition, just like IMEIs in mobile phones.

To interface the modem with a controller/processor, an AT commands is needed to be transmitted through serial communication protocol. The controller/processor initially sends these commands to and receives responses back from the modem. The modem supports various AT commands transmitted by the controller to establish connections with cellular networks of GSM and GPRS.

8.5.5 Adopted Low-Power Modes

In the context of low-power consumption, various low-power modes such as shutdown ALU, power down, and idle modes can be considered:

- The shutdown ALU mode minimize the power consumption. The ALU is shutdown while the controller is reading a port address from ROM or while reading the data from RAM. This helps to decrease the static power consumption of ALU.
- The power down mode is automatically enabled, if there is no activity for, say, 12 h or more. It helps to reduce power consumption while the device is not in use. This is similar to the sleep mode in other devices such as computers.
- In the case of idle mode, while for most of the time, the device will remain in idle mode, it will go into active state only when an operation is needed to be done through an interrupt. This helps to decrease the excessive power dissipation in the device.

Other powers optimizing techniques may also be used, e.g., Clock gating to decreasing the dynamic current; and Power gating to limit the leakage of current.

8.6 Summary and Future Directions

In this chapter, the low-power microcontroller using VHDL in Xilinx software is presented for AmI-based IoT applications. Ambient Intelligence (AmI) has an important role to play in the design of intelligent low-power microcontroller. The study of various architectures is also presented to make the devices, application specific.

Available architectures are reported with the aim to make a comparative study to determine which of these could provide better performance, low-cost and still work satisfactorily for the IoT applications. Different qualities of these are considered and integrated. It is noted that a microcontroller like MSP430 has the highest number of modes some of which are embedded within. It is found that AVR and Arduino consume the least power and so the proposed architecture is based on these. The VHDL code is developed for the proposed architecture and main modules are successfully developed including ALU, Decoder, RAM, ROM, and the Controller. After the implementation of design on Xilinx, the static and dynamic power is calculated using an in-built tool power analyzer which came approximately to the least power consumption in the individual components. In the study, an IoT application is also implemented on the designed microcontroller. As overall integration showed up some real issues, we reworked the hardware and developed further IoT applications for Remote Temperature Monitoring.

These days, microcontrollers are so common in everyday life that any electronic gadget is incomplete without them and therefore our work covers a wide range of areas of scope for smart IoT applications embedded with AmI. For further studies, this work has some wide-ranging applications in the industry, especially in case of Industrial IoT (IIoT), e.g., in terms of remotely monitoring the temperature of the laboratory rooms and manufacturing unit where temperature plays an important part. Also, this application can be used in greenhouses, which can help the farmers to control the ambient environment within and outside the garden houses. Furthermore, our study can be helpful to record the temperature of remote areas like Antarctica7 and Bermuda Triangle.

References

1. Štuikys V, Damaševičius R (2003) Metaprogramming techniques for designing embedded components for ambient intelligence. In: Ambient intelligence: impact on embedded system design. Springer, US
2. Acampora G, Cook DJ, Rashidi P, Vasilakos AV (2013) A survey on ambient intelligence in healthcare. In: Proceedings of the IEEE
3. Bibri SE (2015) Ethical implications of ami and the Iot: risks to privacy, security, and trust, and prospective technological safeguards. In: The Shaping of ambient intelligence and the internet of things, Atlantis Press, Paris
4. Gaglio S, Re GL (2014) Advances onto the internet of things. Springer
5. Sui H, Wang H, Lu MS, Lee WJ (2009) An AMI system for the deregulated electricity markets. IEEE Trans Indus Appl
6. Benini L, Farella E, Guiducci C (2006) Wireless sensor networks: enabling technology for ambient intelligence. Microelectron J
7. Mazidi MA, McKinlay RD, Causey D (2008) Microcontroller, P.I.C., embedded systems. Pearson, New Jersey
8. Sun AC, Chen CL, Lee CH (1999) In-circuit programming architecture with ROM and flash memory. Macronix International Co Ltd, US Patent
9. Rao YR (2017) Automatic smart parking system using Internet of Things (IOT). Int J Eng Technol Sci Res

10. Navghane SS, Killedar MS, Rohokale DV (2016) IoT based smart garbage and waste collection bin. Int J Adv Res Electron Commun Eng
11. Tan L, Wang N (2010) Future internet: the internet of things. In: 2010 3rd International conference on advanced computer theory and engineering (ICACTE), IEEE
12. Bing K, Fu L, Zhuo Y, Yanlei L (2011) Design of an internet of things-based smart home system. In: 2011 2nd International conference intelligent control and information processing (ICICIP), IEEE
13. Wang M, Zhang G, Zhang C, Zhang J, Li C (2013) An IoT-based appliance control system for smart homes. In: 2013 Fourth international conference intelligent control and information processing (ICICIP), IEEE
14. Darianian M, Michael MP (2008) Smart home mobile RFID-based Internet-of-Things systems and services. In: ICACTE'08. international conference advanced computer theory and engineering, IEEE
15. Aarts E, Wichert R (2009) Ambient intelligence, technology guide. Springer, Berlin, Heidelberg
16. Cook DJ, Augusto JC, Jakkula VR (2009) Ambient intelligence: technologies, applications, and opportunities. Pervasive Mobile Comput
17. Eisenhauer M, Rosengren P, Antolin P (2009) A development platform for integrating wireless devices and sensors into ambient intelligence systems. In: Sensor, Mesh and Ad Hoc communications and networks workshops. SECON workshops' 09. 6th Annual IEEE communications society conference, IEEE
18. Nakashima H, Aghajan H, Augusto JC (2009) Handbook of ambient intelligence and smart environments. Springer Science & Business Media
19. Strom SrO, Eieland A, Flodell SrH (2018) How to pick the best 8-or 32-Bit microcontroller for your next design
20. Petre VC (2006) Microcontroller based measurements: how to take out the best we can of them. In: Proceedings of the 8th WSEAS international conference on Mathematical methods and computational techniques in electrical engineering, World Scientific and Engineering Academy and Society (WSEAS)
21. Li M, Li Z, Vasilakos AV (2013) A survey on topology control in wireless sensor networks: Taxonomy, comparative study, and open issues. In: Proceedings of the IEEE
22. Hutter M, Schwabe P (2013) NaCl on 8-bit AVR microcontrollers. In: International conference on cryptology in Africa. Springer, Berlin, Heidelberg
23. Polastre J, Szewczyk R, Culler D (2005) Telos: enabling ultra-low power wireless research. In: Proceedings of the 4th international symposium on information processing in sensor networks, IEEE Press
24. Chokkalingam S, Arunprasath V, Dinesh KP (2014) Implementation of 8 Bit microcontroller using VLSI. Int J Adv Res Comput Sci Technol
25. Laudon JP, Lenoski DE, Manton J, Anderson ME, Graphics Properties Holdings Inc (2000) High memory capacity DIMM with data and state memory. US Patent 6,049,476
26. Singh J, Raj B (2018) Comparative analysis of memristor models and memories design. J Semicond IOP Sci
27. Gubbi J, Buyya R, Marusic S, Palaniswami M (2013) Internet of Things (IoT): a vision. In: architectural elements, and future directions, Future generation computer systems
28. Kelly SDT, Suryadevara NK, Mukhopadhyay SC (2013) Towards the implementation of IoT for environmental condition monitoring in homes. IEEE Sensors J
29. Bandyopadhyay D, Sen J (2011) Internet of things: applications and challenges in technology and standardization. Wirel Pers Commun
30. Li S, Da XuL, Zhao S (2015) The internet of things: a survey. Inf Syst Front
31. Lewis FL (2004) Wireless sensor networks Smart environments: technologies, protocols, and applications
32. Lynch C, o'Reilly F (2005) Processor choice for wireless sensor networks. In: Proceedings 1st workshop on real-world wireless sensor networks REALWSN
33. Jie L, Ghayvat H, Mukhopadhyay SC (2015) Introducing Intel Galileo as a development platform of smart sensor: evolution, opportunities and challenges. In: Industrial electronics and applications (ICIEA), IEEE 10th conference IEEE

34. Piyare R (2013) Internet of things: ubiquitous home control and monitoring system using android based smart phone. Int J Internet Things
35. Mault J (2002) Remote temperature monitoring system. HealtheTech Inc, U.S. Patent Application
36. Frohn RC, Lopez RD, (2017) Remote sensing for landscape ecology: new metric indicators: monitoring, modeling, and assessment of ecosystems. CRC Press
37. Ker MD, Sung YY (2001) Hardware/firmware co-design in an 8-bits microcontroller to solve the system-level ESD issue on keyboard, Microelectronics Reliability
38. Dhia SB, Sicard E, Mequignon Y, Boyer A, Dienot JM (2007) Thermal influence on 16-bits microcontroller emission. In: Electromagnetic compatibility, 2007. EMC 2007. IEEE international symposium, July 2007, IEEE
39. Brandolese C, Fornaciari W, Salice F, Sciuto D (2000) An instruction-level functionally-based energy estimation model for 32-bits microprocessors. In: Proceedings of the 37th annual design automation conference, ACM
40. Grimmer GG, Rhoades, MW (1998) Microcontroller with security logic circuit which prevents reading of internal memory by external program. Motorola Solutions Inc, U.S. Patent
41. Tsai HJ (1999) Microcontroller with programmable embedded flash memory. Winbond Electronics Corp, U.S. Patent
42. Francillon A, Castelluccia C (2008) Code injection attacks on harvard-architecture devices. In: Proceedings of the 15th ACM conference on computer and communications security, October 2008, ACM
43. Kamal R (2011) Embedded systems: architecture, programming and design. Tata McGraw-Hill Education
44. Mazidi MA, Mazidi JG, Mckinlay RD (2000) The 8051 microcontroller and embedded systems. New Delhi
45. Latif FA, Stevens MD, Moysey JA, Shinkarovsky M, Nguyen H, Dale MZ (1994) Programmable multiple I/O interface controller, Unisys Corp, U.S. Patent
46. Huang HW (2013) The atmel AVR microcontroller mega and Xmega in assembly and C, Cengage Learning
47. Filipowicz W (2005) RNAi: the nuts and bolts of the RISC machine, Cell
48. Madrid N, Boulton R, Knoesen A (2017) Remote monitoring of winery and creamery environments with a wireless sensor system. Build Environ
49. Mainwaring A, Culler D, Polastre J, Szewczyk R, Anderson J (2002) Wireless sensor networks for habitat monitoring. In: Proceedings 1st ACM international workshop on wireless sensor networks and applications
50. Chan YJ, Huang JW (2017) Multiple-point vibration testing with micro-electromechanical accelerometers and micro-controller unit. Mechatronics

Part III
Applications and Use Scenarios

Chapter 9
Tax Services and Tax Service Providers' Changing Role in the IoT and AmI Environment

Güneş Çetin Gerger

Abstract New technology trends including Ambient Intelligence (AmI), Machine Learning (ML), Internet of Things (IoT), Cloud and Edge computing all have an important role to play in further developing the tax administration processes. The main challenge is to effectively reform the Revenue Administration (RA) services in today's electronic age. Business activities have become more global and digitalized; and revenue administrations' traditional services are also developing fast through the use of smart devices and smart software applications in the IoT-distributed computing environment. In this digital world, tax service providers also have a big responsibility, in fact, a requirement, to use and adopt revenue administrations' smart e-services built with some intelligence to improve automation. Understanding the need of the hour, tax service providers are attempting to develop IoT environments with AmI for fulfilling citizens' tax obligations. It is understandable and should be acceptable that in doing so and through increased smart automation, their workload is going to decrease drastically. In this chapter, it is aimed to examine RA tax services provision in the IoT environment in some leading countries and discuss, in some detail, how the role of tax service providers is changing in this improving smart taxation environment.

Keywords Tax services · E-tax services · Digital economy · IoT
Revenue services · Tax service providers · Ambient intelligence · AmI
Cloud computing · Edge computing

9.1 Introduction

In the digitalized world, there are some inherent issues regarding taxation processes, first of "where or who to tax" and then of "what to tax". Determining incomes correctly is the main function of revenue administrations. This is because it is crucial

G. Çetin Gerger (✉)
Manisa Celal Bayar University, Manisa, Turkey
e-mail: gunes.cetin@hotmail.com

© Springer Nature Switzerland AG 2019
Z. Mahmood (ed.), *Guide to Ambient Intelligence in the IoT Environment*, Computer Communications and Networks, https://doi.org/10.1007/978-3-030-04173-1_9

for tax fairness and tax compliance. Governments are trying to use technology at all relevant levels for full charge of tax revenues. A smart taxation system has the key role to play in this process. OECD [1] has presented an initial discussion paper on taxation in digital economy to G20 governments in March 2018. It is an important milestone for the digital economy. The idea of smart tax administration is becoming attractive and will soon be reshaping the current RA scenario in the developed and developing economies.

There are many tax-related projects around the world in which the Internet of things (IoT) can pose an important role. IoT can help to optimize the route of field tax collectors (especially relating to geolocation) and seamlessly link up the smart devices belonging to citizens with smart software taxation applications. Utilizing drones and kiosk systems, using blockchain for the collection of various taxes (e.g., VAT in the UK), and using blockchain for salary payments can help to determine and collect tax-related revenues correctly. Here, we present three examples of tax-related projects:

- Brazil-ID: It is a system based on the use of radio-frequency identification technology [2],
- Blockchain Pilot Project at Singapore Customs: Singapore Customs is developing a pilot project with IBM to modernize the entire customs flow, from source to destination, using blockchain technology [22],
- VAT management in the European Union: this is a mechanism to improve VAT management in the European Union, based on the use of blockchain technology [3].

E-services like digital tax accounts assist the taxpayers to manage relevant information and see their tax requirements and calculations. Using these services, taxpayers can complete the tax processes and fulfill their responsibilities easily without the help of tax service providers. As a result, tax service providers' role will soon change because of ambient intelligence advances in the IoT environment. For instance, the IBM Watsons artificial intelligence technology is already being used to work out expenses for deductions [4]. It will be possible for a computer system to calculate the optimum taxation for an investment in the near future.

The World Economic Forum (WEF) released a report in September 2015 summarizing 21 possible factors [5]. The report includes some interesting points. For example, based on the WEF Report, it is foreseen that by 2021, 30% percent of corporate audits of AmI Tax will be collected for the first time by the government through blockchain. The first AmI machine will be in an institutional board of directors, soon. Furthermore, AmI in tax administration will facilitates compliance. In addition, tax service providers and tax authorities will explore artificial intelligence (AI) to validate and report for compliance purposes [5]. Tax service providers' use of AI will decrease their added value in their work and so they will be able to focus on more important activities rather than low-level tasks that machines can easily perform. In the IoT environment, use of AmI is improving and it affects the role of both revenue administrations and tax service providers (tax professionals).

The study reported in this chapter aims to discuss possible arrangements and applications for taxation in the AmI environment. This chapter comprises three parts. The first refers to the new developments in digital economy, which will be evaluated in terms of the concepts of smart revenue administration and AmI. The second part is on how the role of tax service providers (traditional tax intermediaries) is changing in the digital world. The final section concludes and discusses some future directions for further research.

9.2 Taxation Developments and Issues in Digital Economy

9.2.1 Ambient Intelligence

The tax environment is changing rapidly in digital economies. The related taxation issues in the digital world need to be considered by revenue administration to achieve tax compliance. For this purpose, tax authorities need to increase transparency, share information with each other and use high levels of technology. Using technology in taxation can provide validation for compliance purposes. It can increase efficiency, automation, and openness. Tax authorities and tax providers are in fact starting to explore sophisticated data analytics, artificial intelligence, and machine learning (ML). Using artificial intelligence for taxation processes is a reasonably new phenomenon. Ambient Intelligence (AmI) works in a global computing devices world that is also relatively new in which physical settings interact in an intelligent and unobtrusive manner with people [6].

The growth in the use of digital devices is leading to new practices in supporting tax activities such as payments, form filing, and enquiries. Today, officials are using AmI-based processes and devices as digital assistants distinguishing expenses for deduction, and even assessing the optimum taxation for an investment in many countries.

9.2.2 Internet of Things (IoT)

The Internet of Things (IoT) grows inevitably. The term IoT was coined in 1999 by Kevin Ashton, who referred to a global network of radio-frequency identification connected objects. The term now extends to products, vehicles and even entire buildings that have embedded electronics, sensors, software and network connectivity used to collect and exchange data for monitoring and control purposes [7].

The IoT is based on the use of smart devices and Machine-to-Machine (M2M) communication. Electronics smart devices embedded with processing capabilities, and in some cases storage, communicate with each other via the Internet generally over wireless connections.

The possible market for the IoT vision is increasing at an exponential rate. Based on industry estimates, a total of US$6 trillion will be spent on IoT solutions on a global scale between 2015 and 2020. Businesses, governments, and consumers together will invest approximately US$1.6 trillion in IoT solutions in 2020 through hardware, software, and services [8].

According to Gartner [9], the Internet of Things and online connectivity among devices is expanding exponentially—up 31% from 2016 to 2017 and is expected to reach 2.5 times the current levels by 2020.

IoT is about a major transition between selling products and services that brings out the main tax applications when services are on sale across borders. IoT adoption revolves monetization of data; and creates expanded geographical footprints enabling the companies to contact with different tax authorities in other areas. Tax authorities can access the related data and do their audits from the geographical footprint.

9.2.3 Blockchain

Blockchain refers to a transaction and data management technology that is not centralized. It was developed and used for the first time for the Bitcoin cryptocurrency. The interest in Blockchain technology has increased in a continuous manner since the idea was put forward back in 2008 [10].

Like any new technologies, Blockchain technology also brings new taxation issues. The main issues refer to the place of any transactions and the identity of the taxpayer. Such transactions are recorded in thousands of computers across the world that are not centralized [11].

Tax administrators are also exploring the use of blockchain. This is because blockchain is a strong record technology that can be used to store any type of data including financial transactions by recording when a transaction occurs, as well as the details of transactions. This is a secure method for registration, confirmation of taxpayers, as well as a correct recording of transactions [12].

9.2.4 3D Printing

In recent years, 3D printing has increasingly progressed into the mainstream. Traditional mass manufacturing methods have changed, which also altered the supply chain processes. In the aerospace industry, for example, Boeing is already using 3D printing to create more than 50000 units of over 900 distinct parts for both its aircraft and spacecraft. "Narrow" artificial intelligence may be observed in various fields [12].

3D printing results in another concern for taxation authorities all over the world because nearly all the values that are subject to taxation in 3D printing exist as Intellectual Property (IP); and are not mentioned in its production, transportation, or at sales points. IP has become a major concern in OECD because these countries are developing new models for the taxation of digital services and intellectual properties, i.e., the things included in digital media [13].

9.2.5   Sharing Economy

The sharing economy results from the coexistence of all-inclusive use of the expression and availability of a highly heterogeneous group of online platforms that provide various new and innovative economic and social activities that cannot be easily categorized [14].

The widespread use of smartphones has brought a new concept called the *sharing economy* to electronic commerce. Electronic entities, products, and services like Uber, Airbnb, and BlablaCar are devoid of any legal regulation or taxation of income, and their revenues are based on informal income and control. Therefore, legal measures concentrate on preventing the informality of their transactions [15].

In this changing environment, there are multiple possible routes with many of these being explored in OECDs Going Digital Project. Digitalization is also changing the nature of policy-making itself with a new range of tools available to develop, monitor, and evaluate the effectiveness of a range of different policies and their outcomes. It is, thus, also important to be able to harness technological innovation to support the delivery of more effective and tailored solutions and foster a supportive environment for innovation and growth. For tax matters, this means that policy development and implementation should be designed to allow for the changing environment [16].

9.3   Examples of Tax Services Applications in the IoT and AmI Environments

Appropriate digital transformation can improve tax administration. Working with new technology increases tax administrators' motivation, supports optimization in operations, increases operational efficiency, and enhances openness for taxpayers. Moreover, taxpayers feel more satisfied with their tax services experiences [11].

Mobile applications are crucial in this digital age. Additionally, tax applications have become the main service in many tax administrations. These applications allow taxpayers to file returns and pay taxes due. Tax applications integrated with other intelligent systems are used by taxpayers in their everyday lives. They improve tax compliance.

Smart portals and smart mobile devices in the IoT environment provide real-time online assistance to taxpayers using chat or virtual video assistant facilities. This online support option will allow taxpayers to locate the information they need within the channel they are using or move seamlessly between channels. Taxpayers contact these via a virtual assistant and click to call or click to avail chat options. Virtual assistants supported with an online knowledge base enable taxpayers to immediately find information that they need [12].

Smart portals support and provide guidance to taxpayers and community members and enable related compliance activities more easily. They directly contact taxpayers about the assistance they need. These also allow tax administrations to direct multiple contacts to a lower number of contacts. There are additional benefits relating to smart portals. These include lower administrative costs, more timely revenue flows, increased participation and easy access, and reduction in compliance cost. So, all these can help to improve tax compliance.

AmI, like Chatbot technology, assists taxpayers and uses machine learning to acquire better insight from the data for better decision-making at the strategic level in real time. The scientists at MIT Computer Science and Artificial Intelligence Laboratory (CSAIL) have supported the development of a Simulating Tax Evasion and Law through Heuristics [17]. IoT also connects cash records with taxation authorities to aid the fight against frauds in sales and taxation with a variety of services such as carbon tax [5]. Ambient intelligence and IoT help taxpayers to access and connect to taxation and revenue authorities. Personal relations may aid in building reliable relationships between stakeholders. Some examples of how various countries across the globe are using and deploying connected e-services to audit taxpayers are provided below.

9.3.1 The Australian Tax Office (ATO)

The Australian Tax Office (ATO) have managed to drastically reduce the high volume of call center inquiries. With successful collaboration with voice biometrics that can identify a customer and shorten the call by about 40 s [18], ATO helps clients via a digital virtual assistant named Alex. Alex had over two million conversations, and as of September 2017, ATO has achieved a huge reduction in costs as well. Alex has contributed to 8–10% reduction in contact center call volumes [19].

Alex gives personalized answers to customer requests using natural language processing, conversational dialog, and advanced resolution techniques. Thus, there is the use of AI as well. This digital operation allows contact center agents to spend more time to handle complex demands. ATO have suggested that the virtual assistant will continue to be used to establish contextual conversations with customers.

9.3.2 UK HM Revenue and Customs (HMRC)

The HMRC is the main tax agency in the UK. The virtual assistant applications are an important part of HMRCs current and future plans. Robotics systems check the tax returns of millions of businesses and individual taxpayers. The UK tax agency is also looking at implementing robotics and artificial intelligence (AI) for several other areas of their work, with a view to easing the complexity of tax operations and increasing efficiency with reduced errors. The department is using Artificial

Intelligence (AI) technology for working on compliance, and HMRC is acting on digital transformations. AI can help out with the most complex tasks relating to tax operations and services [20].

Moreover, The Data Protection Act 2018 has also come into force. It creates a new framework for data processing, providing for a separate regime to regulate the processing of personal data by intelligence services. This regime is based on international standards, which will be provided for in an amended Council of Europe "Convention for the Protection of Individuals regarding Automatic Processing of Personal Data" [1].

9.3.3 Mexico Tax Administration Service (SAT)

Mexico's tax administration authority, Servicio de Administración Tributaria (SAT), employs a tax collection system that is cloud based. It is used by individuals and business organizations to file tax returns and pay taxes in an easy and timely manner. Taxpayers log onto the SAT website by using their tax ID numbers. The earnings and deductions are calculated automatically in real time. SAT enables clients to perform the transactions that are related to taxation online and to check the status of their tax returns and any overpaid or underpaid dues. The system enables openness and enhances trust.

Synchronization of all access requests with a single authentication directory that depends on PeopleSoft Customer Relationship System (CRM) and PeopleSoft Human Resources (HR) system that provide improved ease for clients and IT staff as well as transparency. Through physical, virtual and cloud media, Compliance Management Platform aids clients to deal with compliance and security challenges. A 100% ROI was achieved within 18 months when Compliance Management Platform was employed [21].

9.3.4 Brazil Federal Secretariat for Revenue (BFSR)

Brazil Revenue Services has a system based on the use of radio-frequency identification technology (RFID) and other wireless communication technologies, to establish a standard of identification to be used in tax-related products and tax documents circulating in the country [22].

The Brazil-ID system was created by the tax administrations of Brazilian states and the Federal Tax Administration. Refer to Fig. 9.1. The control of the transit of goods for the management of the ICMS (Tax on the circulation of goods and services) is under the responsibility of the states. This system responds to the inherent challenges, eliminating or reducing the time of the stops for the control of trucks, improving the accuracy of control, and lowering the costs for transportation companies [2, 22] as well as introducing ease and effectiveness. The project is proving to be a huge success.

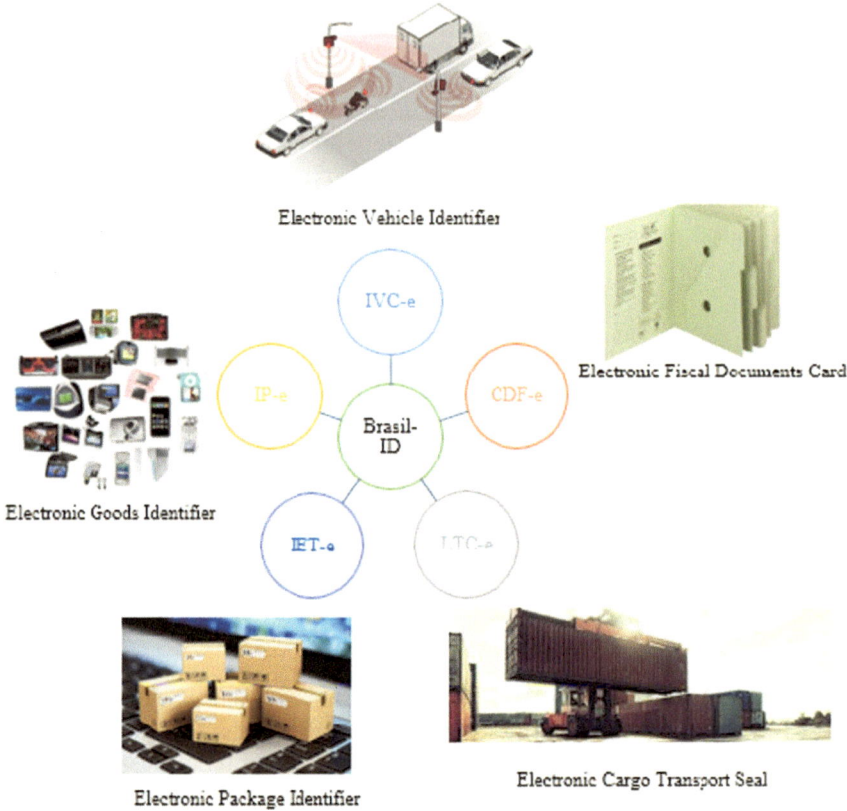

Electronic Vehicle Identifier

Electronic Fiscal Documents Card

Electronic Goods Identifier

Electronic Package Identifier

Electronic Cargo Transport Seal

Fig. 9.1 General scheme of the Brazil-ID system

9.3.5 Inland Revenue Authority of Singapore (IRAS)

Singapore Revenue Authority and Customs Services has introduced a pilot project in association with IBM for the entire customs flow using blockchain technology. The flow is using the typical mechanisms of blockchain for intelligent contracts, consensus determination, etc. The first goal is to reduce the overall time taken and operational costs [23].

The pilot project involves all stakeholders including agents, banks, customers, transporters, and customs administrations. The World Customs Organization participates in the project. The standardized electronic identifiers embedded in or attached to the products and containers would be added to the customs of origin for the case of container description. Automation and streamlining would be further extended, including improved security and fraud reduction.

In Singapore, all citizens over the age of 15 can apply for a SingPass ID to use online government services, including tax services. In Singapore, taxpayers are provided with national digital IDs, allowing them to access a range of public and private digital services. This serves to mitigate identity theft problems and simplify access to services, as well as availability around the clock. In Singapore, in addition to digital ID and password (SingPass), two-factor authentication is required for access to all sensitive government services such as tax return filing and calculation. Citizens can choose two-factor authentication by using a physical token or via their mobile and handheld devices. The platform is being extended to companies and intermediaries this year, significantly strengthening the security of government e-transactions.

9.4 Changing Role of Tax Service Providers

In the age of the Fourth Industrial Revolution, the role of tax professionals is changing. They are free agents, assisting companies for required details and helping with matters relating to tax and revenue. Tax service providers operate in many jurisdictions inputting tax requirements into information systems that are used by taxpayers to run their businesses, manage their bank accounts, and interact with the relevant government departments [23].

Tax service providers perform a variety of tasks. They help with tax returns, provide advice on the application of laws, represent their clients with respect to tax audits and solve disputes and, in general, work as agents on behalf of clients. Tax service providers improve compliance rates, meet service requirements and demands, and support to lower their client's administrative burden.

The Forum on Tax Administration (FTA) has published a report called "Rethinking Tax Services: The Changing Role of Tax Service Providers in SME Tax Compliance" [16]. This report provides an outline of the related technological, business and quality enhancements.

The market for bookkeeping, accounting, and payroll software applications has changed dramatically over the past 5 years. The trend is particularly marked in developed economies with high technology penetration. Use of cloud-related technologies, device connectivity in the IoT paradigm, and embedding AmI in the smart device is becoming the norm. Thus, the role of tax service providers is now necessarily requiring them to use and provide services that are the latest technology based. As an example, the Danish Tax Administration has reported [23] that more than half of the business population now use cloud-based accounting software to manage their bookkeeping and tax affairs. New Technologies such as Fog/Edge Computing and AmI have now begun to be considered for the possible uses for blockchain in tax administration [23].

9.5 Benefits of Employing AmI and IoT for Tax Administration

Using ambient intelligence in smart and IoT-based devices can drastically reduce taxpayers' compliance costs and increase the effectiveness of systems due to automation and embedded intelligence. Using edge computing environment in the IoT infrastructure can help to increase the security of private and sensitive data and thus increase clients' trust. In this context, taxpayers can fulfill their tax and revenue obligations faster and more easily and securely with fewer inherent issues relating to tax services. Taxpayers' tax processes become more simplified. Transparency and data management get improved with AmI, IoT and context-aware autonomous systems. Taxpayers gain the ability to comply voluntarily with the required systems. They will have trust in the knowledge that taxation offices and relevant departments follow the right level of privacy and confidentiality with respect to client data and circumstances.

Cognitive computing, blockchain technology, artificial intelligence, and robotics are important examples of technologies for tax administrations. Some of these are already in use. These new technologies offer tax administrations the opportunity to further improve efficiency and effectiveness [23].

Some ordinary routine tasks can be enabled by AI. These include: voice recognition, ATM check amount recognition, and text spell checking [24]. On upper levels of tax functions, tax applications become more complex in human judgment tasks like answering subtle legal and particular taxation-related questions from legal documents, thereby possibly assisting government oversight. However, machine learning and leaning algorithms can prove useful, though, there is a need for further research on these topics.

In the following sections, we highlight the use of the latest technology in some well-known tax-related applications and scenarios.

9.5.1 Blockchain Technology in VAT Applications

Value-Added Tax (VAT) is the main tax revenue source for tax administrations and the largest contribution to the government treasury. Therefore, tax authorities are always trying to establish more effective VAT calculation and collection methods. Recently, advanced solutions have been put into practice in Brazil, as mentioned in an earlier section. Using blockchain technology, electronic invoices become obligatory and are received by tax authorities in real time. At the European Union level, the EU VAT system will be reformed in the next few years [25].

Real-time tax calculation determines the amount of due VAT correctly and submits it to the tax authorities. There are two main reasons for this: tax returns and payments are calculated over a fixed period, for example, monthly or quarterly; and the calculations are not always based on actual transactions. In an international context, controlling the VAT data is problematic as each country maintains their own ledgers, making it difficult to obtain wholesome data on VAT movements between different jurisdictions [26].

9.5.2 IBM Watson AI Technology to File Tax Returns

IBM Watson is an AI-based platform for businesses. This platform uses cognitive computing techniques for data analysis. Tax preparation, involving huge amounts of data, can be performed with the Watson Tax Return solutions. For this, IBM has partnered with H&R Block who already have deep insights into this field; and are one of the largest tax service providers. The data for tax preparation purposes may be used by H&R Block to absorb, understand, correlate, and analyze and move it to a certain context, e.g., to suggest the most beneficial tax filing options for a specific situation of a client [4].

The underlying AI-based technology is designed for possible tax implications that are based on personalized reasoning interviews with clients. IBM has a great experience in applying Watson to data-rich settings and using cognitive features to derive insights and support human decision-making process in a better manner. This then provides the basis for efficiency and ease of tax return preparation.

The customer may observe the progress of tax-filing process on their display monitors under the guidance of the Tax Pro. The system provides many options for clients to choose to complete a tax return in an acceptable form. Based on its past experiences, knowledge, and judgment, the Tax Pro evaluates the options and determines the best solution with respect to tax requirements and services [4].

9.5.3 IoT for Geolocation of Tax Field Operations

In the Philippines, the Bureau of Internal Revenue (BIR) is modernized with the Internet of Things (IoT) infrastructure for improving the efficiency of tax collection. BIR is utilizing IoT to enhance the Mobile Revenue Collection System. Real-time data are sent by BIR officers to the central database, which allows geolocation and helps to optimize the working of field tax collectors [3].

Through this technology, BIR is able to extend their databases to include the geolocation data such as where the companies manufacture their goods and deploy services. This happens with a stamp system. BIR requests the cooperation of other enterprises to enhance IoT vision by using such technologies and smart interconnected devices for collecting data from external smart devices such as fuel pumps, sensors, actuators, automatic vending machines, etc., to obtain real-time retail information, that will immediately help to calculate tax returns for clients, more correctly [3].

9.5.4 Drone Technology for Map Location to Reduce Tax Avoidance

Indonesia is one of the world's largest producers of tree products, i.e., rubber, palm oil, coffee, cocoa, and spices. The areas that are remote in this country are difficult to reach and the government cannot provide a satellite or helicopter service because of expense [3].

In 2015, drone technology was used to catch frauds like underreported plantation size or mineral extraction volumes [3]. Drones flew over plantations and took pictures every five seconds. The data collected in this way were transferred onto a location map. Analyses were then performed on the data in terms of the location measurements to verify tax returns from the businesses relating to aforementioned tree products.

9.6 Conclusion

Tax administration and services continue to develop and change. The rise of new technologies and novel business approaches will accelerate the transformation of revenue administration and tax service providers' role in the next decade. Cognitive computing, blockchain technology, artificial intelligence, cloud/edge computing, and ambient intelligence (AmI) in the IoT environment are already being used or explored by these administrations, for the development and deployment of tax-related services.

To completely understand and apply technology developments and innovations in the business environment, tax administration should continue to follow these technology trends closely. Transformational change is not simple and has inherent issues, both for administrations and taxpayers, but advantages outweigh any negatives. However, adoption needs to be proper.

In all business processes, including the tax-related processes, there are many important developments related to AmI. These include project design, implementation, investment planning, data access, regulations, law, and government policy; as well as the use of smart intelligent devices interlinked through the IoT and distributed environments. Benefits are enormous, especially in terms of efficiency, effectiveness, ease of use, and availability of services.

Digitalization brings productivity and financial advantages as well in terms of costs. The OECD BEPS Action Report [14] published in 2015 emphasized the impact of digitalization on the economy and society. Data collection, and therefore the amount of data, is increasing every year. Narrow artificial intelligence (AI) is already organized and growing; and broad AI will arrive soon. Nine of the world's top 20 businesses are now digitalized by market capitalization. The worlds' businesses are providing new consumer experiences.

The challenge is in terms of regulating the tax administration services properly in the digitalized economic world. In the current Fourth Industrial Era, many new technologies and approaches have emerged. In the IoT and AmI environments, the roles of both tax administration and tax service providers are changing fast, in line with development in technologies. Nowadays, most of the tax dues are dealt with by AmI within the IoT, resulting in improvements in tax administration and tax-related services.

References

1. Depatment for Digital, Culture, Media and Sport (2018) Data Protection act 2018 factsheet—intelligence services processing. https://assets.publishing.service.gov.uk/government/uploads/system/uploads/attachment_data/file/711233/2018-05-23_Factsheet_4_-_Intelligence_services_processing.pdf. Accessed 23 May 2018
2. Brazil federal revenue and customs administration (2018) http://idg.receita.fazenda.gov.br/. Accessed 20 April 2018
3. Ernest H (2017) How the internet of things is improving tax services in ASEAN? Accessed 01 April 2018
4. HR Block (2018) https://www.hrblock.com/lp/fy17/hrblock-and-watson.html#video. Accessed 1 April 2018
5. KPMG (2017) Technology in tax embracing the now & thinking the future. www.kpmg.com.au. Accessed 01 April 2018
6. AJCSD Ramos C (2008) 3 ambient intelligence, the next step for artificial intelligence, intelligent systems IEEE
7. HW (2016) Tax notes, Accessed 18 April 2018
8. BI Intelligence (2015) The internet of things 2015: examining how The Iot will affect the world, November, Accessed 20 April 2018
9. Gartner (2017) Gartner newsroom website. https://www.gartner.com/newsroom/id/3598917. Accessed 23 April 2018
10. Yli-Huumo J, Ko D, Choi S, Park S, Smolander K (2016) Where is current research on blockchain technology? a systematic review. PLoS ONE 11(10):e0163477. Accessed 05 May 2018
11. PWC (2017) Digital transformation of tax administration [Çevrimiçi]. www.pwc.com. Accessed 02 May 2018
12. EY (2017) https://betterworkingworld.ey.com/trust/in-a-world-of-3d-printing-how-will-you-be-taxed. Accessed 4 April 2018
13. European Parliament (2017) Tax challenges in the digital economy. http://www.europarl.europa.eu/RegData/etudes/STUD/2016/579002/IPOL_STU(2016)579002_EN.pdf. directorate general internal policies, policy department a: economic and scientific policy. Accessed 02 May 2018
14. OECD (2018) Tax challenges arising from digitalisation—interim report 2018: inclusive framework on BEPS, OECD/G20 base erosion and profit shifting project. Accessed 02 May 2018
15. Çetin Gerger ve G, Bozdoğanoğlu B (2017) Evaluation of the problems regarding taxation of electronic commerce in the context of informal economy and tax evasion, Issues in Public Sector Economics, Peter Lang
16. OECD (2016) Technologies for better tax administration: a practical guide for revenue bodies. OECD Publishing, Paris
17. Legal information institute (2015) https://blog.law.cornell.edu/blog/2015/11/06/artificial-intelligence-and-predicting-tax-evasion/. Accessed 05 May 2018

18. CXCENTRAL (2018) https://cxcentral.com.au/technology/ai/virtual-assistant-to-improve-self-service/. Accessed 10 April 2018
19. ATO (2018) https://www.ato.gov.au/About-ATO/About-us/In-detail/Key-documents/A-better-online-experience. Accessed 20 April 2018
20. HMRC (2018) https://www.gov.uk/government/organisations/hm-revenue-customs. Accessed 10 April 2018
21. NetIQ (2017) Mexico's tax administration service. https://www.netiq.com/docrep/documents/toiseh4eaf/mexicos_tax_administration_service_ss.pdf. Accessed 01 April 2018
22. CIAT (2018) Inter- american center of tax administration. https://www.ciat.org/the-internet-of-things-and-tax-administrations-concepts-challenges-and-opportunities-i/?lang=en#_ftn9. Accessed 12 April 2018
23. OECD (2017) Tax Administration 2017: comparative information on OECD and other advanced and emerging economies, Paris
24. BB Milner C (2017) PWC advanced tax analytics & innovation tax analytics, artificial intelligence and machine learning–level 5. www.pwc.org. Accessed 01 April 2018
25. Rikken O (2017) Blockchain real time tax (is this the future?) https://www.darwinrecruitment.com/blog/2017/08/blockchain-real-time-tax-is-this-the-future. Accessed 01 April 2018
26. DELOITTE (2018) Artificial intelligence–entering the world of tax. https://www2.deloitte.com/global/en/pages/tax/articles/artificial-intelligence-in-tax.html. Accessed 01 April 2018

Chapter 10
Ambient Intelligence in Systems to Support Wellbeing of Drivers

Nova Ahmed, Rahat Jahangir Rony, Md. Tanvir Mushfique,
Md. Majedur Rahman, Nur E. Saba Tahsin, Sarika Azad, Sheikh Raiyan,
Shahed Al Hasan, Syeda Shabnam Khan, Partho Anthony D'Costa
and Saad Azmeen Ur Rahman

Abstract The possibilities of ambient intelligence in the healthcare sector are multifaceted, ranging from supporting physical to mental wellbeing in various ways. Ambient intelligence can play an important role in supporting emotional wellbeing and reducing discomfort. Real-time capability in systems to provide support during discomfort can be useful in scenarios which are traditionally neglected. Absence of concern about wellbeing among commercial vehicle drivers during stressful driving situations may lead to accidents and poor lifestyle. Ambient intelligence can play a role in determining such situations to support the drivers when it is required. The availability of low-cost Internet of Thing (IoT) based components has opened up opportunities in areas where resources are constrained. In the current chapter, the focus is on improving the wellbeing of commercial vehicle drivers in a low-income setting. The chapter focuses on understanding the concepts of discomfort and wellbeing through a detailed qualitative study followed by a possible solution approach to address the ongoing challenges. A low-cost wearable IoT-enabled system along with a long-term analytic support is proposed to improve the wellbeing of drivers using ambient intelligence. The entire system is built up using a connectivity framework. The low-cost IoT device would enable support for discomfort for community who traditionally do not receive such support. Wellbeing of drivers is important for improved driving quality and better traffic management. A system in place to support drivers in real time, named Bap re Bap is presented here in the context of Bangladesh.

Keywords Ambient intelligence · AmI · Resource constraint deployment · IoT
Low-cost IoT · Wellbeing · Drivers · Bangladesh

N. Ahmed (✉) · R. J. Rony · Md. T. Mushfique · Md. M. Rahman · N. E. S. Tahsin · S. Azad
S. Raiyan · S. Al Hasan · S. S. Khan · P. A. D'Costa · S. A. U. Rahman
Department of Electrical and Computer Engineering, North South University,
Dhaka, Bangladesh
e-mail: nova.ahmed@northsouth.edu

© Springer Nature Switzerland AG 2019
Z. Mahmood (ed.), *Guide to Ambient Intelligence in the IoT Environment*, Computer
Communications and Networks, https://doi.org/10.1007/978-3-030-04173-1_10

10.1 Introduction

In the middle of busy traffic, congestion, and other ongoing activities, how fast can we know about a sick person driving a motor vehicle? If that person is in charge of driving a vehicle, there is an urgent requirement to immediately support the person. An incident in an economically less developed country asks for a low-cost approach to explore the situation. The abovementioned scenario takes a developing country approach of asking for the ambient intelligence to be strengthened by ubiquitous computing [1–3]. The ability to generate alerts in an emergency scenario; disseminate the information; and finally, being able to respond to it, calls for deployment of a system with ambient intelligence to meet current requirements.

Ambient intelligence creates the conundrum of elevated support with higher risk of privacy concerns of individuals. A system that is able to support an emergency situation requires adequate measure to ensure device-level and user-level authentication. Existing research to support ambient intelligence looks at privacy concerns, ethical concerns, and deployment challenges [2]. There are numerous visionary research projects considering future deployment of a system to ensure embedded ambient intelligence. However, the proposed research envisions deployment of a system under current scenarios to best support real-time response systems. On one hand, automated intelligent system can support best responses; on the other hand, low-cost IoT (Internet of Things) devices would be able to ensure knowledge of its surroundings which can be applicable even in a resource-constrained setup.

The drivers in the context of Bangladesh face challenging traffic scenarios. When it comes to the traffic system, our challenges are multifaceted. From the traffic viewpoint, we have a unique system that is multimodal (often referred to as heterogeneous); cars, buses, rickshaws, CNG-powered auto-rickshaws, bicycles, and taxis all use the same road with their very diverse speeds and maneuvering capabilities. The drivers of these vehicles also greatly vary in their level of education, and knowledge of traffic rules [4] and roadside safety. A rickshaw-puller, for example, often visits from a village for extra seasonal income and gets started on the road without much training. In many cases, their driving behavior is opportunistic or greedy, as opposed to, cautious or defensive.

Furthermore, traffic rules are frequently violated by both, drivers and pedestrians, making the situation even more complex. For instance, it is a commonplace for pedestrians to walk along the vehicular right-of-way, or to cross in a mid-block section where there are no pedestrian crossings. Similarly, it is not uncommon for drivers to violate lane discipline, overtake aggressively, violate traffic signals, and in extreme cases, drive on the wrong side of the road. Heavy traffic is a prevalent problem in many metropolitan cities; the situation of Dhaka exacerbates due to poor road conditions across the city [5]. The multimodal traffic with unruly vehicle drivers calls for very alert driving due to high probabilities of unanticipated maneuvers in the midst of congested conditions. The combination of these factors puts a driver in a unique position, and we want to portray the driver's experiences, which is crucial for studying and ensuring road safety, as well as reducing driving-induced stress that has a significant impact on health and overall wellbeing.

We have contributed in developing a minimalistic IoT tool that is able to monitor our drivers along with a software application that allows us to take a comprehensive look at the experience. The hardware–software system would enable us to look at our journey in a comprehensive manner, leading to a better lifestyle and wellbeing using real time as well as long-term analysis of sensing data. A person having a rough day must be advised to take some extra rest, relax, so that a new day begins in a better way. We have conducted predevelopment studies and designed and developed our IoT sensing system. We have worked closely with a group of drivers to allow user-centric design that has lead us to design a user-defined intervention tool where the user is in charge of defining intervals that would match their daily routines.

We present an initial deployment of an emergency alert system using a unified communication framework assisting ambient intelligence. The work is designed, deployed, and tested in the context of a developing country, Bangladesh.

The rest of the chapter is organized as follows: the related work is presented followed by a background section illustrating the current traffic conditions of Bangladesh. A qualitative study is conducted in the next sections followed by a proposed system and finally, a conclusions section is presented.

10.2 Related Work

It is interesting to see how our work is related to various research efforts from different dimensions. The related work section introduces concepts of ambient intelligence, driving experience, i.e., the comprehensive encounter with the driving task and the resultant impression on the driver, we have looked into work done in similar realms such as driver behavior and mood.

10.2.1 Ambient Intelligence

Ambient intelligence is defined to have ubiquitous computing, ubiquitous communication, and intelligence interface [5, 6].

Bohn et al. [2] raise important topics of ambient intelligence from social, economic, and ethical concerns. These concerns are gaining importance as amount of sensor-related data and capabilities to infer information from data is becoming easier. The ambient economy of interest discussed by Bohn et al. focuses mainly on technologically advanced nations. However, this technology can open up great opportunities for other nations as explored in our work. Bohn and team discuss about privacy and ethical concerns of ambient intelligence. We have considered IoT device authentication and secured data transfer protocols to protect personalized context-sensitive data.

Ambient intelligence is used to provide personalized telemedicine support as discussed by Rubel et al. [7]. It refers to the EPI–MEDICS project. A wider range of discussions from healthcare perspective is presented by Acampora et al. [1]. There is a focus on the perception of intrusive monitoring as data collecting points are becoming pervasive [8, 9].

Emotional wellbeing support system using ambient intelligence is being studied by numerous researchers. There are concepts of wearable devices or environmental sensors to support people for emotional wellbeing [10]. Our current focus is on emotional wellbeing using wearable devices along with other support system elements. We particularly emphasize on a low-cost solution to reach out to economically less advanced communities.

10.2.2   Driving Experience and Behavior

Driving experience and behavior is a comprehensive topic covering various aspects of drivers which are presented here.

Driving and Related Factors

Driving behavior is an inherently complex process, with driving decisions affected by various factors, including network topography (e.g., type of road, number of lanes, curvature, gradient, etc.), traffic conditions (e.g., average speed, density, etc.), surrounding conditions (e.g., position and speed of the adjacent vehicles), path-plan of the driver (e.g., location of the next exit), features of the vehicle (e.g., acceleration and deceleration capabilities) and characteristics of the driver (e.g., age, experience, aggressiveness, impatience, stress level, and numerous others). Driving behavior has been investigated and modeled over decades, primarily in the context of microscopic simulation (e.g., [11, 12]), emission analysis (e.g., [4, 13]), and safety research (e.g., [14–16]). However, most of these researches have focused on distinctly measurable traffic attributes and driver characteristics and more subtle factors (like effect of attention level, mood, stress levels, etc.), have remained a relatively less researched area.

Driving and Attention

A driver's attention level and the type of attention he or she needs to project at a given time could be a very close indicator of the driving experience of the driver at that point of time, and ultimately plays a major role in driving safety. A huge array of work exists on the driver's attention.

Research presented in [17] breaks down the types of attention-related disturbances, which could lead to bad driving experience, into three categories: (1) inattention, (2) distraction, and (3) competition for attention. Inattention denotes the instance when the driver is focusing on thoughts in his/her mind unrelated to the driving task. Competition for attention corresponds to the instance when cognitive abilities are managing two competing driving-related tasks, such as taking a corner while trying to avoid an oncoming vehicle. Distraction points to elements outside driving that are actively hampering the focus on driving task. Driver distraction is itself a highly studied area of research with numerous subtopics.

Research conducted by Iqbal et al. [5, 18] provides an in-depth study on the correlation of phone calls and driving distraction. It is evident from their study that conversations that require cognitive involvement reduces the driver's concentration significantly, which is in line with findings of other researchers [19, 20]. Outside factors like roadside advertisement can be a source of distraction for drivers as well [21].

Work focused on driver attention of particular demographic group exists too; [22] discusses attention-related problems in younger drivers, their risk-taking attitude, and rash personalities, while studies such as [23–25] seek to identify factors affecting decreasing ability to maintain attention in older drivers. The direct and indirect effect of driver's attention has also been studied by Salvucci [26] and have been found to contribute to as much as 25% of unsafe driving [27]. Measuring driver attention is thus a crucial element in the study of driver experiences.

Driving and Mood

In terms of the driver's mood, mood states are found to have a significant effect on driving behavior, particularly on risk perception as well as risk attitude [10, 28, 29]. This has prompted research on developing Road Frustration Indices e.g., [30] and more recently [31]. Interestingly, it has also been observed that the mood of the driver can be positively transformed by interventions such as music. Wider research on psychology also demonstrates the potential to manipulate moods through designed interventions e.g., [32, 33].

Driving and Experience

As already mentioned, the focus of this study goes beyond the temporary mental states. Rather, we focus on the overall experience which is an aggregation of day-to-day encounters and hence evolves over a longer time span. This definition is different from previous researches exploring effect of driving experience on driving behavior where driving experience has been defined by the hours/months/years the drivers have been engaged in driving (e.g., [32, 34, 35]) and often the number of accidents they have been involved in the past (e.g., [35]), rather than the overall quality associated with the hours driven.

The effect of driving stress on day-to-day activities and vice versa have been investigated by Gulian et al. [36] using a diary survey where both have been found to be strongly correlated. But as acknowledged in the research, the diary-based survey is prone to data errors like missed incidents, missed or incorrect details, etc. Also, all these studies have been conducted in the context of developed countries where the triggers for frustration that shapes up the driving experience are very different from that of the unique situations in Dhaka as mentioned in the previous section. On the other hand, there is a previous study particularly concerned about the effect of traffic conditions on driving rule violation probabilities [29] in the context of Dhaka but that is based on stated preference data which is likely to have experimental bias.

There has been some interesting research focusing on various factors of young drivers that may correlate to accidents. A study conducted by Clarke et al. [27] show that age, experience, and time of the day have a great influence on young drivers (aged 17–20). The amount of injury is high among young male drivers of the UK showing an average of 440 injury accidents per 100 km driven. The behavior varies among male and female drivers. A detailed meta-analysis conducted by Caired et al. [37] shows existing studies considering cell phone conversations and driving performances. Similar meta-analysis is conducted from the perspective of texting while driving [38]. The studies presented here consider developed country perspectives.

Drivers and Resource Constraint Setup

The challenge of our research is mainly focused on measuring the parameters of interest in the least intrusive yet comprehensive manner in the midst of extreme resource constraints. We have therefore conducted real-time analysis using custom sensors as real-time data collection and dissemination is expensive due to the data cost of internet in Bangladesh. We have used mobile phone based video sensing for post-driving analysis. To get more detailed insights, we have done qualitative study with various user groups along with sensor-based study. We must keep in mind about the effect of user bias through the interviews, especially when we are discussing issues with low literacy population as discussed by Dell [39]. We have conducted semi-structured or unstructured long (an hour or more) interviews to address such bias.

Zafiroglu et al. [31] discuss recordkeeping studies along with interviews in Brazil, China, and Germany and gives us perspective of developing country challenges in some of the regions of interest. The social problems of mugging, carjacking, and car theft were a reality in Brazil, as discussed in the paper.

On the broader aspect, technology intervention design requires a closer look at ongoing challenges and opportunities in the context of a developing country. Developing and underdeveloped countries suffer from adequate technology infrastructure, support, and acceptance in one hand as sudden intervention is often a reason behind technology-based abuse in such countries [40]. On the other hand, technology is often considered as a weapon to fight all the existing problems in a utopian view [41], which brings in new challenges on its own merit. Postcolonial computing study in the context of Bangladesh is comparatively recent exploring the technology-imposed challenges in different aspects [40, 42–44]—mostly in an urban setup. Our work focusing on technology intervention must closely consider ongoing concerns relating to ongoing concerns.

A driver's experience is impacted by its surrounding environment greatly. The following section presents the snapshot of the current traffic related scenarios of Bangladesh along with a case study illustrating social implications of such conditions.

10.3 Background to the Current Study

An illustration of the current traffic system and road conditions is required to analyze the challenges faced by drivers that impact their wellbeing. The various challenges of the traffic in the metropolitan city of Dhaka in Bangladesh is presented in Fig. 10.1 where Fig. 10.1a shows multimodal traffic where various vehicles share the same road; Fig. 10.1b shows the scenario of traffic congestion; and the contrast of roads with acceptable conditions and uneven roads are presented in Fig. 10.1c, d respectively.

10.3.1 Multimodal Traffic and Road Conditions

Majority of roads (with a few exceptions) allow vehicles of various speeds to coexist on the same road which is referred to as multimodal traffic. There are public buses (50 seats), CNG-driven auto-rickshaw, private cars, microbuses, trucks (after 9 pm), motorcycles, and often pedestrians sharing the same road. The multimodal traffic has a large impact on the traffic system as each vehicle has its own driving

Fig. 10.1 Road Conditions: (1a) Congested road on one lane (1b) Multimodal traffic and heavy congestion (1c) Clear road condition (1d) Broken road

behavior (apart from its different speed). The public buses are unruly in nature; often compete with each other to attract passengers. It is the most rebellious vehicle that takes a greedy approach in driving, changing lanes whenever one seems lighter. This behavior imposes a heavy burden on the other lightweight vehicles.

On top of our heterogeneous traffic, we have challenging road conditions. Only 7.5% of the total land area in Dhaka is dedicated for roads [45] whereas in well-functioning megacities it is typically higher than 15%. On top of that, 30% of this 7.5% road is also occupied by the street vendors and makeshift shops. The roads are not very well constructed and very often have poor riding quality.

10.3.2 Traffic Conditions

Due to the issues mentioned above, there is often severe traffic congestion in Dhaka, resulting in a wastage of 3.2 million working hours [46]. The underlying reasons behind and how the drivers behave at a time of congestion make the situation here quite unique and different from other places. For example, to reduce travel time and to minimize driving efforts, both the motorized and nonmotorized vehicles tend to violate the traffic rules and even resort to driving through the wrong side of the roads.

The roads are heavily congested during certain periods of the day considering the office going and returning traffic at certain times. The number of rickshaws significantly increase during the school drop off and pick up times. The congested traffic has a direct and indirect impact on wellbeing as mentioned by several participants that is shared later. The traffic condition is challenging to describe as there are many real-time scenarios that require immediate attention. For example, the situation where an ambulance is stuck in the traffic congestion which is constantly playing a siren and the relatives of the patient are out anxiously trying to make way for the vehicle or a scenario when there is a sudden aggression among drivers where one vehicle has hit another and many others like this.

10.3.3 A Case Study

A particular incident is presented here to clarify the current condition of our unruly traffic and inefficient law enforcement system. There are existing rules to enforce traffic rules but the implementation mechanism is weak as it is in many economically less developed countries.

The very first court case that charged a company owned vehicle, that hit and ran a journalist M H Montu [13, 47, 48], continued for more than 24 years. It was a Pepsi Cola truck that hit the journalist in 1989; the truck was coming from *the wrong side*. The case lingered on as the company denied its responsibilities, asking the truck-driver to be punished only. After a long-running case, the company was charged on April 13th, 2016 [49]. This incident shows a clear picture of the ineffective law enforcement system. A journalist's family had to struggle for 24 years which could be worse for many others. The entire process was economically and emotionally draining (one of the coauthors of this contribution is closely related to this incident personally) and at times seemed impossible to carry on for many reasons such as long-time frame, lack of evidence, etc.

The car drivers, particularly, commercial car drivers are of the least concern from the authorities' perspective. The adequate training, monitoring and lack of ability to deal with emotional conditions (that may occur after incidents like a major accident) are not practiced in Bangladesh. We hope to draw attention to authorities through our work.

Current scenario explores how an intelligent system in place can usefully provide real-time responses in case of an immediate need.

10.4 Quantitative Study Exploring Drivers Experiences

We have initially focused on understanding driver's experience to define the solution space through a quantitative study approach. The following subsections contain details on the study setup and findings from the study including demography and lifestyle of drivers, traffic-induced discomfort impacting drivers' wellbeing and driving, and then analysis of driver's experiences.

10.4.1 Study Setup

The goal of this study is to complement the previous studies concerning drivers. Here, we are concerned in finding out more about how the drivers feel in terms of wellbeing and are interested in looking at the possible impacts it may have caused in terms of driving behavior. Our updated qualitative study involved 88 detailed interviews (10.2% of participants did not want to share their demographic data), each involving 40 questions which took around one hour of time to complete each questionnaire. Our analysis and discussion of the previous study have been used as a baseline to design this study. One major change we considered was regarding the persons who have taken the interviews. This time male volunteers were chosen to interview drivers rather than the female researcher as there was concern that the drivers might not share their true experiences (especially if these are negative) with a female personnel along with making sure that none of the drivers were previously known to the persons taking interviews. We have chosen an entirely different set of drivers this time focusing mainly on commercial drivers who drive to earn their livings of off cars and CNG-driven taxi only. We have spoken to drivers who were waiting on particular parking spots at two supermarket parking places, two hospital parking places, two educational institute parking places, and parking places of apartment buildings where many drivers wait in between their duty. It must be noted that all our drivers were male, women drivers in commercial activities are still very limited in Bangladesh. We have offered the drivers a choice of a mobile phone card (equivalent to BDT 100 TK) or a monetary incentive of BDT 100 after each interview. It must be noted that all the participants preferred money over the phone cards.

We have used several different sets of questions. The first set of questions concerned demography along with two lifestyle questions where we have asked whether the driver lives with his family or the number of people the driver shares his room with. These lifestyle parameters could indicate extra burden on individuals. The second set of questions were regarding stress and designed around the first set of interviews we had conducted. This set asked the drivers to rate their level of annoyance of previously mentioned events on a scale of 1–5—the questions on annoyance referred to congestion; sudden presence of another vehicle or person in front of the driver; commands from passengers and hunger. It also included the question about whether the vehicle had Air Conditioning (Air Conditioners are not very common in commercial vehicles). We added an indirect question regarding the situations when the drivers thought it was OK to break the traffic rules, or when they were forced to break them. We ended this set of questions with a scenario generation question to find out more about the personality of the driver to determine whether or not they got agitated easily. The final set of questions were formed around the six wellbeing parameters: autonomy, positive relationship with others, personal growth, purpose in life, environmental mastery, and self-acceptance with respect to driving. All the questions were asked in Bengali and the interview questionnaires were filled up by the persons taking the interview using pen and paper. The findings of the survey are presented in subsequent discussions.

10.4.2 Demography and Lifestyle of Drivers

It is observed that the majority of the commercial drivers we talked to fall into the age group of 18–30 years as shown in Fig. 10.2. It must be mentioned that some of the drivers looked younger than they have mentioned and some appeared older than their reported ages and some have mentioned that they are not aware of their age although there is an age associated with each driver's driving license. It shows the lack of awareness in the lower income community.

The working hours of a sample of 88 drivers (P1–P88) show long hours to remain on duty (including the intervals to take rest) as shown in Fig. 10.3. Some columns remain zero, as 10.2% of participants did not feel comfortable sharing their demography-related data to the unknown interviewers. This represents an average of 11 h of duty time with 2 h of average downtime. The long hours often include overtimes where drivers earn extra money for their living.

The basic lifestyle shows the struggle where 35.22% of the drivers do not live with their families and some did not prefer to share this information having 54.54% of the drivers living along with family members. The highest number of members

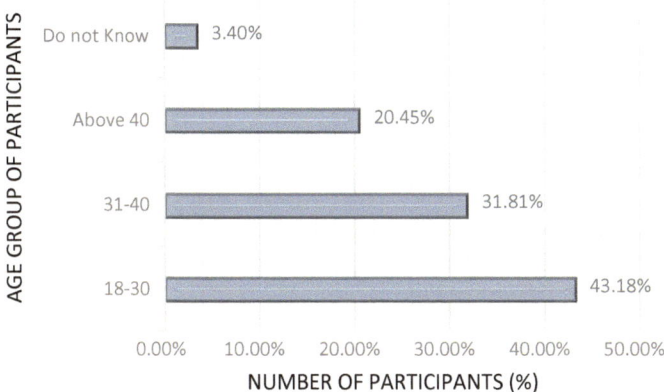

Fig. 10.2 Age groups of participating drivers

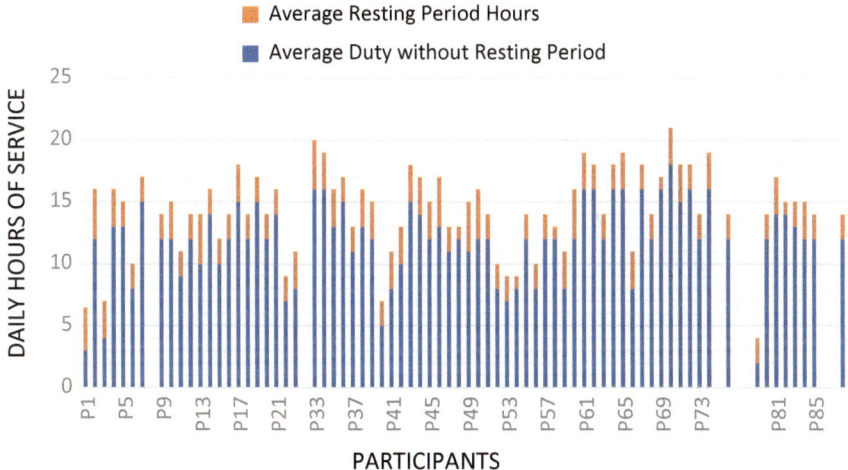

Fig. 10.3 Duty hours of participating drivers

sharing the same room is 17 which shows the level of struggle present in the life of these citizens apart from long hours on driving duties.

10.4.3 *Traffic-Induced Discomfort Impacting Drivers' Wellbeing and Driving*

The source of discomfort considers previously discussed elements such as traffic congestion, sudden presence of a person or vehicle in front of the driving vehicle, hunger, advice on how to drive, etc. However, we have considered two open-ended questions along with the mentioned ones.

One open-ended question was regarding any particular incidents or situations that make the drivers feel annoyed and uncomfortable, referring to the questions discussed. It is evident that the presence of air conditioning system is desirable in the hot and humid weather of Bangladesh. On the other hand, traffic congestion/jam, hunger, nosy passengers, and owner's advice/misbehavior are highly responsible for drivers' discomfort which is illustrated in Fig. 10.4.

To determine results, we divided the drivers into three groups according to their feelings of wellbeing, rating as below:

- Highest level of wellbeing was considered felt when drivers were satisfied in all of the parameters, except two
- Medium level of wellbeing was considered felt when drivers were satisfied in all parameters considered, but four
- Lowest level (regular) of wellbeing was considered felt when drivers were satisfied only with respect to less than half of the parameters considered.

Few drivers shared their worst experiences and so we identified their wellbeing as negative. Refer to Fig. 10.5.

Another open-ended question concerned around breaking of traffic rules. We indirectly asked the drivers about conditions when they felt they had to break the traffic rules. Here, we found some interesting results. A number of drivers shared that they had to break rules in emergency situations, some said they followed other drivers who broke rules for whatever reasons, some other drivers said that sometimes this was due to car/cab owners dictates, some mentioned empty roads or road without police officers as reasons behind breaking rules. From the data, we could also see that there were people who break the rules as a result of frustration and anger relating to their discomfort and lack of wellbeing as shown in Fig. 10.6a, b respectively.

Fig. 10.4 Causes of drivers' discomfort

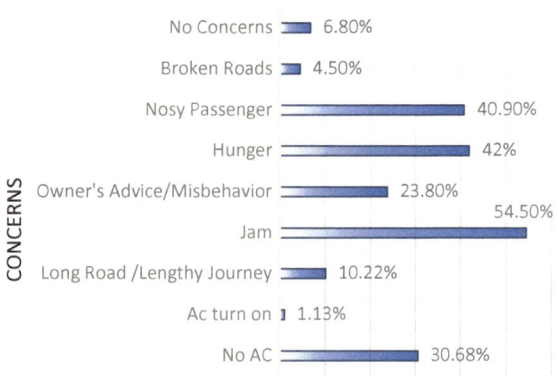

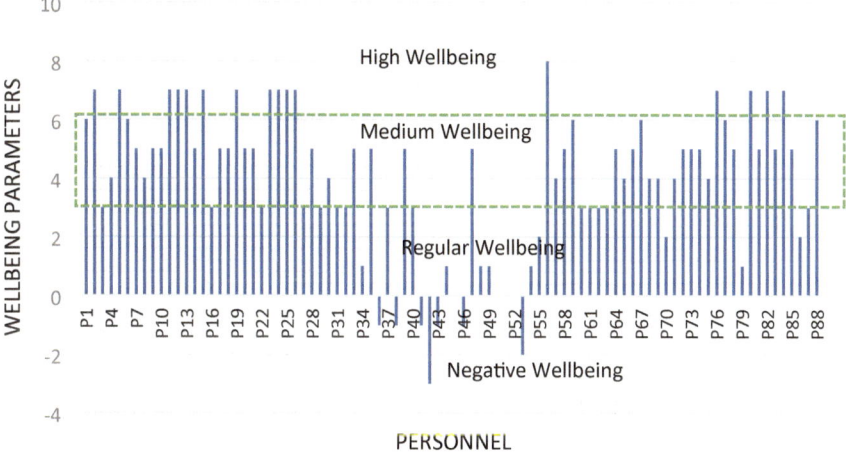

Fig. 10.5 Categories of drivers' wellbeing

10.4.4 Analyzing Drivers' Experience

We have used learning algorithms on our collected quantitative data to find out factors related to breaking traffic regulations. To find out how many groups or categories the drivers should be divided into, we used the k-means clustering algorithm on our scatter plots. The results for each factor are given in the following images, with k values ranging from 2 to 5, which means we tested via our graphs with 2, 3, 4 and 5 groups. We used Google's open-sourced Tensor Flow to run the clustering algorithm as well as designed our own machine learning algorithms.

We have looked at the relationship of wellbeing and likelihood to break law as can be seen in Fig. 10.7a, b with four different cluster sizes ranging from 2–3 and 4–5, respectively. The various cluster sizes allow us to understand how wellbeing is related to likeliness to break the law. The wellbeing cluster is interesting as the cluster size of 2 creates two regions based on wellbeing but both are equally likely to break the rules. The increased number of clusters create zones based on wellbeing related to likeliness to break rules.

The behavior pattern of aggressive drivers is studied against likeliness to break rules as can be seen in Fig. 10.8a, b based on cluster sizes 2–3 and 4–5, respectively. The cluster size of 2 shows that more aggressive drivers are less likely to break the rules. However, the larger number of clusters creates separate zones identifying various behaviors.

We have used the same clustering algorithm on many other pairs of behaviors. Apart from that, we co-related the discomfort, aggressiveness, and wellbeing. We noted that discomfort positively relates to aggressiveness, and wellbeing negatively related to aggressiveness which is shown in Fig. 10.9a, b respectively.

Fig. 10.6 **a** Likelihood to break traffic laws: relating to Discomfort. **b** Likelihood to break traffic laws: relating to Wellbeing

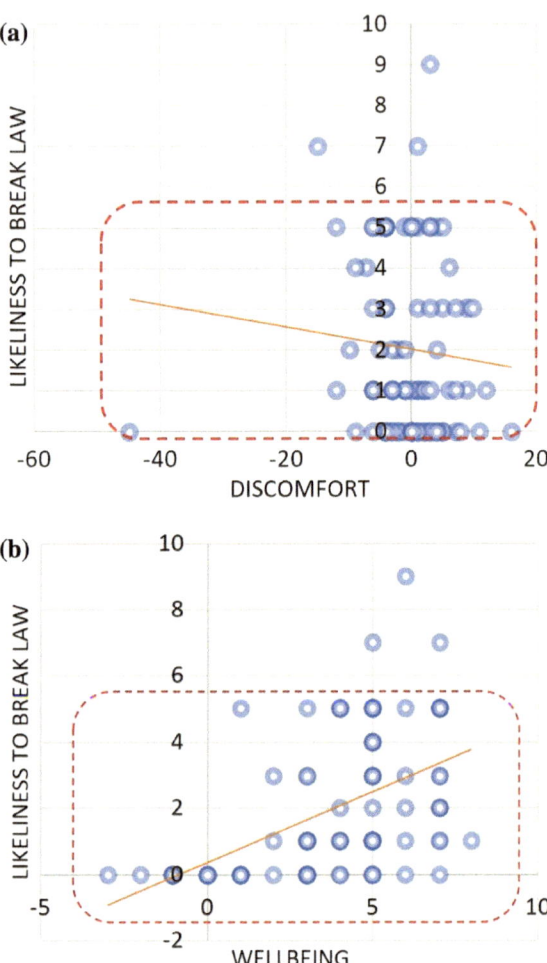

The studies on drivers' experiences present the current status of drivers which shows that the struggle of drivers come from various external and internal factors. There are certain conditions that are not avoidable (e.g., congestion) and drivers must find ways to comfort themselves in such situations knowing how they might be affected otherwise. We hope to work on technology-based intervention that may provide users with better management of discomfort.

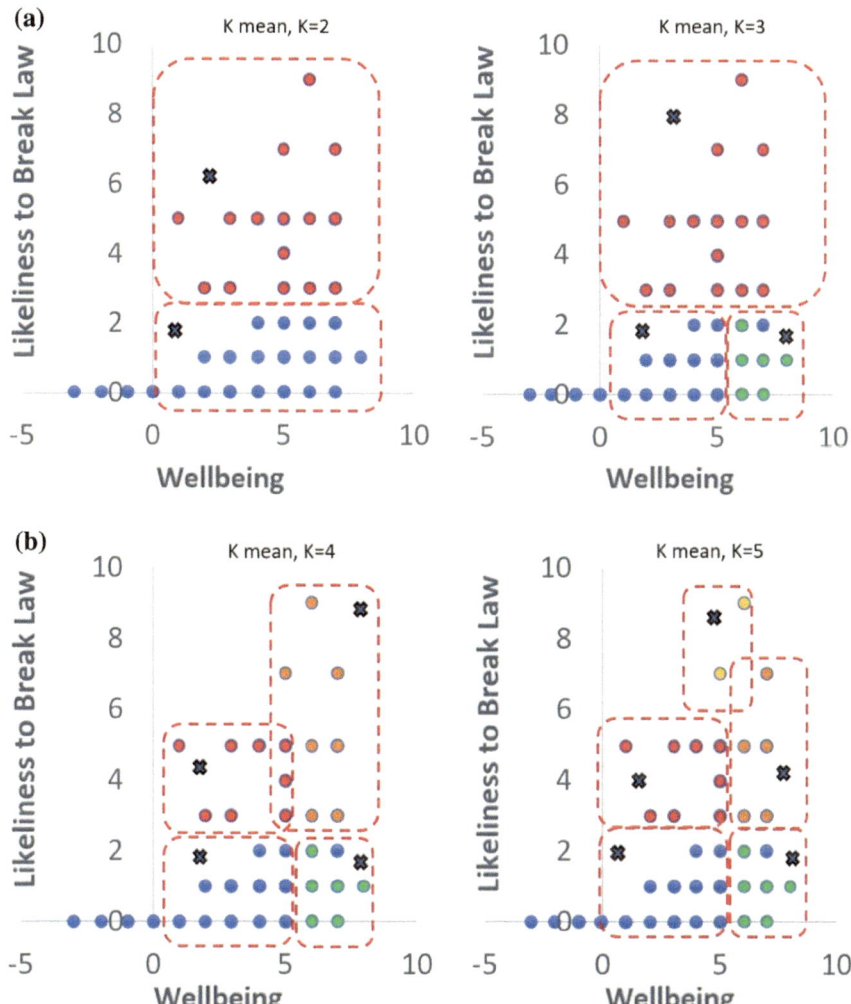

Fig. 10.7 a Factors that impact drivers' wellbeing (K mean = 2 and 3). **b** Factors that impact drivers' wellbeing (K mean = 4 and 5)

10.5   Proposed System Architecture of 'Bap re Bap'

The proposed architecture of our system, called Bap re Bap, is presented in Fig. 10.10. Bap re Bap system comprises four major components along with the wearable sensing devices. We now discuss these briefly in the following sections.

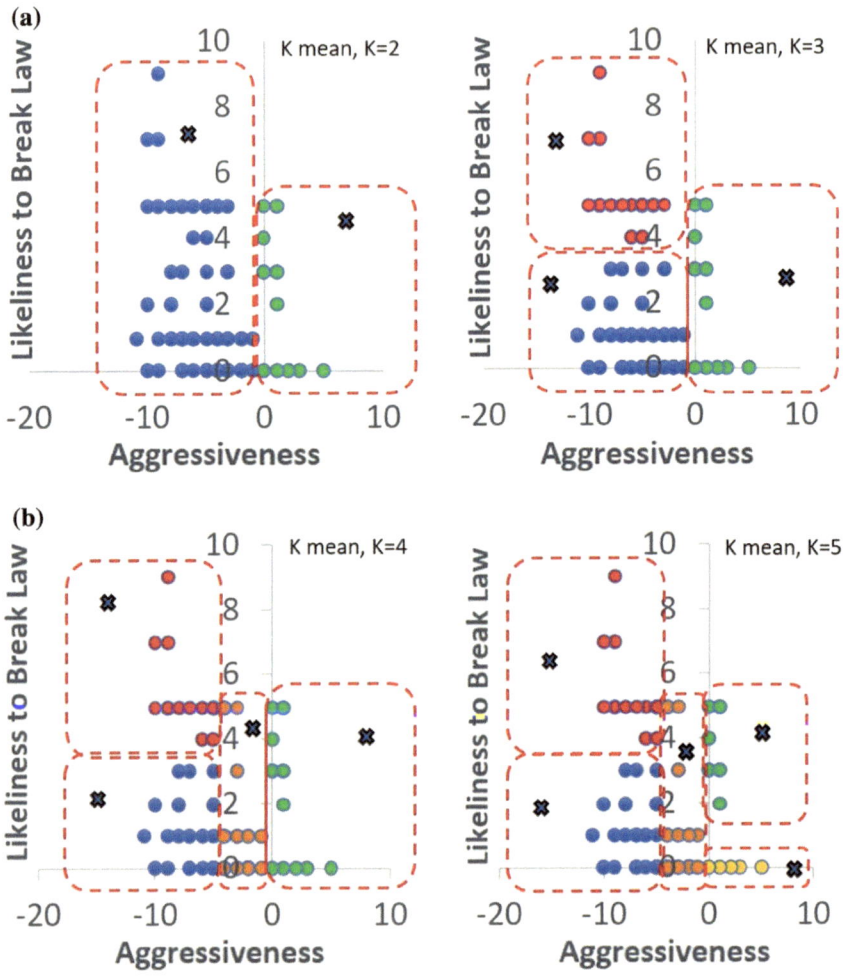

Fig. 10.8 **a** Factors that contribute to traffic regulations violations (K mean = 2 and 3). **b** Factors that contribute to traffic regulations violations (K mean = 4 and 5)

10.5.1 Sensor Data Collection Unit

Sensor data collection unit is in charge of collecting data from sensor devices which can be a commercial off the shelf devices or custom build sensing device. The only requirement we emphasize on is to have 'time-stamped data' so that various inputs can be aligned, synchronized and processed properly.

We discuss our custom designed low-cost sensor data collection device. It is a simple sensing mechanism that collects pulse (heart rate) information with a timestamp and locally processes the data and sends it to the nearby mobile device

Fig. 10.9 a Correlation of discomfort with respect to aggressiveness. **b** Correlation of wellbeing with respect to aggressiveness

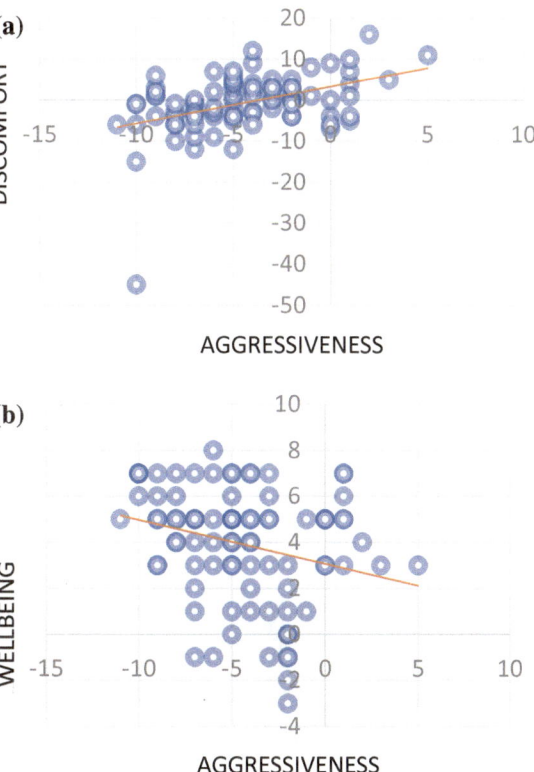

periodically. The frequency of data update can be custom changed as we have designed the components from low level.

IoT Wearable Sensing Module

We developed a low-cost IoT device to detect the pulse (i.e., heart rate) in real time and store the measurements for post-processing. The device is depicted in Fig. 10.11. This low-cost system sensor follows the basic mechanism of the heart. When our heart pumps, oxy–hemoglobin blood flows throughout the human body and de–oxy–hemoglobin blood flows from the body toward the heart. This system detects the reflectivity changes in our blood vessels and gives the real-time data of the pulse. Pulse can be obtained from several points of the body such as fingers, ears, etc. We have considered pulse information extraction from the ear to minimize driver distraction. Figure 10.11 is our updated version, but initially, we started with a large prototype having several components. This system contains an MCU and a pulse sensor where the MCU-centric system uses an Arduino open-source prototyping platform. The miniaturized pulse sensor deals with a phototransistor and highly sensitive photodetector, and it is set to the ear with 120 cm long with an amplifier box via a 3.5 mm jack. This amplifier unit (inset, Fig. 10.11) amplifies the detected

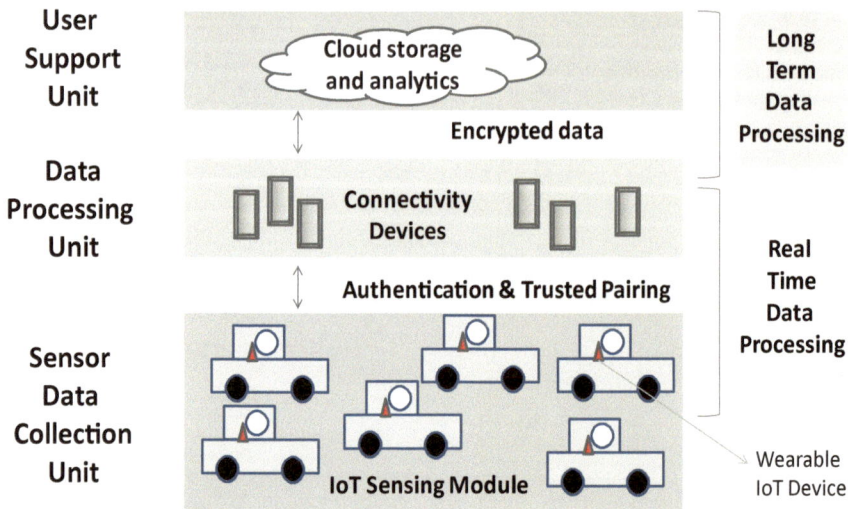

Fig. 10.10 Proposed system architecture

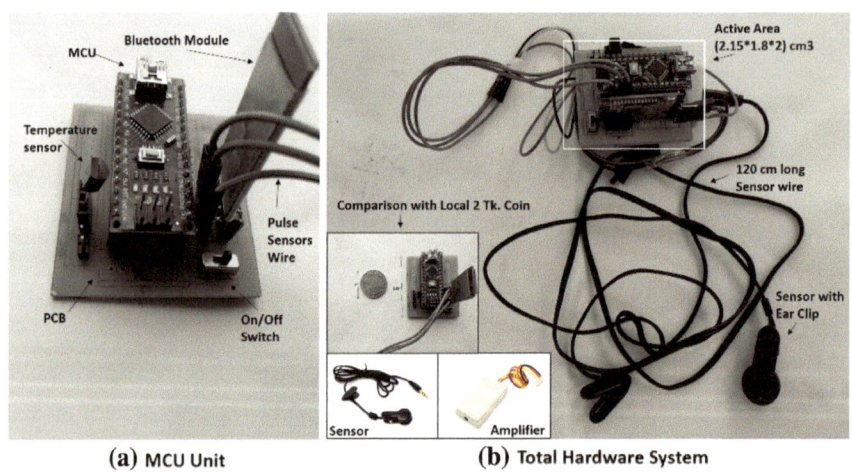

(a) MCU Unit   (b) Total Hardware System

Fig. 10.11 Low-cost IoT wearable device (11a) and MCU unit (11b)

pulse and sends the data to MCU. A temperature sensor is added with the system, to measure the temperature of the environment because temperature has a significant impact on driver's wellbeing, as we have already mentioned in our study. All the measured data is sent to a smartphone application via Bluetooth communication, using HC-06 Bluetooth module. So, the overall system is small enough that, it is possible to mount on a 7.74 cm^3 small box, as shown in the inset in Fig. 10.11. In another inset in Fig. 10.11, we have compared the device size to that of a Tk 2 coin.

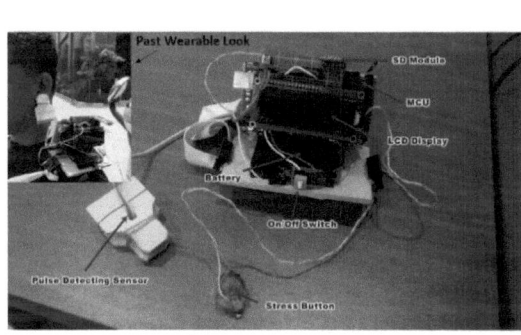

(a) Initial Prototype

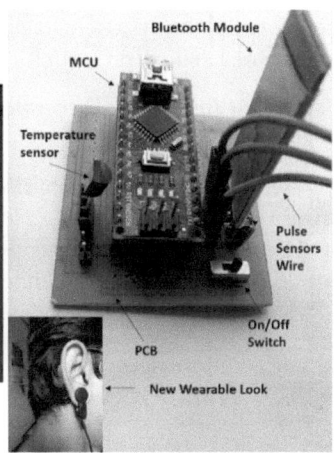

(b) Updated Prototype

Fig. 10.12 Prototype: initial version versus updated

The initial device prototype was significantly large, and it consisted of an MCU (Arduino Platform), a pulse sensor which was bright green (515~nm InGaN Sapphire), LED and a phototransistor sensitive to wavelengths of 800–1000~nm. A 16 × 2 LCD showed the real-time pulse and an SD card stored the raw temporal data for post-analysis. We also included a push button, referred to as stress button, for the user to voluntarily press when they felt anxiety. This prototype consumed a lot of power as it had many components. But now, we minimized the device considerably and also reduced the power consumption and weight, and rest of the operations are now done by the smartphone app.

Initial prototype costed around $27, which was pretty high compared to the updated system that costs around $17. We have reduced the sensor cost and size of the basic sensing unit by printing it to a basic PCB where the active area is 2.15 cm * 1.8 cm. The minimalistic devices price tag is really low, and we ended up having seven devices below the cost of one commercial device.

There are commercial similar devices available that are of much higher cost. An internationally available device (e.g., Fitbit) that costs around $100 is almost 5 times more expensive than our custom-built system. The unavailability of international transactions, and general availability of these products in the local market along with foreign tax make the price even higher in Bangladesh.

We also successfully changed our test run environment. Initial prototype was comparatively big and it was set to drivers' shoulder which made the test run situation seem creepy, we were escorted by the traffic police three times as they were concerned about the huge device hanging to the shoulder (inset Fig. 10.12a). Now, the new system gets connected to the driver's ear only by the sensors 120 cm long connecting wire to the main device, and the main device is not seen from the outside; hence it seems like drivers are wearing an earphone on one ear (inset Fig. 10.12b).

Currently, the system is using Firebase Database supported by Google which is open source and supports multiple platforms.

User Input Regarding Discomfort

In the initial implementation, we had a stress button, as mentioned above, where the users could report his/her discomfort at any time, such as feeling stressed while driving, for whatever reasons. However, the stress button turned out to be underused by the drivers as no one remembered to press it during a situation of turmoil. Only one driver pressed it twice deliberately during traffic congestion.

10.5.2 Data Processing Unit

This unit is responsible for data cleanup (e.g., elimination of duplicate, sudden spikes in data input, etc.) and real-time data analysis. We have used a simple algorithm for window-based learning in real time.

In our current implementation, the data processing unit is coupled with data analytic unit. Our current system involves including sensor parameters from the mobile phone such as temperature (environmental temperature), GPS location and accelerometer information giving a current snapshot of the user along with how it is related to the environment. The sensing system sends periodic data to the real-time analytic component allowing aggregated information available for decision making.

10.5.3 Intelligent User Support Unit

The user support system consists of two software components—one for short-term real-time decision making and the other for longer term decision making. We also have two relevant analytic systems.

Short-Term Analytic System

This component can suggest the user to take a short break, listen to a song or look out of the vehicle window analyzing and experiencing simple clues such as cars stuck in congestion, a sudden bump or hard break. We use our simple real-time stream processing algorithm after data clean up and synchronization step and work on data analytic.

We have developed a mobile phone application connected via Bluetooth to our hardware component along with a web-based system. The mobile phone application aims to represent real-time data while the web-based system is used for long-term data analysis.

The mobile phone based application, as can be seen in Fig. 10.13, is able to keep detailed information regarding time (time of the day, day of the week and whether it is a special holiday or not) in the mobile phone. It also offers a list of songs to play

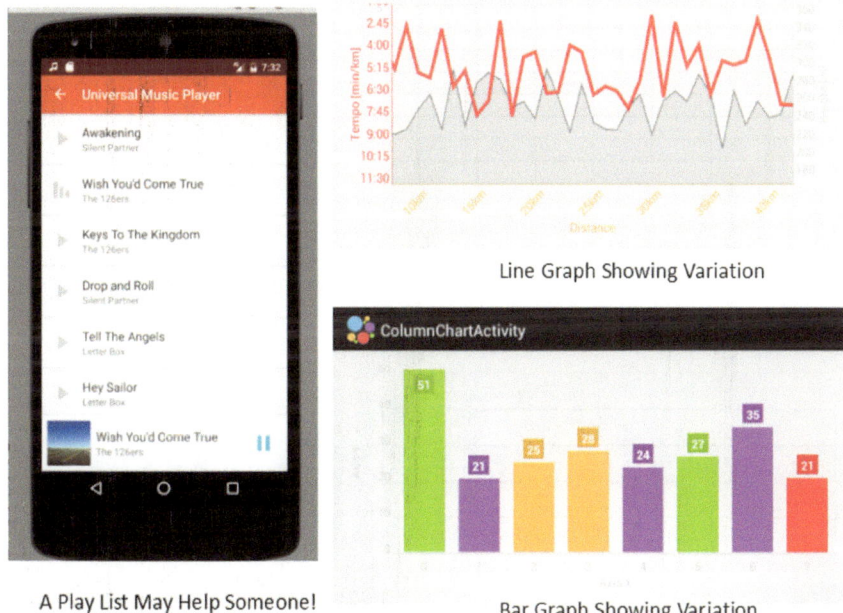

Line Graph Showing Variation

A Play List May Help Someone!   Bar Graph Showing Variation

Fig. 10.13 Mobile phone application to show short-term real-time indicators

based on users predefined playlist. This feature of songs was strongly suggested by the student volunteers and it was received with positive feedback.

Long-Term Analytic System

The long-term analytic system can be compared to a standard healthcare analytic system that is able to monitor health-related information. This is an independent web-based tool with a database. This is a proof of a concept system that is able to serve only educated drivers pool, unlike general commercial drivers.

The current system uses several different representations of visualization and was usable on mobile devices as well. The high internet cost in Bangladesh was a reason for many of the commercial drivers to prefer an application as a web-based program.

Our initial development included several different visual ways to represent daily indicators of how the users/drivers felt throughout the day. We used generic graphs (line and bar), color representations (red, green, and yellow), and buttons that look similar to emoji (smile, indifferent, and sad) to show the daily summary. Two graph representations are shown in Fig. 10.14. The detailed developmental information is archived as a project report at North South University.

The information is also provided in the form of a comprehensive report that provides the health status along with suggestions for the users/drivers. The set of suggestions are in fact user-defined, as stress release options vary from person to person.

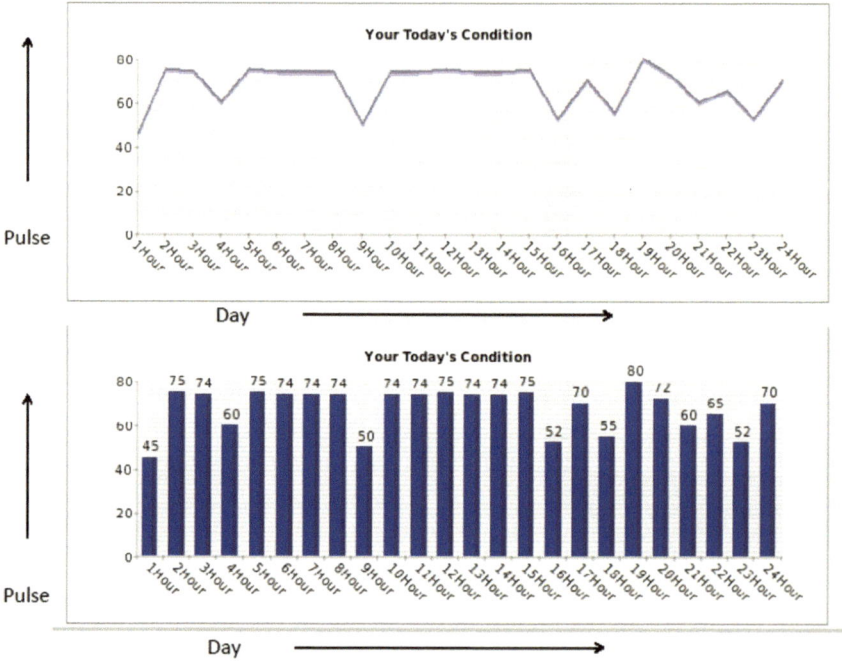

Fig. 10.14 Graphical representations of user experience: line graph and bar chart

10.6 System Evaluation

The evaluation of Bap re Bap system is presented in this section. We have covered, in some detail, the Internet of things (IoT) sensor accuracy along with field evaluation.

10.6.1 IoT Sensor Accuracy Compared to Commercial Sensors

We have conducted a series of experiments to find out the accuracy of our pulse sensor vs the accuracy of a commercially available pulse sensor and manual pulse measurement (using the fingers over blood vessels to measure the pulse). The experiments took place over a period of a week, and we studied persons of various ages, genders, and occupation. Two cases are presented here.

Figure 10.15a shows the pulse readings of a student (male, age 20) through his various activities at the university campus, from morning and ending at night. Figure 10.15b shows the pulse data of a professional commercial driver (male, age 41) over a day. The sensor data deviations clearly show how pulse varies during driv-

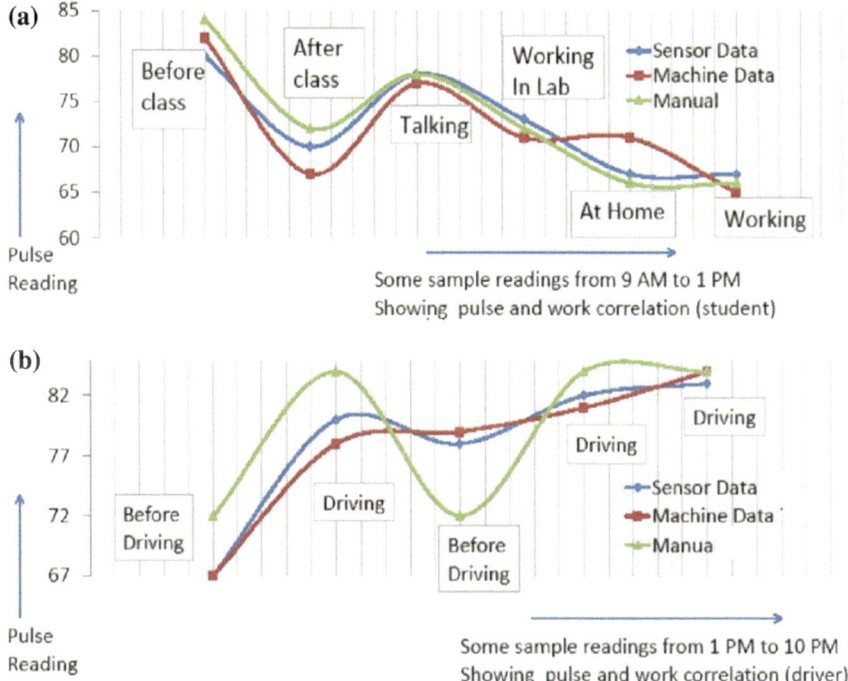

Fig. 10.15 **a** Pulse sensor data (discrete values) from Bap re Bap device compared with manual reading and with a commercial device. Pulse readings taken of student. **b** Pulse sensor data (discrete values) from Bap re Bap device compared with manual reading and with a commercial device. Pulse readings taken of driver

ing compared to the steady state of resting, labeled 'Before Driving' in the figure. It can be noted that the manual measurement varied significantly. This could be due to measurement errors, particularly as it was difficult to count pulses while the subject was driving. It must be noted that in all cases, the standard deviation of error for custom sensor was less than 3% compared to the industrial machine data. Figure 10.16a–e represents the standard deviation of errors of numerous activities data for five different subjects.

Our initial testing revealed the device to be reliable, with pulses easy to measure using the available low-cost hardware. We have therefore designed our wearable system using the above pulse sensing components, with data-logging mechanism.

10.6.2   Field Evaluation of Wearable Sensors

We have conducted ten short driving trips that took one to two hours of time per trip, covering various routes to conduct our testing. Four drivers participated in ten test

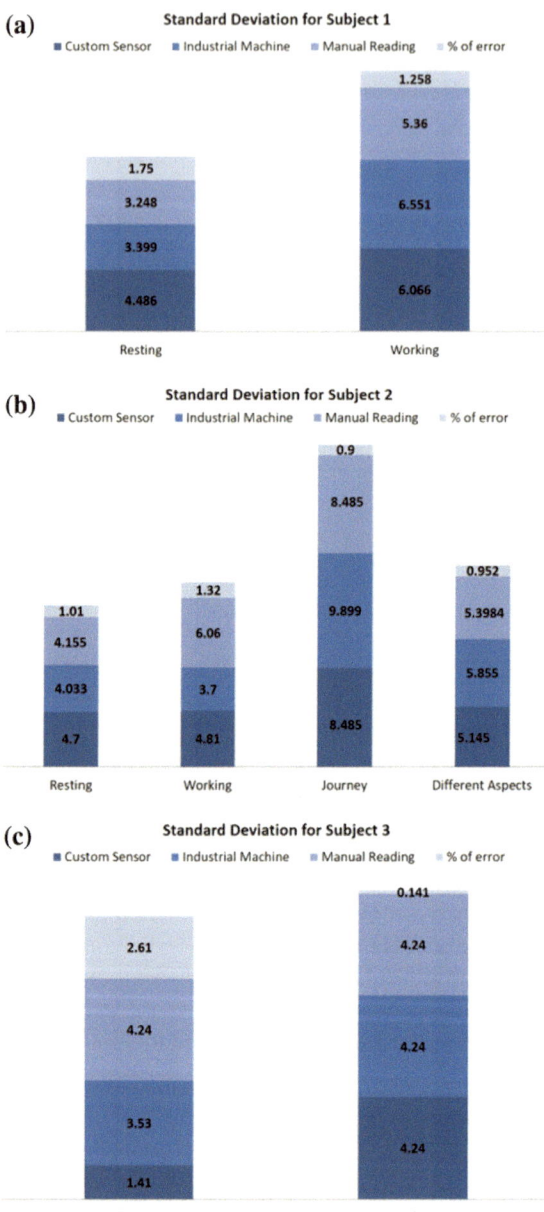

Fig. 10.16 a Standard deviation of errors for five subjects (subject 1). **b.** Standard deviation of errors for five subjects (subject 2). **c** Standard deviation of errors for five subjects (subject 3). **d** Standard deviation of errors for five subjects (subject 4). **e.** Standard deviation of errors for five subjects (subject 5)

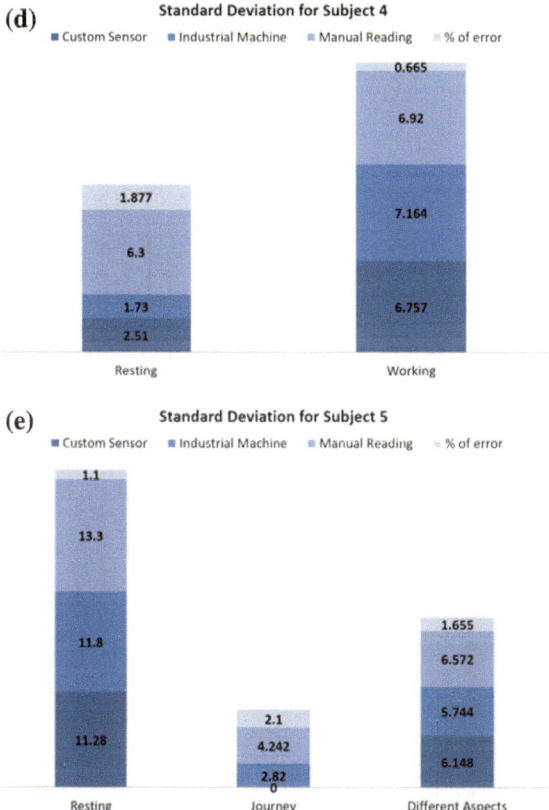

Fig. 10.16 (continued)

runs. We have combined four distinct subject's average pulse values during different traffic and road conditions. Subjects 1 and 3 were two commercial drivers of CNG-powered auto-rickshaws; Subject 2 was a commercial driver who drives a Microbus affiliated with a university; Subject 4 was a professional who drives his personal car. We have offered BDT 300 as a token of appreciation to each of the participants in our test drives. The auto-rickshaw drivers were approached in parking lots, and the two other drivers were contacted through personal links of the researchers. The pulse measurements of the drivers under various scenarios are shown in Fig. 10.17.

It is challenging to conclusively define driving behavior with a limited number of observations. Furthermore, pulse varies from person to person, and from scenario to scenario (such as taking turn, driving normally, driving fast, etc.). However, a trend in the relative change in pulse from the base pulse (i.e., normal driving) under various driving or traffic conditions may be noticed for each of the four subjects in our preliminary tests. In the case of traffic congestion or jam, the average pulse of the private driver (Subject 4) is seen to have markedly increased, whereas the pulse

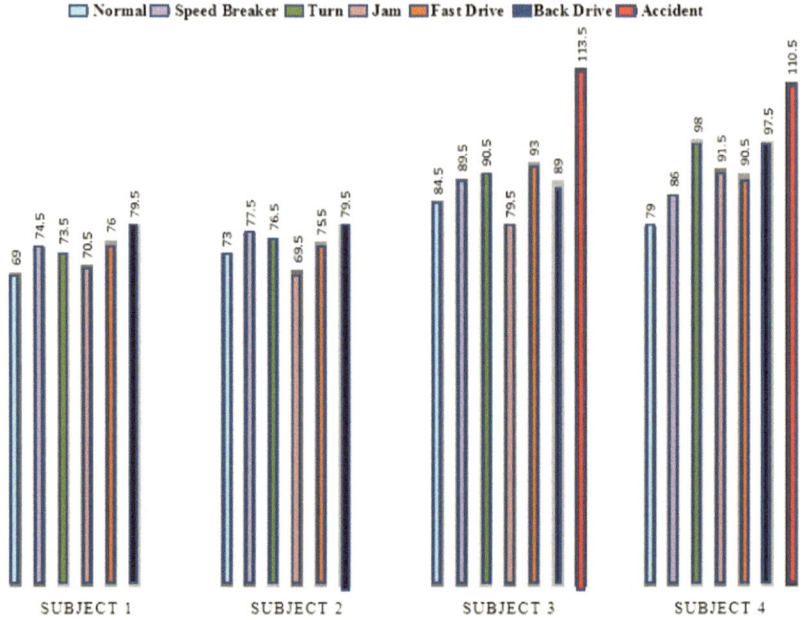

Fig. 10.17 Average pulse rate of four drivers during different traffic conditions

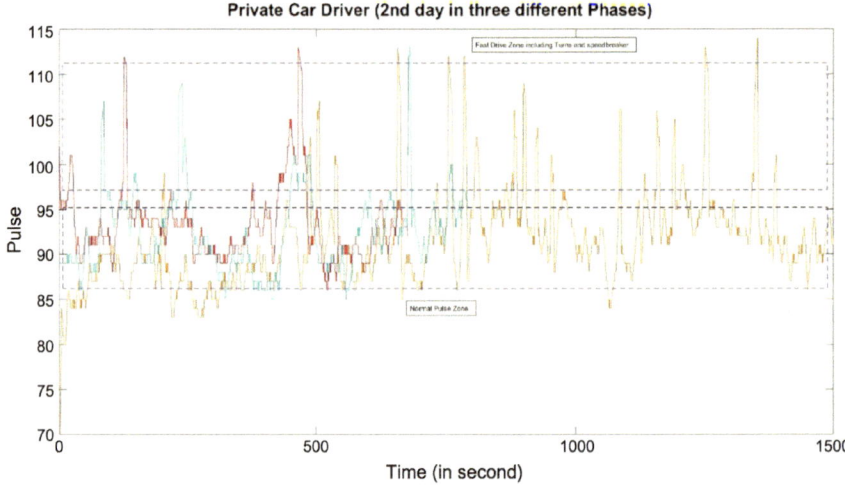

Fig. 10.18 Raw pulse rate data of a private car driver

has remained quite uniform compared to the normal driving pulse of the other three commercial drivers. In all other conditions (speed breaker, turn, etc.), the average pulse can be seen to have increased for each of the drivers compared to their normal driving pulse. Unexpected conditions, such as the accidents of Subjects 3 and 4,

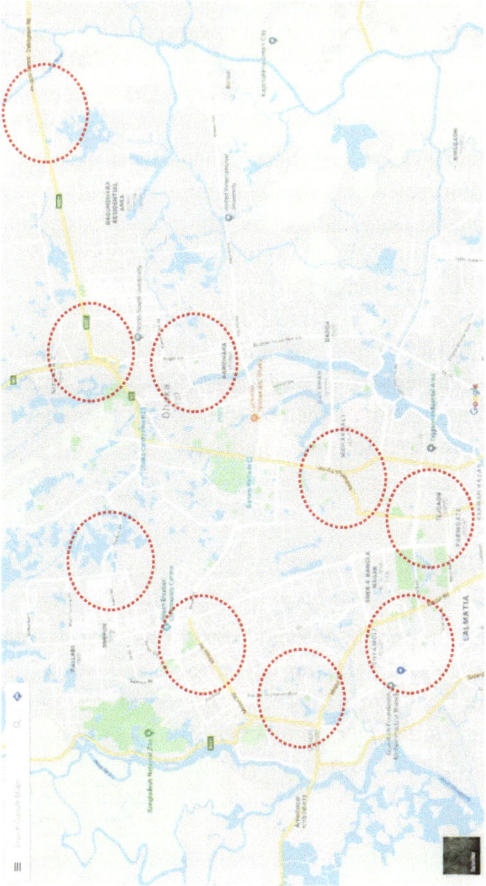

Fig. 10.19 Routes chosen to capture driver experiences

show that the pulse rises considerably in unexpected situations. Figure 10.18 shows a private car driver's raw data at three different phases for a single day. His normal pulse ranged in between (87–93) bpm, but when he faced challenges or unexpected conditions on road, the pulse raised considerably.

The routes were chosen carefully to cover some parts without and some with heavy traffic congestion, including highways and local roads. This was done in order to study driving behavior in various traffic conditions. The overall sample routes around Dhaka city are circled in Fig. 10.19.

The system showed minimal communication overhead when the wearable device was connecting to the mobile application and webserver. Over an average of 10,000 continuous transactions, the server connection required an average of 1.5 s of access time while the wearable device to application connection time was within the range of 2 s.

Here unexpected situations refer to vehicles and pedestrian coming from the wrong side of the road, motorcycles coming from any corner of the road, a honk from anywhere on the road, beggars and sellers who normally stay on the road during congestion and suddenly run when the vehicles start to move. We have listed a few, here, but the list can be much longer.

The presence of a high number of unexpected behaviors can be a good indicator of unmindful conduct and lack of concentration during driving, where our intervention system may advise the driver to relax on that particular occasion. But it must be noted that real-time intervention if implemented, can itself be another factor that may or may not be a trigger for increased stress in drivers.

10.7 Conclusion

Emotional wellbeing is important in today's fast-paced world. Ambient intelligence has been studied to support wellbeing of physical and mental health perspective as an emerging support system. However, the level of awareness and support for wellbeing is low among marginal communities. The focus of our work is on enabling opportunities for such economically less fortunate communities for enhanced support and their wellbeing in challenging situations; the community under discussion is drivers of vehicles.

The focus of the current work is in the context of a developing country, Bangladesh. The country faces challenging traffic scenarios which directly and indirectly impact the vehicle drivers. Commercial vehicle drivers facing challenging traffic scenario require attention and support in real time. Negligence in supporting the commercial vehicle drivers may lead to subsequent problems of poor vehicle driving performances and increased number of traffic accidents. We have conducted extensive interviews of drivers to understand the current conditions of commercial drivers of Bangladesh and realized the importance of a support system for this community.

The recent advancement of Internet of Things (IoT) have opened up opportunities to provide interesting application scenarios. We have proposed a wearable system that is cheap and made with basic components, coupled with other relevant devices embedded with AmI. In this chapters, we have discussed our system and analyzed the usability of the system. The idea is to support drivers on the road for better efficiency. This ambient intelligence can show how drivers are disturbed or stressed on busy roads, asking for real-time support for immediate actions. We can also map areas that are difficult and stressful for the drives and help to make their road design better. With the help of the system that we have proposed, we may be able to advise at the policy level to a developing country, aiming at happier drivers leading to a safer traffic system.

References

1. Acampora G, Cook DJ, Rashidi P (2013) A survey on ambient intelligence in healthcare. In: Proceedings of the IEEE 101.12, pp 2470–2494. https://doi.org/10.1109/jproc.2013.2262913
2. Bohn J, Coroama V, Langheinrich M, Mattern F, Rohs M (2005) Social, economic, and ethical implications of ambient intelligence and ubiquitous computing. Ambient intelligence. Springer, Berlin, Heidelberg, pp 5–29
3. Memon M, Wagner SR, Pedersen CF, Beevi FHA, Hansen FO (2014) Ambient assisted living healthcare frameworks, platforms, standards, and quality attributes. Sensors 14(3):4312–4341
4. Ericsson E (2001) Independent driving pattern factors and their influence on fuel-use and exhaust emission. Journal on Transportation Research Part D: Transport and Environment, Elsevier, Vol 6, Issue 5, pp 325–345. https://doi.org/10.1016/s1361-9209(01)00003-7
5. Iqbal ST, Horvitz E, Ju YC, Mathews E (2011) Hang on a Sec!: effects of proactive mediation of phone conversations while driving. In: Proceedings of the SIGCHI conference on human factors in computing systems (CHI '11). ACM, New York, NY, USA, pp 463–472. http://dx.doi.org/10.1145/1978942.1979008
6. Jonsson I-M, Nass C, Endo J, Reaves B, Harris H, Le Ta J, Chan N, Kerkenbush NL, Lasome CE (2003) The emerging role of electronic diaries in the management of diabetes mellitus. AACN Adv Crit Care 14(30:371–378
7. Mahmud K, Gope K, Chowdhury SMR (2012).Possible causes & solutions of traffic jam and their impact on the economy of City. J Manag Sustain 2(2):112
8. Ahmed N (2010) Reliable framework for unreliable RFID devices. In: 2010 8th IEEE international conference on pervasive computing and communications workshops (PERCOM Workshops), IEEE
9. Demiris G, Oliver DP, Dickey G, Skubic M, Rantz M (2008) Findings from a participatory evaluation of a smart home application for older adults. Technol Health Care16(2):111–118
10. Hu TY, Xie X, Li J (2013) Negative or positive? The effect of emotion and mood on risky driving. Transp Rese Part F Traffic Psychol Behav 16(2013):29–40
11. Choudhury CF, Ben-Akiva ME (2013) Modelling driving decisions: a latent plan approach. Transportmetrica A Transp Sci 9(6):546–566
12. Miyajima C, Angkititrakul P, Takeda K (2013) Behavior signal processing for vehicle applications. APSIPA Trans Signal Inf Process 2(2013):e2
13. Abedin J (2008) Accidents or Murders? http://www.daily-sun.com/home/printnews/304913. Accessed 2018
14. Bonsall P, Liu R, Young W (2005) Modelling safety-related driving behaviour-impact of parameter values. Transp Res Part A Policy Pract 39(5):425–444
15. Ranney TA (1994) Models of driving behavior: a review of their evolution. Accident Anal Prev 26(6):733–750
16. Toledo T, Koutsopoulos HN, Ben-Akiva M (2009) Estimation of an integrated driving behavior model. Transp Res Part C: Emerg Technol 17(4):365–380

17. Mierlo JV, Maggetto G, Burgwal EV, Gense R, (2004). Driving style and traffic measures-influence on vehicle emissions and fuel consumption. Proceedings of the Institution of Mechanical Engineers, Part D: Journal of Automobile Engineering 218, 1 (2004), 43–50
18. Iqbal ST, Ju YC, Horvitz E, (2010) Cars, calls, and cognition: investigating driving and divided attention. In: Proceedings of the SIGCHI conference on human factors in computing systems (CHI '10). ACM, New York, NY, USA, pp 1281–1290. DOI:http://dx.doi.org/10.1145/1753326.1753518
19. World Health Organization and others (2011) Mobile phone use: a growing problem of driver distraction. http://www.who.int/violence_injury_prevention/publications/road_traffic/distracted_driving/en/index.html
20. Strayer DL, Drews FA, (2007). Cell-phone–induced driver distraction. Current Dir Psychol Sci 16(3):128–131
21. Crundall D, Loon EV, Underwood G (2006) Attraction and distraction of attention with roadside advertisements. Accident Anal Prev 38(4):671–677
22. Pollatsek A, Fisher DL, Pradhan A (2006) Identifying and remedying failures of selective attention in younger drivers. Curr Dir Psychol Sci 15(5):255–259. DOI:http://dx.doi.org/10.1111/j.1467-8721.2006.00447.x
23. Mathias JL, Lucas LK (2009) Cognitive predictors of unsafe driving in older drivers: a meta-analysis. IPG Int Psychogeriatr 21(4):637–653. DOI:http://dx.doi.org/10.1017/S1041610209009119
24. Okonkwo OC, Crowe M, Wadley VG, Ball K (2008) Visual attention and selfregulation of driving among older adults. Int Psychogeriatr 20(2):162–173. Issue 01. DOI:http://dx.doi.org/10.1017/S104161020700539X
25. Pollatsek A, Romoser MRE, Fisher DL (2012) Identifying and remediating failures of selective attention in older drivers. Curr Dir Psychol Sci 21(1):3–7
26. Salvucci DD (2013) Distraction beyond the driver: predicting the effects of in-vehicle interaction on surrounding traffic. In: Proceedings of the SIGCHI conference on human factors in computing systems. ACM, pp 3131–3134
27. Clarke DD, Ward P, Bartle C, Truman W (2006) Young driver accidents in the UK: the influence of age, experience, and time of day. Accident Anal Prev 38(5):871–878
28. Abdu R, Shinar D, Meiran N (2012) Situational (state) anger and driving. Transp Rese Part F Traffic Psychol Behav 15(5):575–580
29. Islam MM, Choudhury CF, (2012) A violation behavior model for non-motorized vehicle drivers in heterogeneous traffic streams. In: Transportation research board 91st annual meeting
30. Gunatillake T, Cairney P, Akcelik R (2000) Traffic management performance: development of a traffic frustration index (2000)
31. Zafiroglu A, Healey J, Plowman T (2012) Navigation to multiple local transportation futures: cross-interrogating remembered and recorded drives. In: Proceedings of the 4th international conference on automotive user interfaces and interactive vehicular applications. ACM, pp 139–146
32. Valck ED, Groot ED, Cluydts R (2003) Effects of slow-release caffeine and a nap on driving simulator performance after partial sleep deprivation. Percept Motor Skills 96(1):67–78
33. Webb TL, Sheeran P, Totterdell P, Miles E, Mansell W, Baker S (2012) Using implementation intentions to overcome the effect of mood on risky behaviour. Br J Soc Psychol 51(2):330–345
34. Lajunen T, Summala H (1995) Driving experience, personality, and skill and safety-motive dimensions\in drivers' self-assessments. Personal Individ Diff 19(3):307–318
35. McCartt AT, Shabanova VI, Leaf WA (2003) Driving experience, crashes and traffic citations of teenage beginning drivers. Accident Anal Prev 35(3):311–320
36. Gulian E, Glendon AI, Matthews G, Davies DR, Debney LM (1990) The stress of driving: a diary study. Work Stress 4(1):7–16. http://dx.doi.org/10.1080/02678379008256960
37. Caird JK, Willness CR, Steel P, Scialfa C (2008) A meta-analysis of the effects of cell phones on driver performance. Accid Anal Prev 40(4):1282–1293
38. Caird JK, Johnston KA, Willness CR, Asbridge M, Steel P (2014) A meta-analysis of the effects of texting on driving. Accid Anal Prev 71(2014):311–318

39. Dell N, Vaidyanathan V, Medhi I, Cutrell E, Thies W (2012) Yours is better!: participant response bias in HCI. In: Proceedings of the SIGCHI conference on human factors in computing systems. ACM, pp 1321–1330
40. Ahmed SI, Jackson SJ, Ahmed N, Ferdous HS, Rifat MR, Rizvi ASM, Ahmed S, Mansur RS (2014) Protibadi: a platform for fighting sexual harassment in urban bangladesh. In: Proceedings of the 32nd annual ACM conference on human factors in computing systems. ACM, pp 2695–2704
41. Toyama K (2015) Geek heresy: rescuing social change from the cult of technology. PublicAffairs
42. Ahmed SI, Ahmed N, Hussain F, Kumar N (2016) Computing beyond gender-imposed limits. In: Proceedings of the second workshop on computing within limits. ACM, p 6
43. Ahmed SI, Jackson SJ, Rifat MR (2015a) Learning to fix: knowledge, collaboration and mobile phone repair. In: Proceedings of the seventh international conference on information and communication technologies and development. ACM, p 4
44. Ahmed SI, Mim NJ, Jackson SJ (2015b). Residual mobilities: infrastructural displacement and post-colonial computing. In: Proceedings of the 33rd annual ACM conference on human factors in computing systems. ACM, pp 437–446
45. Assignmentpoint.com (2015) Traffic Jam in City. http://www.assignmentpoint.com/arts/modern-civilization/traffic-jam-dhaka-city.html. Accessed July 2015
46. Dhakatribune.com (2013) Congestion costs Tk200bn every year: Survey http://www.dhakatribune.com/bangladesh/2013/jul/14/congestion-costs-tk200bn-every-year-survey. Accessed July 2015
47. Prothomalo.com (2016) http://www.prothomalo.com/bangladesh/article/831007/. Accessed July 2017
48. The Daily Observer. SC orders company to compensate victim's family. http://www.observerbd.com/2016/04/16/146874.php
49. Dhakatribune.com (2015) Bus driver remanded for killing CNG drivers. http://www.dhakatribune.com/bangladesh/crime/2015/12/23/bus-driver-remanded-for-killing-cng-driver/

Chapter 11
A Vision-Based Posture Monitoring System for the Elderly Using Intelligent Fall Detection Technique

E. Ramanujam and S. Padmavathi

Abstract Elderly monitoring systems are the major applications of care for elderly and the disabled who live alone. Falls are the leading factor to be detected in the elderly monitoring system to avoid serious injuries and even death. The detection systems often use ambient sensors, wearable sensor, and vision-based technologies. In case of sensor-based devices, the elderly are required to wear the detection devices, however, quite often, they forget to wear these or do not wear them correctly. Moreover, the sensors need to be charged and maintained regularly. Also, the ambient sensors need to be installed in all the rooms to cover the whole actuation. The additional difficulty is that they are complex in circuitry and sensitive to temperature. Vision-based devices are the only plausible solution that can replace the aforementioned sensors. Besides, the cost of vision-based implementation is much lower and related devices are better than wearable devices in activity recognition. Much like Ambient sensors, cameras can also be installed in all the rooms; the cost and maintenance of these are less as compared to ambient sensors. This chapter proposes a vision-based posture monitoring system using infrared cameras connected to a digital video recorder and a fall detection mechanism to classify the falls. In the chapter, we observe the behavior of the elderly through the specially designed clothing fabricated with retroreflective radium tape (red in color) for posture identification. The proposed fall detection technique comprises various modules of operations such as image segmentation, rescaling, and classification. The infrared cameras observe the movement of the elderly people and signals are transmitted to a digital video recorder. The digital video recorder snaps only the motion frames from the signal. The motion images are segmented to red band using image segmentation and further rescaled for better classification using k-Nearest Neighbor and decision tree classifiers. The tests have been conducted on 10 different subjects to identify the falls during various motions such as supine, sitting, sitting with knee extension, and standing. We have shown a detection rate of 94% for the proposed model with k-nearest neighbor classifier.

E. Ramanujam (✉) · S. Padmavathi
Department of Information Technology, Thiagarajar College of Engineering,
Madurai, Tamil Nadu, India
e-mail: erit@tce.edu

© Springer Nature Switzerland AG 2019
Z. Mahmood (ed.), *Guide to Ambient Intelligence in the IoT Environment*, Computer
Communications and Networks, https://doi.org/10.1007/978-3-030-04173-1_11

Keywords Ambience intelligence · AmI · Falls · Posture · Image segmentation
Ambient assistance living · AAL · K-Nearest neighbor · Decision tree
Classification · Reflective tape

11.1 Introduction

Fall is one of the major common risks faced by the elderly and disabled individuals.
World Health Organization (WHO), in a study conducted in 2007 [1], estimates that
28–35% of people, aged 60 years or above, fall at least once in a year. Fall incidents
increase by up to 42% among the people over 70 years of age in which, 50% of
elderly patients are hospitalized and remaining half suffer nonnatural mortalities
due to fall. In developed countries, mortality of the elderly people who live alone is
majorly caused by falls because a significant amount of time can pass before they
receive assistance. In European countries, about one-third of the elderly aged over
65 years stay alone, and this figure is expected to increase significantly over the next
20 years [2]. According to Census report, in India, more than 15 million live all
alone and close to three-forth of them are women. In some states like Tamil Nadu,
the proportion of such "single elders" is even higher with one in eleven of those aged
above 60 years living alone.

In the perspective of elderly monitoring system, ambient assistance living [3, 4]
plays a significant and vital role in research and development. The core element
of ambient assistance living (AAL) is to reduce the assistance time after the fall.
AAL aims at applying technology relating to sensors, sensor networks, pervasive
computing, and artificial intelligence techniques to make elderly people feel safer
and secured to live in their preferred environment.

The availability of AAL optimizes the anticipation of emergencies which can have
a practical effect on public and private health services relating to emergencies such as
cardiac arrest, fainting, seizures, falls, immobilization, or helplessness. When these
emergencies remain unnoticed for a longer time, they lead to severe complications
and even death. With the benefits of generous assistance services, chronic diseases
like dementia, arthritis, Alzheimer's disease, stroke, and epilepsy can be handled in
a proactive and preventive manner.

In view of AAL, emergency conditions and chronic diseases need to be differ-
entiated while monitoring by using automatic systems. The essential requirement
of emergency conditions, which may occur unknowingly in a very short period of
time is timely notice and detection. In contrast, anomaly or deviation in the normal
behavior is the symptom that helps to detect chronic diseases. Research works as in
[5–8] suggest wearable and implantable devices for the earlier detection of emergency situations, e.g., by pressing buttons on wearable alarm devices. The success
of those devices depends on the user's understanding and involvement in the usage
of the devices.

In the case of chronic diseases, the users may lose their consciousness or have poor decision-making capacity. In these scenarios, wearable devices are producing good results in activity recognition [9] but there are many drawbacks as represented by [10] in terms of handling wearable devices. The elderly people have to wear the devices continuously, though, due to aging or age-related diseases like dementia, they may forget to wear them. As a result, any crucial incidents that occur at that time cannot be intimated to the caretakers. Also, the devices need to be charged regularly. Wearing or injecting such devices may cause inconvenience to older people. Hence a noninvasive device [11] is needed to monitor the daily activities of such persons.

Surveillance camera is one plausible solution that can replace many numbers of sensors whose cost of implementation is also much less. Hence, surveillance camera is better than wearable devices in activity recognition. Research papers [5, 12] have suggested a number of automatic vision-based monitoring systems for the elderly. The objective of these studies is to monitor the elderly people in controlled environment and in case of falls, the system will send alert messages to the caretaker or an emergency center. Vibration, location, acoustic, and video sensors are used in these systems.

This chapter aims to develop a vision-based fall detection system to monitor the patient posture behaviors effectively. Unusual behavior is defined as anomaly behavior of the elderly in case of any medical assistance. Various postures such as standing, sitting, supine, and sitting with knee extension from the bed are monitored through the proposed approach. The approach monitors the elderly behaviors at all times (day and night) at low cost by using an infrared camera and digital video recorder with special retroreflective radium red tape [13] which is fabricated within the clothing of the elderly for detecting their movements.

The objective of the proposed chapter is twofold, which are given as follows:

- to use low-cost retroreflective radium red tape for posture detection instead of scanning the entire human body
- to use low-cost infrared cameras with digital video recorder to snap the motion image of the elderly on motion.

The rest of the proposed chapter has been organized as follows. The state of the art in fall detection is discussed in Sect. 11.2. Section 11.3 covers the system setup relating to our study. Section 11.4 describes the methodology of the system; Sect. 11.5 describes the experimental results and Sect. 11.6 concludes of the chapter.

11.2   State of the Art

The fall detection techniques are classified broadly in terms of wearable sensors, ambient sensors, and vision-based technologies. These are now discussed in the following sections.

11.2.1 Wearable Sensors

Accelerometers and gyroscopes are the most commonly used wearable sensors. These sensors are easy to wear. But, they have some drawbacks such as sensitivity to the body movement, power consumption; also the devices rely on user's ability to wear them and manually activate the alarm during the occurrence of fall. Furthermore, these devices have more number of false positives, even if they incorporate fully automated system. In the context of commercial devices, these sensors are fabricated into a pendant, belt or watch. Bagala et al. [5] have collected data for real-world falls among a patient population using accelerometers. They have presented the performance of 13 published fall detection algorithms applied to the database of 29 real-world falls. They have reported an average detection rate of 83% and a fall detection rate of 98% for the well-performing algorithms. The drawback of the typical mechanisms is that the accelerometers generate more false fall warnings in case of abrupt movement of the sensor. To overcome the issues reported in [5], Wang et al. [14] have proposed a technique by placing the accelerometer at the head level to provide highly reliable and sensitive fall detection rates.

Mathie et al. [15] have proposed an advanced wearable device which incorporates multiple sensors with multiple fusion technologies. In this work, they have used gyroscopes and accelerometers in a single waist-mounted system to acquire data about the inclination angle and movement of the subject. The system successfully distinguishes between activity and rest positions. Bianchi et al. in [16] have extended the work presented in [15] by adding barometric sensors to sense the height variations caused by falls and reported 71% success rate. The major advantage of wearable sensors in the context of biometric is that wearables have greater rehabilitation and fall detection.

Ghasemzadeh et al. [17] have presented an array of sensors that can read a patient's posture and obtain muscular activity readings using electromyography (EMG) sensors with a fall detection rate of 98%. The development of Mobile Technologies with the incorporation of sensors implies a very interesting option for home-based fall detection system. Abbate et al. [18] have proposed accelerometer-based algorithm on mobile and reported 100% fall detection rate. Combination of wearable sensors and mobile phones have been considered in [19, 20]. It uses sensors including tri-axis-based accelerometer, gyroscope, and magnetometer with a mobile phone for data processing, fall detection, and messaging.

Many research works suggest that wearable devices are appropriate in case of emergency situations and health monitoring system. Wearable devices are producing good results in activity recognition but there are also many drawbacks in terms of handling wearable devices as follows:

- The elderly are required to wear the devices continuously
- Due to aging or age-related diseases, they may forget to wear the devices
- Devices need to be charged and maintained regularly
- Degree of success depends on the wearers' appropriate use of these, e.g., pressing the button on the device on emergencies

- Costs of the wearable devices are generally high.
- In case of crucial incidents occurring, the time of fall cannot be intimated to the care staff.

Table 11.1 represents the sensors-based fall detection techniques with various considerations of each research work.

Table 11.1 Sensor-based fall detection techniques

Article	Basis	Fall types	Subjects	Declared performance	Positions	Elderly (Yes/No)
Bagala et al. [43]	Triaxial Acceleration Sensor fixed by a belt	9	9 subjects (7 women, 2 men, age: 66.4+6.2)	Average of 13 Algorithms SP: 83.0%	Lower back	Yes
			15 subjects	SP: 83.0±30.3%		
			29 subjects	SE: 57.0%		
			1 subject	SE: 57.0±27.3%		
Wang et al. [13]	6G Triaxial accelerometer	3	5 volunteers	SE: 70.48%	Head	No
Mathie et al. [45]	Triaxial accelerometer	4	NA	NA	Waist	CHF & COPD Patient
Bianchi et al. [14]	Accelerometer-based systems with a barometric pressure sensor	3	20 subjects (12 male, 8 female; mean age: 23.7)	SP: 96.5%	Waist	No
			5 subjects (2 male, 3 female; mean age: 24)	SE: 97.5%		
		2	5 subjects (5 male, mean age: 26.4)	SE: 98.2%		
Ghasemzadeh et al. [15]	Electromyography Sensors + Triaxial accelerometer	9	5 male (Age 25–32)	F—Measure: 0.63	Tibialis anterior, gastroc-nemius muscles	No
Abbate et al. [16]	Triaxial Accelerometer	3	7 volunteers (5 male, 2 female, ages 20–67)	SP: 100%, SE: 100%	Waist	No

11.2.2 Ambient Devices

Ambient devices measure various parameters in the environment of the subject under protection using groups of infrared sensing devices, sound, vibrators, and so on. The major drawbacks of the ambient devices are that they need to be installed in several rooms to cover the movement of the elderly to cater for all situations. In commercial fall detection devices, ambient technologies use presence and pressure sensors associated with wearable sensors [21–23]. Zhuang et al. [24] have used acoustics sampling and shown high failure detection rates using an ensemble of machine learning algorithms. Alwan et al. [25] have proposed an automatic system using interesting vibration sensors embedded into the flooring. In this work, authors have reported 100% fall detection ratio with the ability to distinguish activities through vibrations. Rimminen et al. [26] have reported 91% success rate using electromagnetic sensors in the floor plates that create an image of objects which touch the floor. Infrared ceiling sensors are proposed in [27] to know the existence/nonexistence of the persons under the sensor, and consequently it detects falls, if the person remains too long in the same position. It is also interesting to note the system that activates an airbag when a fall is detected, as mentioned in [28]. The proposed airbag inflates to trigger using acceleration and angular velocity signals. Table 11.2 represents the Ambient sensors based fall detection techniques. Ambient sensors also have some major drawbacks, which are given as follows:

Table 11.2 Ambient sensor based fall detection techniques

Article	Basis	Fall types	Subjects	Declared performance	Features	Elderly (Yes/No)
Zhuang et al. [22]	Acoustic Sensors	4	13 subjects	F—Measure: 67%	Single far-field microphone	No
Alwan et al. [23]	Floor Vibration sensor	3	NA	SE:100%	Raw Vibration Signal	No
Rimminen et al. [24]	Floor sensor based on near-field imaging	8	10 subjects	SE: 91%, SP: 91%	Near-field image sensor	No
Tao et al. [25]	Infrared Ceiling Sensor	3	5 subjects	F—Measure: 95.14	Binary responses	No
Tamura et al. [26]	Airbag	4	16 subjects	Acc: 93%	Thresholding technique with accelerometer and gyroscope	Yes

- Sensors need to be installed in several rooms to cover the whole actuation
- These require complex circuitry and are sensitive to temperature
- Lifetime of sensors is limited
- They have poor precision and poor signal-to-noise ratio.

11.2.3 Vision-Based Devices

Most commercial fall detection techniques in the market are based on portable devices as discussed in [29]. It is not easy to find any vision-based commercial devices in the market, but their technical advancements and related literature remain promising. Vision-based devices have similar drawbacks as ambient devices, for example,

- Installation of vision-based devices are required in all rooms to cover the whole area of actuation
- Issues regarding privacy on how to deal with images of a real person's life.

For these reasons, the streaming is controversial. As discussed, in this chapter, the proposed system captures the images of an elderly only when movement takes place and transfers the motion image to the server for processing. Advantages of vision-based systems are that they can run on many computers and the algorithms and libraries are mostly implemented as open source. Varieties of algorithms have been developed for fall detection, of which some are designed to analyze only static images and treat each frame individually. The other concerns are explained in the following subsections.

11.2.4 Cameras

In vision-based systems, the camera is the most important device for fall detection systems. Vision-based approaches are focused on real-time execution of algorithms using standard computing platforms and low-cost cameras. There are several methods that are used to obtain semantic information through video analysis. Many of them use 2D and 3D models, and the others are based on the extraction of some features after the video segmentation. Two types of cameras are often used for fall detection 2D cameras and 3D Time of Flight (TOF) cameras as discussed in [30, 31]. The resolution of TOF cameras is lower than 2D cameras and they are much more expensive. The proposed system uses simple 2D Infrared camera along with the digital video recorder to store the motion images for processing.

11.2.5 Processing Units

Most vision-based fall detectors need high computational processing power and expensive hardware. An active vision system for the automatic detection of falls and recognition of several postures of elderly homecare applications deploys a variety of platforms, for example,

- TOF MESA SwissRander SR3000 as used in [30]
- Intel Core i5 processor (Santa Clara, CA, USA) at 2.6 GHz as used by [32].
- Intel Core i7-2600 3.40 GHz processor and 16 GB of RAM clocked at 1333 MHz.
- Heterogeneous platform Zynq-7000 SoC (System on chip) platform as used by [33], which combines ARM Cortex A9 processor and a FPGA (Field-programmable gate array).

The proposed approach uses Intel Core I3 processor with 4 GB of RAM for processing and classification of the falls.

11.2.6 State Classification

Decision systems are typically required in a fall detection technique. Machine learning algorithms are widely used for computer vision fall detection. Data classification algorithms such as k-Nearest Neighbor, Support Vector Machine (SVM), and Decision Trees have been successfully used in computer vision-based fall detection techniques and more advanced machine learning algorithms like artificial Neural Networks, Hidden Markov Model. Deep learning architectures are also used for the classification of falls. Our proposed system uses a simple yet efficient Machine Learning algorithm viz k-Nearest Neighbor and Decision Tree classifier. Machine learning algorithms are often well trained using more number of scenarios as in [34] to reduce the misclassification accuracy, however, certain techniques, e.g., those discussed in [35] use cross-validated results to report the detection ratio.

11.3 Controlled Environment

The proposed vision-based posture monitoring technique uses a controlled environment for fall detection. The schematic layout of the experimental setup is shown in Fig. 11.1. The system uses infrared cameras connected to a digital video recorder (DVR). Infrared cameras [36] operating at 25 frames per second are used to obtain an all-round view of coverage of activities of each subject. The digital video recorder [37] automatically snaps the frames with motion and stores the motion image on a built-in hard disk.

Clothing of the elderly are especially designed at a low cost with ordinary cotton cloth and retroreflective radium red tape (referred further as "tape" in the remainder of the chapter) fabricated especially with lamination in the clothing at four different places. The places being bronchium or arm; hip; thighs; and legs. The tapes are originally used in the vehicles for the reflection of lights during night time. The same tape has been used in the elderly dress to identify the posture during motion. The advantage of the system is that the red tape can be easily segmented from the image for classification and the color of the tape does not fade when washing the clothes for any number of times, as it is specially laminated. Image of a person with tapes placed at different positions in the dress is shown in Fig. 11.2.

Fig. 11.1 Schematic layout
of the elderly living space

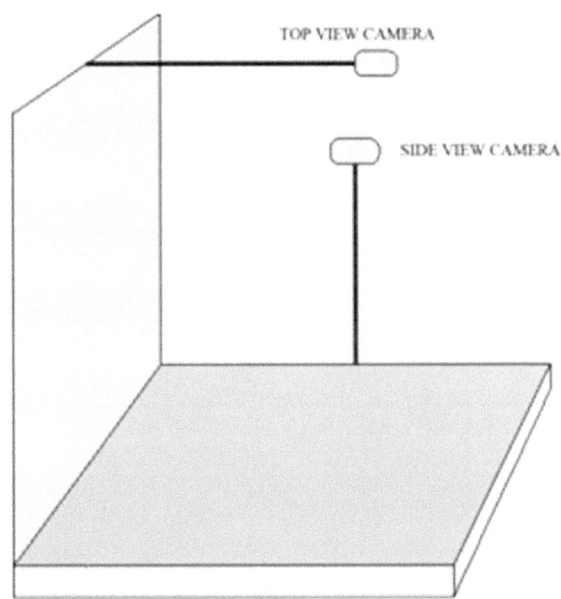

TOP VIEW CAMERA

SIDE VIEW CAMERA

11.4 Proposed Methodology

The architecture of the proposed vision-based posture monitoring system is a three-step process as shown in Fig. 11.3. In step 1, the infrared cameras are used to observe the movement of the elderly and the captured signals are transmitted to the digital video recorder (DVR). It step 2, the DVR snaps the motion images and store them in a built-in hard disk. The motion images are also transferred to the server for the fall detection technique as shown in Fig. 11.4. The fall detection technique undergoes various modules of operations such as image segmentation, rescaling, and classification in the server. The captured motion images are segmented to "Red" band for representing the retroreflective radium tape using k-means clustering technique. The segmented images are further rescaled to a numerical vector to undergo better classification. k–NN and decision tree classifiers are used to group the rescaled images as normal images with any fall. If the system classifies any of the motion as fall, then an alert message is passed to medical experts or care persons for emergency actions as step 3 as shown in Fig. 11.3.

The specific advantages of the proposed system are as follows:

- No interruption in the elderly privacy due to motion capture
- Use of the camera will be less depending on the room size or the space used by the elderly
- Tapes are reflective at the night time also, so there is no need for any other specific monitoring facility in the night time

Fig. 11.2 Subject with the dress

- Less communication delay to the server as the cameras capture only the motion images
- No video processing is carried out, instead only image segmentation is carried out to classify the elderly person's posture.

Various modules of the proposed fall detection technique are now discussed in the following section.

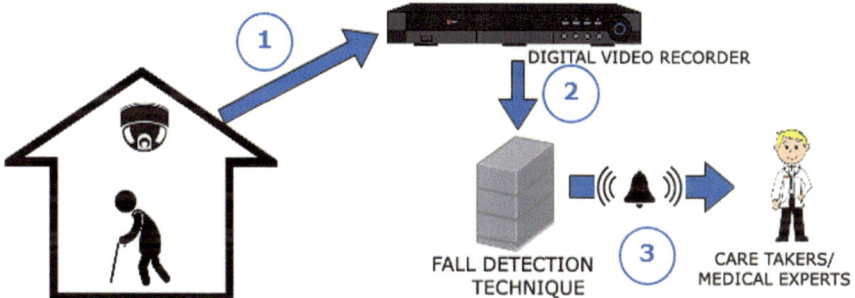

Fig. 11.3 Proposed architecture of the posture monitoring system

Fig. 11.4 Modules of fall
detection technique

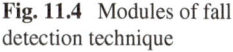

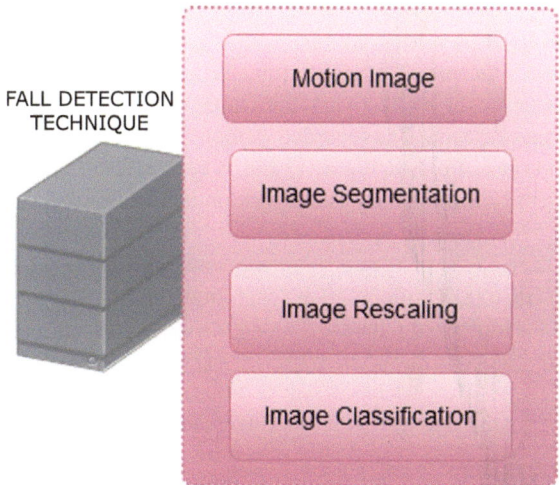

11.4.1   Image Segmentation

Image segmentation is the process of partitioning a digital image into multiple segments (set of pixels or superpixels). Traditionally, numerous image segmentation algorithms have been proposed by various researchers [38, 39]. Here, the clustering is the familiar technique which groups the pixel values into a specific number of clusters. Clustering works on the principles of high intraclass similarity and less interclass similarity among the pixel values. In the last decade, the algorithms such as Fuzzy C-Means (FCM) clustering [40], k-means clustering [41], subtractive clustering [42] and others have been proposed for image segmentation. The proposed system uses simple yet efficient k-means clustering, an unsupervised algorithm for the process of image segmentation. The general definition of a k-means clustering for image segmentation is as follows.

k-means clustering partitions a collection of data (pixel value) into a k-number of clusters ($k = 3$, k represents three color bands) and classifies into k-number of disjoint clusters. k-means clustering works in two phases. In the first phase, the algorithm calculates the k-centroid and in the second phase, it assigns each data point (pixel value) which has the nearest centroid from the respective data point. The distance between the data point and the centroid are calculated by different distance measures [43] such as Euclidean distance, Manhattan distance, and Minoskwi distance. The proposed system uses the Euclidean distance for the distance measure. Once the grouping of all data points to the centroid has been completed, the algorithm recomputes the new centroid for each k-cluster. For the new centroid, again Euclidean distance is calculated between each new centroid and data points and the data points are assigned to the cluster with minimum Euclidean distance. Each cluster is the partition defined by its member objects and by its centroid. Centroid is the point at which the sum of distances from all the objects in that cluster is minimized. k-means is an iterative algorithm, in which the algorithm minimizes the sum of distances from each object to its cluster centroid.

In the proposed algorithm, the k-means clustering process is used to segment only the retroreflective radium tape which is in red color for further process of image classification. The recorded motion image of any resolution (x, y) is grouped into k number of clusters ($k = 3$) and $p(x, y)$ refer to the input pixels to the cluster and c_k is the cluster center. The algorithm is as follows.

Input:

RGB image with (x, y) as resolution
k—Number of Clusters

Output:

Segmented image with color band as Red

Algorithm:

1. Randomly initialize the cluster center as c_k.
2. For each pixel of an image, calculate the Euclidean distance d between the center and each pixel of an image by using the Eq. (11.1).

$$d = ||p(x, y) - c_k|| \tag{11.1}$$

3. Assign the pixels $p(x, y)$ to the nearest center based on distance d
4. After all the pixels have been assigned, recalculate the new position of the center using the Eq. 11.2

$$c_k = \frac{1}{k} \sum_{y \in c_k} \sum_{x \in c_k} p(x, y) \tag{11.2}$$

5. Repeat the process until no new cluster is formed.
6. Reshape the cluster pixels into new image.

Image segmentation is carried out particularly for the tape fabricated in the dress and not for the entire body of the elderly. This may reduce the processing cost on region on interest (ROI) and bounding box issues for the further processing of images. After segmentation, the human posture is displayed through the tape. The segmented image contains only the pixel values of the tape and it is taken for the classification process. Before classification, the segmented image has different resolutions, to avoid the conflicts in classification.

11.4.2   Image Rescaling/Reshaping

The segmented images are rescaled and reshaped into a single numerical vector of size $100X1$ for the classification process. The vector contains numerical value only at the indices of tape and the remaining cells are filled with 0. The target class values are appended to the numerical vector which now becomes $101X1$ for the classification process.

11.4.3   Image Classification

The goal of image classification [44] is to predict the categories of the input image using its features. However, the image classification requires more time to interpret the spectral classes and it may vary across different images. This present chapter proposes the image classification through the traditional data classification, as the image has been rescaled to a numerical vector of size $101X1$. Classification is basically grouped as supervised and unsupervised. Numerous supervised classification algorithms have been proposed for data classification as in [45] such as k-nearest neighbor (k-NN), artificial neural network (ANN), decision tree algorithms, support vector machine (SVM) and the like. This section utilizes k–NN, decision tree algorithm and artificial neural network (ANN) for the classification of motion images.

11.4.4   Decision Tree Algorithm

Decision tree learning uses a decision tree to go from observations about an item to conclusions about the items' target values. In decision trees, leaves represent one or more classes and the branches represent conjunctions of the features that lead to the classification process. In decision analysis, decision trees are used to visually and explicitly represent decisions and decisions making. Decision trees can be learned by splitting the source set into subsets on an attribute test value. The process is repeated on each derived subset in a recursive manner called recursive partitioning. The recursion is completed when the subset at a node has all the same values of the target variable or when splitting no longer adds value to the predictions. Decision tree learning is the construction of a decision tree from class-labeled training tuples. Numerous algorithms have been proposed for decision tree classification [45]. This chapter uses the concept of C4.5 (successor of ID3).

Algorithms for constructing the decision trees usually works as a top-down approach, by choosing a variable at each step that best splits the set of items. Different algorithms use different metrics for measuring the best split. The measures like Gini impurity, information gain and gain ratio, and so on are proposed for the best split measures. C4.5 uses information gain which works on the concept of Entropy and Information Theory. Entropy is defined as in Eq. 11.3 where D represents the tuple to be classified and p_i represents the probability that tuple D belongs to class C_i:

$$Info(D) = -\sum_{i=1}^{m} p_i log_2(p_i) \tag{11.3}$$

11.4.5 k-Nearest Neighbor

In pattern recognition, the k-nearest neighbor algorithm (k-NN) is a nonparametric method used for classification and regression. In both cases, the input consists of the k closest training examples in the feature space. In k-NN classification, the output is a class membership. An object is classified by the majority of votes of its neighbors, with the object being assigned to the class most common among its k nearest neighbors (k is a positive integer, typically small). If $k = 1$, then the object is simply assigned to the class of that single nearest neighbor. k-NN is a type of instance-based learning, or lazy learning, where the function is only approximated locally and all the computations are deferred until classification. The k-NN algorithm is one of the simplest of all machine learning algorithms. Both for classification and regression, a powerful technique can be used to assign weight to the contributions of the neighbors, so that the nearer neighbors contribute more to the average than the more distant neighbors. Closeness is defined in terms of a distance metric such as Euclidean distance between two points or tuples: $X_1 = (x_{11}, x_{12}, \ldots, x_{1n})$ and $X_2 = (x_{21}, x_{22}, \ldots, x_{2n})$, as

$$dist(X_1, X_2) = \sqrt{\sum_{i=1}^{n}(x_{1i} - x_{2i})^2} \tag{11.4}$$

11.5 Results and System Evaluation

This section demonstrates the operations and performance of the proposed vision-based fall detection technique. To analyze the performance of the proposed technique, an experiment has been carried out with various subjects of different ages and both genders as shown in Table 11.3. During the experiment in the controlled environment, the subjects are clothed with a special dress fabricated with a retroreflective radium tape. When recording, the subjects are requested to do a variety of motions like sitting, sitting with knee extension, sleeping, supine, standing and some falls during motion. In Motion 1 in Table 11.3, the subject undergoes from the standing position to a sitting position (Motion 1). In Motion 2, the subject undergoes from the sitting position to standing position. Similarly, from the sitting position to sitting with knee extension is referred to as Motion 3. Motion 4 is from the sitting with knee extension to supine position. From supine position to sitting with knee extension is considered as Motion 5. Finally, Motion 6 considers falls from any position. Motion 6 alone is experimented for less number of motions, due to risk in injuries or other health issues. The numerical entries of each Motion represent the number of time the subject has carried out the particular motion during the experiment. In Table 11.3, subject S.No 1 at age 23 Male performs Motion 1 for 80 times (standing to sitting position during record); Motion 2 records for 70 times, i.e., from sitting position to standing position; and so forth.

Figure 11.5 represents various motions of two different subjects detected along with the Tapes. The motion images are processed with various modules as described earlier, i.e., image segmentation using k-means clustering and rescaling. Figure 11.6 represents the segmented image that shows only the posture of the subject through the tapes. The pixel value of the stickers is indexed with numeric value and the other

Table 11.3 Elderly attributes with their different Motions

S.no	Gender/Age	Height (cm)/Weight (Kg)	Motion 1	Motion 2	Motion 3	Motion 4	Motion 5	Motion 6
1	23/M	176/74	80	70	85	100	80	28
2	30/M	184/90	70	80	90	100	95	20
3	35/M	170/74	65	80	75	80	90	25
4	19/M	172/70	80	70	80	100	80	30
5	28/F	165/52	90	80	100	80	70	25
6	45/F	170/64	88	90	80	80	85	20
7	70/M	168/59	75	60	55	70	66	20
8	62/F	152/58	80	70	80	90	90	30
9	40/M	174/88	85	80	90	70	80	20
10	34/M	156/70	80	75	95	80	90	22
		Total motions	793	755	830	850	826	240

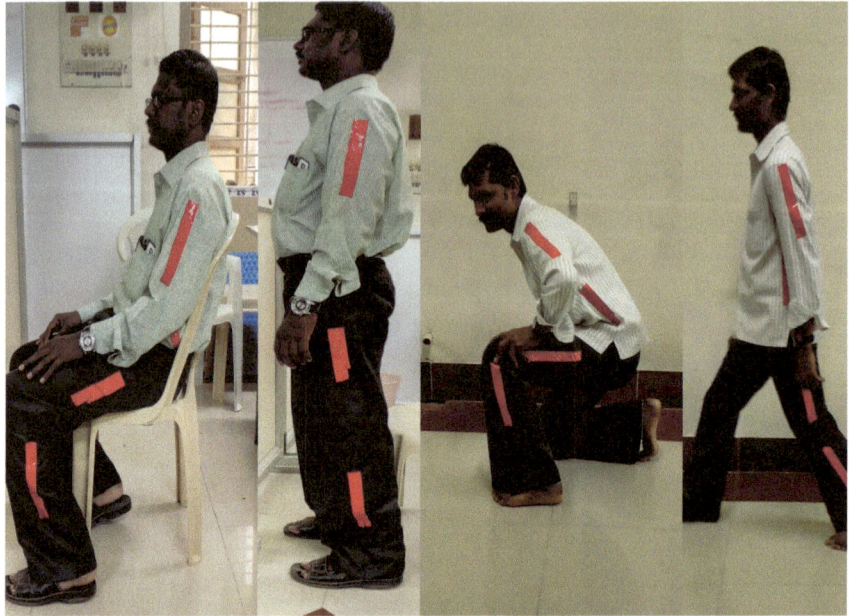

Fig. 11.5 Various subjects on motion detected

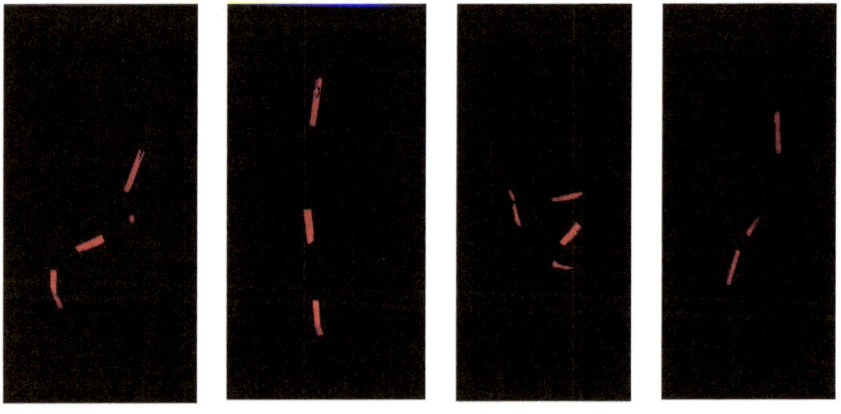

(A) SITTING POSTURE　　(B) STANDING POSTURE　　(C) FALL POSTURE　(D) STANDING POSTURE

Fig. 11.6 Posture of the elderly visible through tapes after image segmentation

value seems to be zero. On performing rescaling and reshaping of the image, the pixel values are rescaled to a numerical vector for easier classification.

The vector value of each image has been considered for classification process with the target label value. Finally, the 10-fold cross validation is used for classifying the motion images using k-NN and decision tree classifier. For assessing the performance

Table 11.4   10-fold cross-validated result of k-NN classifier

Motions	Motion 1	Motion 2	Motion 3	Motion 4	Motion 5	Motion 6
Motion 1	755	7	8	12	8	7
Motion 2	8	716	9	14	6	8
Motion 3	6	9	789	11	4	8
Motion 4	7	8	7	792	7	9
Motion 5	9	6	8	11	794	6
Motion 6	8	9	9	10	7	202
Accuracy (%)	95.21	94.83	95.06	93.18	96.13	84.17

Table 11.5   10-fold cross-validated result of decision tree classifier

Motions	Motion 1	Motion 2	Motion 3	Motion 4	Motion 5	Motion 6
Motion 1	748	7	10	14	9	10
Motion 2	10	706	12	16	8	9
Motion 3	9	12	773	13	6	8
Motion 4	8	10	10	781	7	10
Motion 5	9	9	12	14	788	9
Motion 6	9	11	13	12	8	194
Accuracy (%)	94.33	93.51	93.13	91.88	95.4	80.83

of the proposed methodology, all the motions from 1 to 6 of all the 10 persons are merged. In total, 4294 motions are used for the assessment. The cross-validated results of k-NN and decision tree are shown in Tables 11.4 and 11.5, respectively. The performances of the classifier are measured using the familiar metrics such as Sensitivity (Sen), Specificity (Spe) and Accuracy (Acc) as given in Eqs. 11.5, 11.6 and 11.7 where M_x and M_y are two motions.

$$Sen = \frac{TP}{TP + FN},\tag{11.5}$$

$$Spe = \frac{TN}{TN + FP},\tag{11.6}$$

$$Acc = \frac{TP + TN}{TP + FP + FN + TN},\tag{11.7}$$

where

TP—Motion M_x identified as M_x
TN—Motion M_x identified as M_y
FP—Motion M_y identified as M_y
FN—Motion M_y identified as M_x

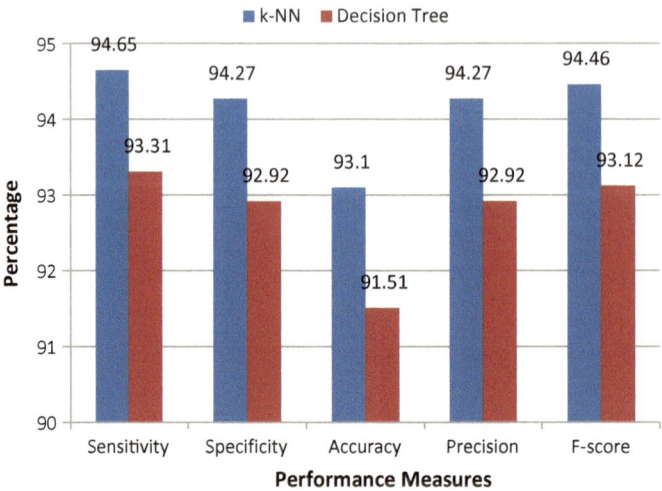

Fig. 11.7 Performance measure of k-NN and decision tree classifier

Tables 11.4 and 11.5 represent the 10-fold cross-validated results (Confusion Matrix) of k-NN classifier and decision tree, respectively, for the Motions from 1 to 6 for all 10 participants. Table 11.4 shows the performance of k-NN classifier, for all Motions 1–6. Accuracy of Motion 1 is 95.21, Motion 2 is 94.83, Motion 3 is 95.06, Motion 4 is 93.18, Motion 5 is 96.13, and Motion 6 is 84.17. The Motions are detected at the scale of 94–95%. However, Motion 6 has the range of 84 due to less number of samples taken for the assessment. The overall average accuracy of k-NN classifier for the posture monitoring of all the motions is 93.1%.

Table 11.5 shows a similar performance of decision tree classifier, for all the Motions from 1 to 6. Accuracy of Motion 1 is 94.33, Motion 2 is 93.51, Motion 3 is 93.13, Motion 4 is 91.88, Motion 5 is 95.40, and Motion 6 is 80.83. The overall average accuracy is 91.51. On comparing the performance of both the classifiers, the performance of k-NN classifier shows better accuracy than the decision tree classifier. Performance of the system/classifier is measured using different measures like accuracy, sensitivity, and specificity. From Eqs. 11.5 and 11.6, sensitivity and specificity only detect the false positive which is the important factor in determining the performance of the posture monitoring system. Figure 11.7 shows the performance measures of k-NN classifier and decision tree. From the performance comparison which is shown in Fig. 11.7, it is clear that sensitivity and specificity show higher value than decision tree in the range of 94–95%. This shows the efficiency of the proposed posture monitoring system. However, in future, the robust monitoring system may be proposed to improve the accuracy, sensitivity and specificity of the system.

11.6 Conclusion

This chapter introduces a low-cost posture monitoring system using retroreflective radium tapes which are fabricated in a loose cotton dress with infrared cameras. The Tapes are fabricated at four different parts of the dress. Although the system presented in this chapter is currently under development, it reliably detects falls in the controlled environments. The system performs with approximately 94% efficiency in the controlled environments. The present system has an enormous advantage in the sense that a person under surveillance is not required to wear any sensory devices. There is no privacy interference for the elderly as it records only the motion images. The cost of the proposed approach is much less when compared to other fall detection techniques, as it requires only a simple infrared camera with digital video recorder. Lifetime of the system is high. The system reliability has been proven with different subjects of different age and genders involved in the experiment. As regards the future, work can be extended in terms of occlusion, state differentiation, and gait recognition.

References

1. World Health Organization (2007) WHO global report on falls prevention in older age. World Health Organization: Geneva, Switzerland, 2007
2. Rodrigues R, Huber M, Lamura G (2012) Facts and figures on healthy ageing and long-term care. European Centre for Social Welfare Policy and Research, Vienna
3. Costa R, Carneiro D, Novais P, Lima L, Machado J, Marques A, Neves J (2009) Ambient assisted living. In: 3rd Symposium of Ubiquitous Computing and Ambient Intelligence 2008. Springer, Berlin, pp 86–94
4. Pieper Michael, Antona Margherita, Cortés Ulises (2011) Ambient assisted living. Ercim News 87(2011):18–19
5. Bagalà F, Becker C, Cappello A, Chiari L, Aminian K, Hausdorff JM, Zijlstra W, Klenk J (2012) Evaluation of accelerometer-based fall detection algorithms on real-world falls. PLoS ONE 2012(7):37062. https://doi.org/10.1371/journal.pone.0037062
6. Wang C-C, Chiang C-Y, Lin P-Y, Chou Y-C, Kuo I-T, Huang C-N, Chan C-T (2008) Development of a fall detecting system for the elderly residents. In: Proceedings of the 2008 2nd International Conference on Bioinformatics and Biomedical Engineering, Shanghai, China, 16–18 May 2008, pp 1359–1362
7. Lindemann U, Hock A, Stuber M, Keck W, Becker C (2008) Evaluation of a fall detector based on accelerometers: a pilot study. Med Biol Eng Comput 2005(43):548–551. https://doi.org/10.1007/BF02351026
8. Mathie MJ, Coster ACF, Lovell NH, Celler BG (2004) Accelerometry: providing an integrated, practical method for long-term, ambulatory monitoring of human movement. Physiol Meas 2004:25. https://doi.org/10.1088/0967-3334/25/2/R01
9. Glaros C, Fotiadis DI (2005) Wearable devices in healthcare. In: Intelligent paradigms for healthcare enterprises. Springer, Berlin, pp 237–264
10. Varshney U (2007) Pervasive healthcare and wireless health monitoring. Mobile Networks and Applications 12(2–3):113–127
11. Rashidi P, Mihailidis A (2013) A survey on ambient-assisted living tools for older adults. IEEE J Biomed Health Inform 17(3):579–590

12. Mubashir M, Shao L, Seed L (2013) A survey on fall detection: principles and approaches. Neurocomputing 2013(100):144–152. https://doi.org/10.1016/j.neucom.2011.09.037
13. Wang C-C, Chiang C-Y, Lin P-Y, Chou Y-C, Kuo I-T, Huang C-N, Chan C-T (2008) Development of a fall detecting system for the elderly residents. In: Proceedings of the 2008 2nd international conference on bioinformatics and biomedical engineering, Shanghai, China, 16–18 May 2008; pp 1359–1362
14. Bianchi F, Redmond SJ, Narayanan MR, Cerutti S, Lovell NH (2010) Barometric pressure and triaxial accelerometry-based falls event detection. IEEE Trans Neural Syst Rehabil Eng 18(6):619–627
15. Ghasemzadeh H, Jafari R, Prabhakaran B (2013) A body sensor network with electromyogram and inertial sensors: multimodal interpretation of muscular activities. IEEE Trans Inf Technol Biomed 2010(14):198–206. https://doi.org/10.1109/TITB.2009.2035050
16. Abbate S, Avvenuti M, Bonatesta F, Cola G, Corsini P, Vecchio A (2012) A smartphone-based fall detection system. Perv Mob Comput J 2012(8):883–899. https://doi.org/10.1016/j.pmcj.2012.08.003
17. Aihua M, Ma X, He Y, Luo J (2017) Highly portable, sensor-based system for human fall monitoring. Sensors 2017:17. https://doi.org/10.3390/s17092096
18. Albert MV, Kording K, Herrmann M, Jayaraman A (2012) Fall classification by machine learning using mobile phones. PLoS ONE 2012(7):e36556. https://doi.org/10.1371/journal.pone.0036556
19. Chaccour K, Darazi R, El Hassani AH, Andrès E (2017) From fall detection to fall prevention: a generic classification of fall-related systems. IEEE Sens J 2017(17):812–822. https://doi.org/10.1109/JSEN.2016.2628099
20. FATE Project (2017). https://fate.webs.upc.edu/project. Accessed 30 Nov 2017
21. Tunstall Products (2017). https://uk.tunstall.com/services/our-products/. Accessed 30 Nov 2017
22. Zhuang X, Huang J, Potamianos G, Hasegawa-Johnson M (2009) Acoustic fall detection using gaussian mixture models and gmm supervectors. In: Proceedings of the IEEE international conference on acoustics, speech and signal processing (2009), Taipei, Taiwan, 19–24 April 2009, pp 69–72
23. Alwan M, Rajendran PJ, Kell S, Mack D, Dalal S, Wolfe M, Felder R (2006) A smart and passive floor-vibration based fall detector for elderly. In: Proceedings of the 2nd information and communication technologies, ICTTA'06, Damascus, Syria, 24–28 April 2006, vol 1, pp 1003–1007
24. Rimminen H, Lindstrom J, Linnavuo M, Sepponen R (2010) Detection of falls among the elderly by a floor sensor using the electric near field. IEEE Trans Inf Technol Biomed 2010(14):1475–1476. https://doi.org/10.1109/TITB.2010.2051956
25. Tao S, Kudo M, Nonaka H (2012) Privacy-preserved behavior analysis and fall detection by an infrared ceiling sensor network. Sensors 2012(12):16920–16936
26. Tamura T, Yoshimura T, Sekine M, Uchida M, Tanaka O (2009) A wearable airbag to prevent fall injuries. IEEE Trans Inf Technol Biomed 2009(13):910–914. https://doi.org/10.1109/TITB.2009.2033673
27. The Top 10 Fall Detectors (2016). http://www.toptenreviews.com. Accessed 30 Nov 2017
28. Diraco G, Leone A, Siciliano P (2010) An active vision system for fall detection and posture recognition in elderly healthcare. In: Proceedings of the 2010 Design, Automation and Test in Europe Conference and Exhibition (DATE 2010), Dresden, Germany, 8–12 Mar 2010, pp 1536–1541
29. Kepski M, Kwolek B (2014). Fall detection using ceiling-mounted 3D depth camera. In: Proceedings of the 2014 international conference on computer vision theory and applications (VISAPP), Lisbon, Portugal, 5–8 Jan 2014, pp 640–647
30. Hsu YW, Perng JW, Liu HL (2015) Development of a vision based pedestrian fall detection system with back propagation neural network. In: Proceedings of the 2015 IEEE/SICE international symposium on system integration (SII), Nagoya, Japan, 11–13 Dec 2015, pp 433–437

31. Nguyen HTK, Fahama H, Belleudy C, Pham TV (2014) Low power architecture exploration for standalone fall detection system based on computer vision. In: Proceedings of the European modelling symposium, Pisa, Italy, 21–23 Oct 2014, pp 169–173

32. de Miguel K, Brunete A, Hernando M, Gambao E (2017) Home camera-based fall detection system for the elderly. Sensors 17(12): 2864

33. Liu CL, Lee CH, Lin PM (2010) A fall detection system using k-nearest neighbor classifier. Expert Syst Appl 37(10):7174–7181

34. Claridge AW, Mifsud G, Dawson J, Saxon MJ (2005) Use of infrared digital cameras to investigate the behaviour of cryptic species. Wildl Res 31(6):645–650

35. Kim M-G, Byun W-G (2006) Digital video recorder. US Patent Application 29/226,437, filed May 2, 2006

36. Haralick RM, Shapiro LG (1985) Image segmentation techniques. Comput Vis Graph Image Process 29(1):100–132

37. Pal NR, Pal SK (1993) A review on image segmentation techniques. Pattern Recognit 26(9):1277–1294

38. Zhang DQ, Chen SC (2004) A novel kernelized fuzzy c-means algorithm with application in medical image segmentation. Artif Intell Med 32(1):37–50

39. Ray S, Turi RH (1999) Determination of number of clusters in k-means clustering and application in colour image segmentation. In: Proceedings of the 4th international conference on advances in pattern recognition and digital techniques, pp. 137–143

40. Dhanachandra N, Manglem K, Chanu YJ (2015) Image segmentation using K-means clustering algorithm and subtractive clustering algorithm. Procedia Comput Sci 54(2015):764–771

41. Davies DL, Bouldin DW (1979) A cluster separation measure. IEEE Trans Pattern Anal Mach Intell 2:224–227

42. Kotsiantis SB, Zaharakis I, Pintelas P (2007) Supervised machine learning: a review of classification techniques. In: Emerging artificial intelligence applications in computer engineering, vol 160, pp 3–24

43. Kotsiantis SB, Zaharakis I, Pintelas P (2007) Supervised machine learning: a review of classification techniques. Emerging artificial intelligence applications in computer engineering 160:3–24

44. Harrison GM (1989) Traffic lane marking device. US Patent 4,875,799, issued October 24, 1989

45. Mathie MJ, Basilakis J, Celler BG (2001) A system for monitoring posture and physical activity using accelerometers. In: Proceedings of the 23rd annual international conference of the Engineering in Medicine and Biology Society, 2001, vol 4. IEEE, pp 3654–3657

Chapter 12
Twenty-First-Century Smart Facilities Management: Ambient Networking in Intelligent Office Buildings

Alea Fairchild

Abstract Ambient Technologies, such as beacons, sensors, and other similar smart devices, can be used in work places such as offices to determine everything from whether an employee is in the building, to where they are located, and whether a booked conference room is actually in use. This is part of a larger smart office strategy involving digital facilities management solutions that respond to modern methods and manners of working, as well as smart building technologies providing digital ecosystems that allow workers empowerment through personalization and automation. This new data-driven environment contributes to energy efficiency, optimized space utilization, enhanced workplace experience and occupants' comfort. However, all of this requires standards for data interoperability and seamless networking. Facilities managers are also now taking on a different role as to how they visualize new smarter office spaces, where it is expected that new environments would support their inhabitants intelligently by promoting easier management, better efficiency, increased productivity, and enabling the buildings to be part of the creation process for design and project development. There are obviously numerous sensitivity issues with respect to gathering, storing, maintaining, and processing of the ambient environment data in terms of user privacy, security, and possibility of potential data misuse. In this chapter, we discuss the new approaches to facilities management in terms of developing smarter office spaces, embedded with devices employing Ambient Intelligence (AmI) . We also articulate cases and examples of ambient technologies implementation.

Keywords Ambient technology · Ambient intelligence · AmI · Inoperability SaaS · Privacy · Strategic design · Co-location · Facilities management Efficiency · Collaboration

A. Fairchild (✉)
Constantia Institute sprl and Faculty of Economics and Business,
KU Leuven, Brussels, Belgium
e-mail: alea.fairchild@kuleuven.be

© Springer Nature Switzerland AG 2019   271
Z. Mahmood (ed.), *Guide to Ambient Intelligence in the IoT Environment*, Computer
Communications and Networks, https://doi.org/10.1007/978-3-030-04173-1_12

12.1   Introduction

To make office spaces more functional, efficient, and productive, facilities managers are now tasked with finding ways to ensure better usage of the available facilities. This suggests the deployment of technological infrastructures that allow employers and businesses to maximize their investment in, not only the buildings, but also the people who work there. It is recognized that ambient technologies, in the form of smart devices that sense, store, and distribute information, are appropriate tools to aid in these activities. In this section, we first define certain related terms used in discussing the enablement of intelligence in office spaces. Then, we discuss new business approaches related to Ambient Intelligence (AmI) and present the chapter organization.

12.1.1   Smart Buildings, Spaces, and Facilities Management

A smart building can be defined as one that enables integration and control of building systems in terms of efficient facilities management, employee well-being, and employee engagement. Smartness is directly proportional to *awareness*, which may be defined as *the state or ability to perceive, to feel, or to be conscious of events, objects, or sensory patterns* [1].

Smart spaces are workspaces within smart buildings that use technology, and ambient intelligence built within the smart technological devices, that allow for monitoring and measurement of occupancy levels, available vacancies, use of amenities, etc. These spaces are designed for optimal use within the smart buildings. These are triggered by sensors and intelligent devices (such as smart watches, mobile devices, and handheld badges) for the identification of office workers, their status, their relationship with office spaces (i.e., occupancy status), and the relevant courses of action for better utilization of spaces, and at the same time, ensuring workers' well-being.

The reality of smart spaces for users is not just how the relevant data is used, but how the optimization of the space is taken care of. In the later sections of this chapter, we further elaborate on this.

The concept of smart buildings, smart spaces, and Smart Facilities Management (SFM) is driven by the need for energy efficiency, environment functionality, and space optimization. The distinguishing features of smart facilities management include interoperable control systems, use of sensory devices, automated building systems diagnostics, and self-commissioning of building systems (sensors and control systems) [2]. In this context, appropriate management of the device connectivity and networked features allow the required levels of control that result in more efficiency, better comfort, and cost reduction. It can also benefit both the users and the owners of the buildings.

Buildings facilities include office workspaces including unassigned work areas, workstations with computers connectivity, meeting rooms with various layouts, privacy enhanced soundproof areas for private telephony or web-based conversations, private rooms for solitary work and public areas for either meetings or group work at a large scale.

Office buildings and spaces within buildings do not only facilitate accomplishment of work tasks, they have also the *potential to improve by contributing toward the provision of the optimum working and business environment* [3]. As many functions are carried out in office spaces, optimized utilization of space in office environments has now become vitally important. Office space has always assumed *the role of being the place and the environment where people work, research, team together, create and document information* [4].

Adding "intelligence" to these flexible spaces adds much more than mere knowledge application. It also adds means for enhanced communication as well as provide better working environments that satisfy user actions as people adapt their working behaviors to match their environment. This embedded intelligence is another driver for smart buildings and smart spaces, so that users can find each other in the building, as well as walk into a meeting room that becomes instantly enabled with the right applications and the correct connectivity of devices and data.

However, to achieve the said aims and benefits, there needs to be an understanding of the underlying infrastructure that makes this happen, including standards and interoperability for data exchange.

12.1.2 Ambient Technologies

Ambient technologies can help control the resources that users have access to. They also help to create a more productive and adaptable environment. These technologies work in the background, aiding in the learning how teams may be formed and how they might work together more effectively. This type of technology is still seen as largely experimental; however, users are keen to accept and use them much more than previously anticipated. One example of how smart spaces may be used differently with ambient technologies can be highlighted by the team from Robin which uses meeting room assistants in their smart meeting rooms [5]. Robin is further discussed in the use case section of this chapter.

Technology is making it easier not just to communicate with distant colleagues about work, but also to collaborate and work together. Technologies instill closer personal interactions with distant colleagues. In 2018, it has become much less of a stigma not to be co-located with co-workers. In fact, hot desking, working from home with flexible hours, and smarter working environments are becoming attractive and popular—all this through the use of smart technologies embedded with Ambient Intelligence (AmI) [6].

12.1.3   New Business Models

Mechanisms to use office space more effectively has also evolved over time. Space has now moved from an item of the commodity to a premium article in terms of commercial value. Managing that value is now part of facilities management. This has led to new business models in office space usage, such as Space-as-a-Service (SaaS) that has evolved as a new model of working. The SaaS idea is gaining popularity. Companies like WeWork, Pure House, Krash, and Common are all aware of the millennial desire for convenience, flexibility, and less liability, and have developed successful business models based on the idea of office SaaS. The traditional notions of "private" and "public" space have changed with a more collaborative service economy and technological advancement. Space is being recognized as a profitable commodity that can be leveraged to further business advantage.

Companies such as WeWork lease space wholesale from landlords and then sublet this space, at a margin, in small blocks of floor space, turning real estate into a technology platform for co-working. WeWork currently manages over 3 million square feet of space. They offer its use on a pay-as-you-go basis. Their *unlimited commons* membership option allows people to use WeWork locations anywhere in the world anytime. They provide tenants with the Internet facility, printing services, and separate spaces to relax when taking breaks during working sessions [7]. The co-working location managers handle all services for the use of their facility in terms of actual office management, from payment of utility bills to replenishing the ink in the printer and the coffee in the coffee machines. Managing a facility as a platform in SaaS for a variety of user organizations is a newer but highly successful approach to facilities management.

12.1.4   Chapter Structure

This Chapter looks at the drivers for the need for smarter buildings and smarter spaces; and discusses the rationale for more technologically engaged intelligent facilities management. We first address the technological factors, such as interoperability and standards, that allow the required intelligence to be leveraged. We then discuss the evolution of facilities management as an organizational enabler. The chapter then moves on to discuss ambient technologies and what role they might be able to play in making the buildings and spaces more intelligent and smarter. We examine some of the downsides of this as well, looking at occupants' privacy and sensitivity as well as their being tracked and sensed. Case studies of ambient technology implementations are also discussed before we conclude with final thoughts on where this concept could lead to in future.

12.2   Interoperability and Standards for SFM

There are numerous technical issues with respect to smart facilities management (SFM). In this section, we briefly discuss the issues of interoperability and standards. For the appropriate use of required smart devices in the smart ambient networks, facilities managers need to first acquire an understanding of the levels of interoperability needed for such networks. Since it can be a huge challenge to seamlessly integrating AmI devices in a technological intelligent environment, a discussion on operability and standards for smooth communication is in order. In this respect, what should be the technological basis of AmI is similar to the layers of the Open Systems Interconnection (OSI) model. In this approach, lower levels of the model aim at the collection of lower level contextual information data with the help of various sensors and other similar devices. The middle and upper layers usually consist of computing nodes (in an edge computing context) with enough computing power to allow for the interpretation of the acquired contextual information and automated decision-making. The computing nodes of the upper levels may provide value-added services such as the collection of statistical data or integration of various business processes.

Sensing is a key function of smart buildings, and therefore, of critical importance for the sensing infrastructure. For this reason, a variety of wireless sensor networks has emerged as enablers for delivering sensor data. A consequence of having hundreds of devices is that these networks can become huge bottlenecks.

In terms of standards, two current standards have become popular for wireless sensor networks in buildings, viz.: VZigBee and 6LowPan. Both standards utilize IEEE 802.15.4 radios and are geared toward low-power wireless networks [8].

Occupancy and vacancy sensors are devices that determine if certain space is unoccupied and, if so then, automatically turn off (or dim) the lights and switch off (or lower) the central heating, thereby saving energy and cost. The sensor devices may also turn the lights on automatically upon detecting the presence of people, providing occupancy convenience and potential security aid. According to the Lawrence Berkeley National Laboratory, occupancy-based strategies can produce average lighting energy savings of around 24% [9].

The precursor to ambient networks in smart buildings was the development of two open communications standards for building automation, viz.: BACnet (for Building Automation) and LonWorks, developed by Echelon Corporation in the US. "Lon" in this case stands for Local Operating Network. The two standards have created possibilities for developing smart building controls and automation [10]. BACnet and LonWorks take different approaches to system integration, as follows:

- BACnet, developed in the mid-1990s, is a communications-only standard developed for a building's mechanical and electrical systems, particularly heating, ventilation, and air conditioning (HVAC). Companies that manufacture such systems are now beginning to make voice-controlled devices. BACnet was specifically designed for building automation systems and was adopted as Standard 16484-5 by the International Organization for Standardization (ISO, Geneva, Switzerland) in January 2003. As of 2018, the BACnet Standard achieved a global market share of over 60 percent, according to the latest analysis by British BSRIA, which also forecasts further growth for the next five years. The

BACnet standard comprises rules for data communication for hardware and software used in building control. It includes 23 virtual object types that together represent much of the functionality a building needs to operate. These virtual objects can be grouped together to represent the functions of real building systems [10]. However, there is a growing need to manage the functionality remotely with better efficiency as building owners are driving the trend to monitor and tweak building functions remotely including lighting, fire safety, security, internal conveyors such as escalators, heating, ventilation, and air-conditioning systems [10].

- Developed in the early 1990s, LonWorks combines a communications standard known as LonTalk, with a piece of hardware called the Neuron Chip. LonWorks is already being actively used in the transportation and utilities industries; and now it is being actively adapted for smart buildings environment. Evidence suggests that LonWorks is installed in more buildings globally than the BACnet standard.

Fortunately, the two standards are not mutually exclusive [10]. Unfortunately, however, even if the systems are based on BACnet or LonWorks or both, manufacturers can still program devices to preclude free-flowing data exchange with another vendors' equipment. LonWorks is a standards technology for many of the global standards organizations including ASHRAE, IEEE, ANSI, and SEMI.

The challenge for applications developers is both the development of industry standards and the integration of APIs, data protocols, and network communication standards. There is also a need for quality middleware. Although the need for middleware is well recognized in the AmI community, current research usually takes a top-down approach focused on the seamless integration of lower level nodes in high-end layers [11]. Given the lack of cooperation between device manufacturers, it would be logical that the integration of devices happens at a higher level in the AmI middleware model. This will be occurring in the Edge Computing aspects of the model which we will discuss below.

IDC, in its November 2017 Futurescape Worldwide IoT Predictions [12], has stated that: *by 2020, IT spend on Edge infrastructure will reach up to 18 percent of the total spend on IoT Infrastructure, driven by deployments of converged IT/IOT systems that reduce the time to value of data collected from their connected devices.* This reduction in time is critical to creating value from the infrastructure, which is part of the new role of facilities management in the current technological age of the Internet of Things (IoT) vision.

12.3 Evolution of Facilities Management (FM)

12.3.1 FM as a Profession

Facilities management has changed as a profession from its original management of physical assets of the building. It is now defined by the International Facilities Management Association (IFMA) as *a profession that encompasses multiple disciplines to ensure functionality of the built environment by integrating people, place, process, and technology* [13]. In the twenty-first century, this has become a profession dealing with state-of-the-art office spaces and developing smart facilities. This is not only managing costs and efficiencies, but also reimagining and measuring the efficiencies of flexible meeting spaces that can be used by individuals and groups to support a wide variety of different tasks.

Facilities management (FM) has played an important role in office space evolution. In the early 1980s, the FM industry developed the concept of service bundling, where companies sought to externalize services as well as soft and hard FM outsourcing. It then moved in the 1990s toward service integration, and then in 2000, toward concepts such as total facility management, sustainability management and now to what we refer to as complete workplace management.

Becker [14] stated that *facilities management is responsible for coordinating all efforts related to planning, designing, and managing buildings and their systems, equipment, and furniture to enhance the organization's ability to compete successfully in a rapidly changing world*. As space as an enabling resource has the potential to facilitate positive change in the organization and provide competitive advantage [15], the objective of facilities managers in any organization is to channel resources to provide the right workplace environment for conducting the core business activities on a cost-effective manner that provides value-for-money.

12.3.2 Evolution of FM for Workspaces

The early approaches to the planning of workspaces were to go big and spacious. This was done in anticipation of an increase in volumes of business and the headcount within organizations. Such approaches have *failed to take into consideration the present requirements of occupants, businesses, and their future requirements* [16]. In terms of both the energy and cost efficiency, space management has become a high priority for most office organizations, mainly due to *the high cost of space, demands for more desirable space, and frequent adjustments required to accommodate the rapid growth or expansion of organizations* [17].

Developing and implementing space standards is one of the key responsibilities of any space management department in any organization as *the development of space standards is a prerequisite to the interior design process* [18]. For facilities management, the efforts toward space management are meant to provide a variety of support services to orchestrate all the organization's functions, putting the efforts toward an integration of primary activities in both strategic and operational levels. This makes facilities management an important element in the process of selecting ambient technologies and making sure that they are implemented in an efficient and profitable manner. For example, Near-Field Communication (NFC) technologies may be combined with IoT networks to create "all access" passes to an office space for ease of facilities management. A user could already be carrying around a wireless device that prepares for the user's arrival as they approach the building. It determines user's credentials, unlocks the front doors if credentials are ok, signs the user in, and starts user's computer before they reach the office. While it is a fact that security is going to be a big issue, there are vulnerabilities that need to be resolved; some such vulnerabilities being the same risks that already exist with the use of key cards [19].

Managing the space effectively requires the use of metrics to determine efficiency and cost savings, such as the energy indices that we discuss in the next section.

12.3.3 Indices for Facilities Energy Management

One new element coming into play for FM is the use of energy indices for buildings' design, selection, and occupancy. Climate Energy Index (CEI) and Building Energy Index (BEI) are both applicable. These were developed as a common basis for comparison of building energy performance and different design strategies in a simple and independent fashion [20]:

- The Climate Energy Index (CEI) provides an indication of the consequences of climate with respect to building performance at an *accepted standard of comfort at a particular geographic location* [20].
- The Building Energy Index (BEI) is designed to be a performance indicator for the overall building design strategy. BEI *comprises the climate related and climate unrelated energy loads, which are, respectively, derived from the CEI and benchmark data for non-space conditioning energy uses. The BEI can be compared directly with simulated or measured energy consumption data of a proposed building to benchmark its energy performance* [20].

A third index has also been developed by CBRE and Maastricht University, called the Green Building Adoption Index (GBAI). Together with the U.S. Green Building Council (USGBC) and CBRE Research, this index shows the growth of ENERGY STAR certified and LEED certified spaces for the 30 largest U.S. office markets, both in aggregate and in individual markets, over the previous 10 years.

It should be noted that the way the ambient technology is selected, measured, and rated in relation to FM is considered part of the evolution of the twenty-first century facilities management and evolving role of facilities managers.

12.4 Ambient Technology Management

12.4.1 Ambient Environments

European Commission Information Society Technology Advisory Group (ISTAG) and Philips organization proposed the Ambient Intelligence (AmI) concept in the late 1990s, and defined it as *eEnvironments that are integrated with sensors and intelligent systems* [21]. To be considered ambient, it was suggested that such environments would have the following characteristics [21]:

- Awareness of the presence of individuals;
- Recognition of individual's identities;
- Awareness of the context (e.g., weather, traffic, news);
- Recognition of activities;
- Adaptation to changing needs of individuals.

In creating these environments, AmI-based devices should be designed to be able to deliver user-specific services automatically and in anticipation of the needs of the inhabitants and visitors, assuming that those visitors can be appropriately categorized and their needs predefined [21].

ISTAG itself did not fully define the specifications for AmI. However, it looked ahead and recognized in advance the need and rapid evolution of the technologies and the markets involved. ISTAG took a more holistic approach and identified what had to happen for the development and realization of AmI in terms of technology, society, and business.

In terms of office space, these ideas can be correctly labeled as ambient facility management (or AmFM) [22]. Although still developing, AmFM allows users to communicate in ways that machines can interpret spoken words and take actions in response. Ambient FM uses tiny sensors, discretely located throughout the surroundings to record our movements and actions, learn our preferences, and then predict our desires and adopt to the required new situation.

Ambient technologies need to be designed to be subtle. Business leaders often talk about *digital disruption,* but the adoption of ambient intelligence seems to be the opposite of being disruptive. Facilities managers will welcome this because it enables more information on patterns of usage and productivity.

The technologies and mechanisms that are enabling efforts toward AmFM include the following:

- Minimal or possibly no user interface, that is the replacement of a computer–screen–and–keyboard combination with machines that respond to voice, touch, movement, and biometrics (e.g., fingerprint and retina recognition);
- Artificial intelligence, that refers to the computer systems that perform tasks that humans would generally do as a routine, such as reading documents, data analysis, decision-making, and language translation;
- Machine learning and machine-to-machine communication, that is the ability of computerized devices to learn and improve the performance of tasks with the use of built-in artificial intelligence;
- Natural language processing, that enables computers to recognize the voice of authenticated users, understand natural language words and phrases, and compose responses [22];
- Edge computing, which improves responsiveness and turnaround time by moving processing from centralized processing centers (most likely in the cloud) to smaller distributed processing centers close to where the information is created, held, or delivered [22];
- Mesh networks (hardware and network), which provide continuous connectivity as computerized device users move from one place of another geographical area.

Fig. 12.1 Layered stack for
AmI model

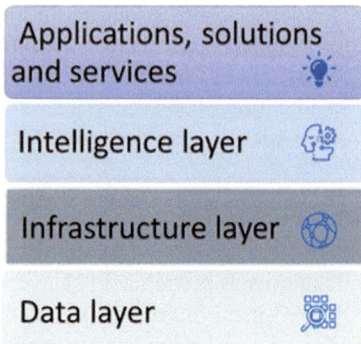

Figure 12.1 suggests an AmI model that has the necessary layers of applications, infrastructure (fog computing, mesh networks), and data tools that might be involved in adding intelligence to data being collected in the workplace.

The two middle layers of Fig. 12.1 are most useful to making the data integration work. Indeed, both edge computing and mesh networks are important to be able to make sense of the data being collected. Without these layers, the intelligence aspect of the infrastructure and applications would not be feasible.

Edge computing paradigm helps to reduce the amount of data, that is often transferred to the cloud or a remote data management site, for processing and analysis. It allows the analysis to happen closer to the point where the data is generated. It also improves data security by directing the data to a close and secure collection point for analysis. The edge computing layer is a way to be more efficient with networking resources to mimic centralized capabilities at the edge of the network and support the quick turnaround of results and decision-making.

Mesh networks, often used to provide mobile services, track our movement as we move across physical areas and ensure device connectivity seamlessly. AmI requires flexible and seamless interoperation across and between networks. It is a primary requirement of hardware platforms or de facto standard that exists to permit this interoperability to take place. In a highly functional mesh network, we would not need to reconnect devices to new networks as we move across physical areas; our presence and, more importantly, the presence of the devices, would be detected by a smart network as we come into the range and would be authorized automatically based on either preconfigured authentication or via authentication on the previous network. With mobile phones set to automatically recognize our home and work networks, and with the use of cell networks while we travel across different zones, we get something akin to what we are currently already used to.

Other aspects of AmFM are also being developed, in particular resolving the interoperability-and connectivity-related issues. Some of us may find it easier to send voice instructions to our computers. Others would value smart furnishings that automatically adjust height and support to accommodate a new occupant whose preferences are already known to a building's knowledge base.

Table 12.1 Technologies and levels of intrusion in smart spaces

Activity	Technology used	Level of intrusion
Checking in, Location, presence	Beacons, WiFi, GPS, automated location tracking (location APIs), geofencing, and activity recognition	Medium—requires personal identification and location, as well as some token from the person as to their whereabouts
Conference room availability	Movement sensors, light detection	Light
Personalized environment	Movement detectors, WiFi, smart key	Medium—involves storage and use of personal data

A well-established business enterprise holds a great deal of information about its customers, their preferences, and their personal data. This data could be usefully employed for managing facilities, applications, and resources, to be beneficial to the company. It could also be beneficial to the employees, suppliers, and new perspective customers.

12.4.2 Ambient Technologies in the Smart Office

As evidenced by the existing literature on this topic, it appears that experience with ambient intelligence in workplace conditions could lead to the ability to reduce unpleasant working and living conditions [23], e.g., room temperature adjustments, etc. Even in today's collaborative business environment, individuals are found heads-down in individually assigned work projects than to be engaged in collaborative work. But the type of work also matters, e.g., if the speed and necessary outcomes of a certain piece of work are fast-paced, or level of the job autonomy is low, then the additional demands created by the distractors will highlight the resulting stress caused by the working environment [24]. Some businesses have chosen to address this by changing wall structures, desk heights, and seating options to make the physical environments adapt to the type of work done, e.g., use of ergonomically designed chairs and work desks.

In relation to the questions such as what technologies are in use in smart spaces, and how intrusive are these on workers' productivity, taking into account the cases reported in the existing literature, Table 12.1 provides a summary of our conclusions.

12.5 User Sensitivity to Profiling, Privacy, and Data Sharing

ISTAG in their initial brief in 2000 on AmI [21] stated that *AmI technologies should support the rights to anonymity/privacy/identity of people and organizations, offering, e.g., relevant combinations of biometrics, digital signature, or genetic-based methods. Also of fundamental importance is research toward safe and dependable large-scale and complex systems (self-testing, self-repairing, and fault tolerant) to underpin the increasing reliance on ICTs implicit in the AmI landscape.* Information security is, therefore, a reasonable concern in a world of ubiquitous, ambient computing.

Ambient Intelligent (AmI) systems are proactive that can assume responsibility and behave proactively, unlike traditional facilities management where the user is in charge and the systems are passive. The shift from passive systems to what could be regarded as a partnership between humans and intelligent artifacts (e.g., light dimmers) creates a demand for a socially adept system [25] in such a way that intelligent systems should show certain abilities that would be traditionally ascribed to humans [26]. To be able to act proactively to mitigate or eliminate undesired situations while regarding user's specific needs, it is important to allow the Ambient FM systems to be extensible, predictive, and to incorporate decision-making capabilities [26].

The term "proactive computing" was first used by Tennehouse, who suggested the following principles for proactive systems [26]:

- They should be closely connected with their surrounding world.
- They should also deliver results to humans before the user action.
- They must operate autonomously.

The challenges in enabling AmI systems are with respect to the following:

- Ambient systems should be able to manage heterogeneous sources (sensors and appliances) to provide high-level information on the state of the environment and what situations are currently available for users [26].
- The ambient systems should be able to process events for detecting situations that are unwanted in the environment; and predict when these situations might arise and proactively manage these situations in advance.
- The ambient systems should be able to determine the policy of actions to consume appropriate services for adapting the environment ahead of the possible new situations [26].
- The ambient system should have expansive capacities to manipulate and adopt to different situations.

If a system is considered ambient intelligent, it should have the ability to build a trust relationship with the users. The system's ability to communicate rationale on its own behavior is one of the most important abilities that such a system can exhibit to gain trust. These explanations are not just a supplement to an ambient intelligent system, but the core requirement of the design and implementation of such a system. Explanations help both for the reasoning process itself and as a means of communicating with the users [27]. However, there is always a risk that information provided, even under a system of trust, can have the potential to be misused. For example, lighting conditions in the user's surrounding convey rich and sensitive data describing users and their behavior. This information could be hijacked and abused, applied to profile the users and perhaps discriminate against them. That is why web standards and APIs are designed and implemented with privacy and trust in mind. It is challenging and interesting to design, create, and analyze products with privacy in mind, as multiple factors need to be considered.

One of the core issues central to technological aspects of privacy engineering is identifiers. Sometimes software is developed in a way that reveals too much about the user's system. This may be formed from tiny, possibly even innocuous data snippets. It turns out that those may introduce interesting consequences when this data is used as identifiers.

But could this go too far? As shown in a recent New York Times article [28], Amazon has two patents for wristbands that track movement and correct employee actions with vibrations to signal more appropriate movement. The use of the Amazon technology could track the employee by means of emitted ultrasonic pulses and radio transmissions to assess, where an employee's hands were in relation to inventory bins and provide "haptic feedback" to steer the worker toward the correct bin. The aim is to streamline time-consuming tasks, like responding to orders and packaging them for speedy delivery. With guidance from a wristband, workers are able to fill in order forms faster. However, these wristbands should raise employee concerns about the privacy of data and could perhaps add additional surveillance layers to the workplace.

The obligation to define such tracking rests with the employer and/or facilities manager when the person tracked is not an employee of the facility. Electronic monitoring of employees is especially intrusive and can lead to unacceptable levels of workplace surveillance. Any oversight of employees in this manner should be narrowly tailored in time, place, and manner, and it must be transparent to employees. The same can be said for visitors to the facility. Thus, a stated policy should be made clearly known to users when entering the facility, e.g., how and why information and location are tracked by sensors.

Rao et al. [29] suggest that location context control policy gets privileges depending on the classification of users. If the user decides to opt-out, the system might give "fake" information, such as the standard location of the user (their office, perhaps) or the reception area of the facility. This would make the Ambient FM system for that user nonfunctional in that it would not be personalized and the user would receive the standard lighting and air condition package as prescribed by company policy.

Tsai et al. [30] observed users' risk and advantage perception associated with the practice of these apparatuses and privacy restrictions on current location-sharing approaches. Their study involved an online survey and found that although most of their respondents had heard of location-sharing approaches, they did not see the probable value of mentioned approaches. They also found that company rules to manage disclosure of user's place only offers a modicum of privacy.

To illustrate what can be achieved by way of benefits of working with ambient technology in the office, the following section discusses some use cases for ambient technology deployment and how the adopters have addressed user concerns.

12.6   Case Studies in Ambient Technology Usage for FM

We have chosen three cases to illustrate device management, user control, and energy efficiency as core drivers for facilities management. These cases focus on the latest developments in the internet of things and always-on communications. They show how the implementation of data-driven mechanisms can help to increase productivity, enhance occupant happiness and well-being, improve sustainability, and optimize service delivery and operations.

12.6.1   *Robin-Powered Conference Room*

Robin is a Boston-based start-up that provides a *software layer for office buildings* [5]. In terms of device management, Robin uses iBeacon and BLE (Bluetooth Low Energy) devices to detect the presence of nearby people and things in a conference room context. It can automate conference room bookings for users just by the action of the users walking into a room. And, after the user enters the conference room, Robin gives the user control to update the screens in the room and also gives them control of the nearby devices.

Their web scheduling product allows companies to determine what facilities and items each room has; and the availability or otherwise of these items. Companies like to customize their spaces with respect to products based on their requirements and interests. In this context, Blue Apron names their rooms after exotic spices and fancy cheeses; Casper names their rooms after everyone's favorite meal to wake up to, e.g., breakfast; Foursquare uses their badges as room names, decorated with colors and props. Robin's desk products allow employees to find a place to sit for the day. Employees know as to what is happening around them and are in control of their workspace, as seen in Fig. 12.2.

To get started, users would need to install the Robin software app on their iOS or Android phone so the spaces can properly identify them, and they can identify what information they would want to share with the app. As for the ambient hardware, only beacons are required; each beacon covering a zone of up to 30 m. An accompanying dashboard provides office-wide overview and analytics for the Robin-powered rooms. And, as new connected devices are added to the workplace, Robin can be customized to control them as well, including things such as Chromecast, smart thermostats, lights, and more [5]. Lighting is important for creating mood, as well as for productivity, as discussed in the following case.

Fig. 12.2 A user-aware workspace with flexible desk scheduling

12.6.2 Intelligent Lighting with Igor Using PoE

With the wired Ethernet lighting control opportunities, market leader Philips Lighting has been embedding ZigBee in lights and luminaires and selling it as a service. Users can control lights wirelessly using ZigBee from phones or tablets. Firms such as Igor [31] have focused on Power over Ethernet (PoE) and found that, with the availability of low-cost ultra-miniature LEDs, sensors, and communications protocols, it makes it possible to embed Internet connectivity into every lighting fixture and many low-wattage sensors. By using a standard Ethernet connection and PoE ports, network-enabled PoE devices can provide any user immediate access to building automation control throughout an entire lighting system.

LEDs are used to indicate occupancy, adjust mood, conserve energy, and remotely control the building automation systems. LEDs can expand their role by combining intelligence with movement or ambient light sensors and interconnecting the PoE nodes in a programmable network.

Igor's open PoE platform provides direct API access to setup, control, and real-time data streams from embedded sensors and analytics, delivering a simple dash-board for energy savings, space utilization, security, and more. The intelligent PoE lighting system can increase staff productivity and well-being by optimizing light and temperature to create comfortable and desired work environments.

Central Iowa Power Cooperative (CIPCO), the largest cooperative energy provider in Iowa, distributes power to nearly 300,000 residents [31]. It used its Cedar Rapids office as a showcase for next-generation lighting, using solutions provided by Igor. CIPCO wanted precise control over lighting through a Crestron audio–video control system, using network switches that could provide 48 Power over Ethernet (PoE) ports and deliver 60 watts per port.

Through these mechanisms, CIPCO has successfully managed to reduce lighting energy use by 75 percent in areas where the lighting solutions have been deployed. The company expects this saving to increase to 85% with its smart control strategies, e.g., by using daylight harvesting and occupancy sensors to automatically dim and shut off lights, if not required. As lights are now considered just another IP endpoint, CIPCO can control scheduling and policies through its audio–video control system. CIPCO have now built their "nervous system" in place to support the adoption of ambient intelligence and continued efficiencies [31].

The Igor system also optimizes space utilization by identifying occupied and unoccupied spaces for energy savings, safety, and security. These long-lasting and low-energy LED lights drastically reduce costs and maintenance by identifying high/low use areas. But we should also focus on space usage in other ways, e.g., by space developments by design with the user comfort in mind.

12.6.3   The Edge Office Space in Amsterdam

The Edge is an example of a groundbreaking new office space venture opened in 2015 in Amsterdam [32]. This has inspirational workspaces developed throughout the building as places to reflect, think, collaborate, and innovate. This innovative landmark building in the Zuidas business district of Amsterdam was developed by OVG Real Estate [32]. It is a 40,000 m^2, multi-tenant office building, which embraces leading-edge smart technology to support flexible and activity-based working. It exhibits the highest standards of sustainability and innovative data-driven insights to enable the most efficient facilities management. The architects' concept included intelligent floorplans to enhance employee comfort and efficiency, flexible workspaces, and the use of environmentally friendly materials. Integrated sensors capture data on room occupancy, temperature, and humidity, which the building owners can use to precisely target the delivery of lighting and other resources, such as heating/cooling and cleaning, to maximize energy efficiency. Light levels and cleaning can be reduced in low-occupancy areas resulting in the saving of time, cost, and energy.

The idea that employees use all aspects and areas of the building is fundamental to a smart building design. Open spaces can be located around a vast atrium bathed in natural lighting. As its hub can be a coffee bar and other utility points where meetings and discussions can take place, working at tables in a bar rather than a traditional desk is highly possible.

To enable personalization within the edge environment, personal comfort has been enhanced further with the ability to control and flex the lighting and temperature, even in open-place spaces, via the Philips Personal Control iPhone app, especially designed for use in the edge. The connected lighting system makes it possible for employees to locate colleagues within the building in real time, check on rooms availability, and find their way easily from place to place.

The system also offers building managers rich real-time and historical data on systems' operations and activities. This data gives them the insights they need to create a premier experience for employees, maximize operational efficiency, and reduce the building's CO2 footprint [32].

Designers kept three key objectives in mind when defining the connected lighting system. The system had to seamlessly integrate with the building as a whole, enable customized solutions purpose-built for the unique environment of the edge, and offer smart interfaces that allow individual users of the building to control the environment effortlessly.

The system's 6,500 connected luminaires over 15 floors share data about their status and operations with Philips Envision lighting management software, running in the IT environment. Facility managers can use the software to capture, visualize, and analyze this data, allowing them to track energy consumption and streamline maintenance operations. The expected energy savings at the edge is around €100,000 and €1.5 m in space utilization costs [32].

12.7   Conclusion

Although the drivers for AmI still suffer from resource and energy constraints, there is a movement toward personalized comfort and intelligent design. Offices have been moving away from the idea that time spent at a desk and at a fixed location are measures of productivity. This departure from tradition can be seen from some of the more ambitious innovations at corporate headquarters, e.g., the GooglePlex. With the inclusion of sensors, beacons, WiFi, and other enabling smart technologies embedded with intelligence, companies are now measuring the effectiveness of both the workspaces and employees through the use of ambient intelligence. The company which gives the employees the choice at any given moment to optimize the space effectiveness is going to benefit many times over, in terms of both profitability and worker satisfaction.

With the use of ambient technologies, the role of facilities management has expanded dramatically over the past decade, as businesses start to understand the true impact of smart working and smart environments. Gone are the days when management teams were responsible for little more than checking boxes and ticking off safety requirements. The role is now integrated with enterprise asset management (EAM) processes and energy efficiencies for better reliability and cost savings.

Given its role in the operations of the business, we find that facilities management has become more sensitive to social and cultural changes. The facilities managers now need to ensure that social constraints are effectively balanced with productivity, development, and worker satisfaction. Not an easy task but effective use of appropriate technology is the secret to success.

The nature of ambient technologies and their acceptance into the workplace will further evolve over time and so the employee tracking will become more commonplace. However, this will need a more careful policy on how privacy and security of user information are kept and maintained in the company.

References

1. Gurgen L, Gunalp O, Benazzouz Y, Gallissot M (2013) Self-aware cyber-physical systems and applications in smart buildings and cities. In: Proceedings of the conference on design, automation and test in Europe, pp. 1149–1154. EDA Consortium
2. Armstrong P, Brambley MR, Pratt RG, Chassin DP (2000) Building controls and facilities management in the 21st century, summer study of energy efficiency in buildings. http://aceee.org/files/proceedings/2000/data/
3. Atkin B, Brooks A (2000) Total facilities management. Blackwell Science, UK
4. Hassanain MA (2010) Analysis of factors influencing office workplace planning and design in corporate facilities. J Build Apprais 6(4):183–197
5. Robin (2018). https://robinpowered.com/blog/case-studies/
6. Softdb (2016). What is sound masking and how does it work? https://www.softdb.com/sound-masking/technology/#comfort
7. Grozdanic L (2016) Space as a service: business models that change how we live and work. http://archipreneur.com/space-as-a-service-business-models-that-change-how-we-live-and-work/
8. Weng T, Agarwal Y (2012) From buildings to smart buildings—sensing and actuation to improve energy efficiency. IEEE Design Test Comput. 29(4): 36–44. (Cearley DW (2010) Cloud computing: key initiative overview. Gartner Report)
9. Dilouie C (2017) All about occupancy and vacancy sensors. http://lightingcontrolsassociation.org/2017/08/21/all-about-occupancy-and-vacancy-sensors/
10. Snoonian D (2003) Smart buildings. IEEE Spectr 40(8):18–23
11. Anastasopoulos M, Niebuhr D, Bartelt C, Koch J, Rausch A (2005, October) Towards a reference middleware architecture for ambient intelligence systems. In: ACM conference on object-oriented programming, systems, languages, and applications
12. Corner S (2018, March 31) Edge computing will be key to IoT, say experts. https://www.iothub.com.au/news/edge-computing-will-be-key-to-iot-say-experts-488021
13. IFMA (2018) What is facility management? https://www.ifma.org/about/what-is-facility-management
14. Becker F (1990) The total workplace. Van Nostrand Reinhold, New York
15. Langston C, Lauge-Kristensen R (2002) Strategic management of built facilities. Butterworth-Heinemann, UK
16. McGregor W (2000) The future of workspace management. Facilities 18(3/4):138–143
17. Brauer RL (1992) Facilities planning: the user requirement method, 2nd edn. American Management Association, New York
18. Owen DD (1993) Facilities planning and relocation. RS Means, USA
19. SelectHub (2017) The future of facilities management: AI, robots, and outer space. https://selecthub.com/cmms/future-facilities-management/
20. McLean D, Roderick Y, Quincey R, McEwan D (2011) Climate energy index and building energy index: New indices to assess and benchmark building energy performance. In: Proceedings of the Building Simulation 12th Conference of International Building Performance Simulation Association, Sydney, Australia, vol 1416, p 792799
21. Lee F (2017) Ambient intelligence—the ultimate IoT use cases. https://medium.com/iotforall/ambient-intelligence-the-ultimate-iot-use-cases-5e854485e1e7

22. Karpook D (2018). Ambient intelligence for buildings. https://facilityexecutive.com/2018/02/ambient-intelligence-for-buildings/
23. Veitch JA, Newsham GR (2000) Exercised control, lighting choices, and energy use: an office simulation experiment. J Environ Psychol 20(3): 219–237
24. Veitch JA (2018) How and why to assess workplace design: facilities management supports human resources. Organizational Dynamics
25. Marsh S (1995) Exploring the socially adept agent. In: Proceedings of the first international workshop on decentralized intelligent multi agent systems (DIMAS 1995), p 308
26. Machado A, Lichtnow D, Pernas AM, Wives LK, de Oliveira JPM (2014) A reactive and proactive approach for ambient intelligence. In: ICEIS, vol 2, pp 501–512
27. Cassens J, Kofod-Petersen A (2007) Explanations and case-based reasoning in ambient intelligent systems. In: CaCoA
28. Yeginsu C (2018) If workers slack off, the wristband will know. (And Amazon Has a Patent for It.). Retrieved from the New York Times online from https://www.nytimes.com/2018/02/01/technology/amazon-wristband-tracking-privacy.html
29. Rao J, Pattewar R, Chhallani R (2018) A privacy-preserving approach to secure location-based data. In: Bhalla S, Bhateja V, Chandavale A, Hiwale A, Satapathy S (eds) Intelligent computing and information and communication. Advances in intelligent systems and computing, vol 673. Springer, Singapore
30. Tsai JY, Kelley PG, Cranor LF, Sadeh N (2010) Location-sharing technologies: privacy risks and controls. ISJLP 6:119
31. CIPCO case study (2018). https://www.igor-tech.com/filesimages/Case%20Studies/cipco-case-study.pdf
32. Philips (2018). http://www.lighting.philips.com/main/systems/system-areas/office-and-industry/offices/futureoffice/connectivity/connecting-with-the-future-of-office-design

Index

© Springer Nature Switzerland AG 2019
Z. Mahmood (ed.), *Guide to Ambient Intelligence in the IoT Environment*, Computer
Communications and Networks, https://doi.org/10.1007/978-3-030-04173-1